Springer Series in
OPTICAL SCIENCES 119

founded by H.K.V. Lotsch

Editor-in-Chief: W. T. Rhodes, Atlanta

Editorial Board: A. Adibi, Atlanta
T. Asakura, Sapporo
T. W. Hänsch, Garching
T. Kamiya, Tokyo
F. Krausz, Garching
B. Monemar, Linköping
H. Venghaus, Berlin
H. Weber, Berlin
H. Weinfurter, München

Springer Series in
OPTICAL SCIENCES

The Springer Series in Optical Sciences, under the leadership of Editor-in-Chief *William T. Rhodes*, Georgia Institute of Technology, USA, provides an expanding selection of research monographs in all major areas of optics: lasers and quantum optics, ultrafast phenomena, optical spectroscopy techniques, optoelectronics, quantum information, information optics, applied laser technology, industrial applications, and other topics of contemporary interest.
With this broad coverage of topics, the series is of use to all research scientists and engineers who need up-to-date reference books.

The editors encourage prospective authors to correspond with them in advance of submitting a manuscript. Submission of manuscripts should be made to the Editor-in-Chief or one of the Editors. See also www.springeronline.com/series/624

L. Pavesi   G. Guillot
(Eds.)

Optical Interconnects

The Silicon Approach

With 265 Figures (5 color)

 Springer

Seplae
phys

Professor Lorenzo Pavesi
Universitá di Trento, INFM and Dipartimento di Fisica
via Sommarive 14, 38050 Povo, Italy
E-mail: pavesi@science.unitn.it

Professor Gérard Guillot
Laboratoire de Physique de la Matière, INSA Lyon, Bât Blaise Pascal
7 avenue Jean Capelle, 69621 Villeurbanne Cedex, France
E-mail: gerard.guillot@insa-lyon.fr

ISSN 0342-4111

ISBN-10 3-540-28910-0 Springer Berlin Heidelberg New York

ISBN-13 978-3-540-28910-4 Springer Berlin Heidelberg New York

Library of Congress Control Number: 2005936104

Springer is a part of Springer Science+Business Media.

springer.com

© Springer-Verlag Berlin Heidelberg 2006
Printed in The Netherlands

Typesetting: SPI Publisher Services
Cover concept by eStudio Calamar Steinen using a background picture from The Optics Project. Courtesy of John T. Foley, Professor, Department of Physics and Astronomy, Mississippi State University, USA.
Cover production: *design & production* GmbH, Heidelberg

Printed on acid-free paper   SPIN: 11018469   57/3100/SPI   5 4 3 2 1 0

4/11/06
GL

4/24/2006
MMC

To our parents
Annalisa, Vittorio
Marie-Thérèse, Maurice

and to our families
Anna, Maria Chiara, Matteo, Michele, Tommaso
Blandine, Antoine, Marie, Laure

Preface

This book is the brainchild of the third Optoelectronic and Photonic Winter School on "Optical Interconnects" which took place from the February 27 to March 4, 2005 in Sardagna, a small village on the mountains around Trento in Italy. This school, held every two years, has been promoted in Trento to trace the very fast developing technologies and the tremendous progress that has been and will occur in the near future. It is a common view that its current explosive development will lead to deep paradigm shifts in the near future. Identifying the plausible scenario for the future evolution of Photonics presents an opportunity for constructive actions and for scouting killer technologies.

In analogy with electronics, the term *Photonics* was coined in 1967 by Pierre Aigrain, a French scientist who worked on semiconductor lasers. While electronics is the science and technology of electron motion, photonics is the science of the mastery of light made up by photons and the technology of using it. The term photonics is used broadly to encompass: (1) the generation of light, e. g. by LED and lasers; (2) the transmission of light in free space through conventional optic systems or guided in a material through optical fibres and dielectric waveguides; (3) the processing of light (modulation, switching, scanning, computing); (4) the amplification and frequency conversion of light by the use of non-linear materials; (5) the detection of light and of images; (6) the use of light for data storage (optical discs and holography); etc.

Optical interconnects have revolutionized telecommunications over the last few decades. Single frequency lasers, fibre optics, fast detectors, dense wavelength division multiplexing (DWDM) have been used for decades in long-distance application such as telephony and wide area networks to overcome the limitations imposed by standard electronic communication. This technology faces a rapid increase, where its performances double every nine months. Note that microelectronics is also increasing with a very fast rate, although the famous Moore's law predicts that the doubling occurs every 18 months.

The evolution of Moore's law is finding red-brick walls. In fact, electronic transmission over copper is currently limited to distances around $100\,\mathrm{m}$ for data rates exceeding $1\,\mathrm{Gb\,s^{-1}}$ and this distance will certainly shrink as data rates rise. So there is a clear need to create a radically new communication landscape using optical interconnects for the computing and communication industries to face increasing challenges to deliver more data faster and at lower power.

After having moved from long-haul backbones to the metropolitan area networks (MANs) and local area networks (LANs) during the last decade, optical communication technologies are and will certainly become increasingly more important for the rack-to-rack, board-to-board, chip-to-chip and even component-to-component within a single chip (on-chip interconnect).

On the other hand, semiconductor productivity and performance have increased at exponential rates in the last 40 years. A dramatic scaling down in feature sizes (the next node is fixed at $65\,\mathrm{nm}$ for the field effect transistor (FET) gate length by the ITRS Road Map) has been the main driving force of the microelectronics industry illustrated by Moore's law over 30 years.[1]

Smaller devices on larger wafers led to larger yield, lower cost and faster circuits. For example, in about 5 years complementary metal oxide semiconductor (CMOS) transistors will be fast enough to operate at clock speeds of roughly $15\,\mathrm{GHz}$, fast enough to support data-transfer rates of $20\,\mathrm{Gb\,s^{-1}}$. This will certainly lead to an interconnect bottleneck due to the current copper-trace base technology, which worsens in terms of both speed and power because the important scaling in sizes has a negative impact on the resistance and inductance of metal interconnects. Transfer information lengths in a single chip are nowadays of the order of $10\,\mathrm{km}$, while in 10 years they will approach $100\,\mathrm{km}$. So interconnects have become the primary limit on gigascale integration causing significant propagation delays for clock signals, overheating, information latency and electromagnetic interference. Up to now the strategy of the microelectronic industry is to try to release this bottleneck by improving design and material performances. However, all the potential improvements can be costly in the future.

Optical interconnect technology is therefore an increasingly attractive alternative and has become more and more essential for some industrial leaders of microelectronics who are very active in this field of research. Optical interconnects can provide much greater bandwidth, lower power consumption, decreased interconnect delays, resistance to electromagnetic interference and reduced signal crosstalk. Photonic materials in which light can be generated, guided, modulated, amplified and detected need to be integrated with standard electronic circuits to combine information processing capabilities of electronics with data handling of photonics.

However to be widely adopted, these converging technologies must provide significant performance breakthrough within a cost-effective engineering.

---

[1] International Technology Roadmap for Semiconductor (ITRS) published by International SEMATECH, Austin, TX. http://public.itrs.net/

An optical solution could replace electrical interconnect when it has higher performance at lower cost and strong manufacturability in high volume.

Today many commercial photonic devices are made from exotic materials such as InP and GaAs, which make them difficult to manufacture and assemble and, consequently, expensive. This limits their applications to special fields such as long-haul telecommunications and wide area networks.

Silicon has remained the microelectronic industry's semiconductor of choice from the very beginning. Integration and economy of scale are two key ingredients in the technological success of silicon. Because of this history, silicon could be an ideal material to integrate both optical and electronic functionalities and to achieve a monolithically integrated silicon microphotonics leading to optics on a silicon chip.

Photonizing silicon is a major challenge for the microelectronic industry to produce inexpensive and performing photonic devices out of silicon. On the other hand, *siliconizing* photonics is a major challenge for the photonic industry to introduce in the production processes of photonic devices the same concepts (standardization, economy of scale, integration and roadmap) which make the success of the microelectronic industry.

Even if some very important enabling devices, like the injection silicon laser, have not yet been demonstrated, many important building blocks of silicon microphotonics exist and certainly silicon will be the fastest growing and highest volume market segment for photonic data link in the next decade. Roadmapping is beginning to be considered very seriously for the photonics industry. [2]

This book is aimed at present the state-of-the-art in optical interconnects in silicon and out of silicon, and all the challenges that need to be addressed before this technology can be successfully used for data communications at low power, high bandwidth, high speed at short distances and to resolve current microelectronic bottlenecks by bringing optics closer to and around the microprocessor.

The chapter by Moussavi gives an in-depth introduction on multi-level advanced interconnect networks in microelectronics and related problems that have become the primary limit in both the performance and the energy dissipations of gigascale integration.

Then follow a cluster of chapters on discrete photonic components. The chapter by Pavesi gives an introduction to the optical properties of silicon and to all the strategies developed to fabricate efficient light-emitting devices paving the road for an injection laser. The chapter by Boyraz presents recent advances in the application of stimulated Raman scattering in silicon-to-silicon Raman laser, optical amplifiers and wavelength converters. The principle of light modulation with silicon devices by using electro-optic and thermo-optic effects is presented in the chapter by Libertino and Scinto. The chapter by Zimmermann gives a review of the state-of-the-art silicon photoreceivers integrated in standard CMOS technology. The advantages of SiGe alloys for

---

[2] http://mph-roadmap.mit.edu/about_ctr/report2005

photodetectors and optical modulators are reported in the chapter by Cassan and coworkers.

Integrated waveguides are the common words of the next few chapters. The chapter by Reed presents the basics of guided optics and passive devices on silicon-on-insulator (SOI). The chapter by Van Thourhout presents the challenges to realize silicon waveguides and other passive components in silicon wafers. The chapter by Krauss discusses the mastering of light at extremely short length scales by photonic crystals to obtain photonic integrated circuits. The fundamental challenge of in/out coupling in silicon submicronic waveguides and the different approaches for light coupling on silicon are presented in the chapter by Orobtchouk.

The last set of chapters deals with the system approach. The chapter by Wada and coworkers is related to new Ge photodetector, modulator and on-chip wavelength MUX/DEMUX integrated on a silicon CMOS platform. An industrial perspective on silicon photonics and a vision of potential future applications are given in the chapter by Paniccia and coworkers which also details recent key components obtained at Intel, like high-frequency CMOS modulator and CW silicon Raman laser. The basic possibilities and limits of free space optical interconnects are reported in the chapter by Kirk.

At the end, we would like to thank all the authors who produced very interesting state-of-the art lectures and chapters. Last but not the least, we also express our gratitude to all those who have contributed to the success of the Third Optoelectronic and Photonics Winter School on Optical Interconnects: the staff of the University of Trento and all the students whose participation was lively and stimulating. Support from the University of Trento, the Physics Department, ITC-irst within the PROFILL project, INSA-Lyon, Hamamatsu-Italia, Laser Optronic and Spectra-Physics, Crisel-Instruments, Advanced Technologies and Raith GmbH is gratefully acknowledged.

Trento, Lyon,   *Lorenzo Pavesi*
November 2005   *Gérard Guillot*

Contents

8 Submicron Silicon Strip Waveguides

12 Silicon-Integrated Optics

13 Free-Space Optical Interconnects

List of Contributors

D.H. Ahn
Department of Materials Science
and Engineering
Massachusetts Institute
of Technology
Cambridge, MA 02139, USA

S. Akiyama
Department of Materials Science
and Engineering
Massachusetts Institute
of Technology
Cambridge, MA 02139, USA

O. Boyraz
Department of Electrical Engineering
and Computer Science
University of California
Irvine, CA 92697
oboyraz@uci.edu

D.D. Cannon
Department of Materials Science
and Engineering
Massachusetts Institute
of Technology
Cambridge, MA 02139, USA

E. Cassan
Institut d'Electronique
Fondamentale
UMR CNRS 8622
Bâtiment 220
91405 Orsay, France
eric.cassan@ief.u-psud.fr

D.T. Danielson
Department of Materials Science
and Engineering
Massachusetts Institute
of Technology
Cambridge, MA 02139, USA

X. Duan
Department of Materials Science
and Engineering
Massachusetts Institute
of Technology
Cambridge, MA 02139, USA

M. Halbwax
Institut d'Electronique
Fondamentale
UMR CNRS 8622
Bâtiment 220
91405 Orsay, France

H.A. Haus
Department of Electrical Engineering
and Computer Science
Massachusetts Institute
of Technology
Cambridge, MA 02139, USA

Y. Ishikawa
Department of Materials Science
and Engineering
Massachusetts Institute
of Technology
Cambridge, MA 02139, USA

S. Jongthammanurak
Department of Materials Science
and Engineering
Massachusetts Institute
of Technology
Cambridge, MA 02139, USA

L.C. Kimerling
Department of Materials Science
and Engineering
Massachusetts Institute
of Technology
Cambridge, MA 02139, USA

A.G. Kirk
Department of Electrical
and Computer Engineering
McGill University
3480 University St.
H3A 2A7 Montreal,
Canada
andrew.kirk@mcgill.ca

S. Koehl
Intel Corporation
Santa Clara CA 95054, USA
sean.m.koehl@intel.com

T.F. Krauss
The Ultrafast Photonics
Collaboration
School of Physics and Astronomy
University of St. Andrews
St. Andrews, KY16 9SS
Scotland, UK
tfk@st-andrews.ac.uk*

S. Laval
Université Paris-Sud
Institut d'Electronique
Fondamentale
UMR CNRS 8622
Bâtiment 220
91405 Orsay, France

K.K. Lee
Department of Materials Science
and Engineering
Massachusetts Institute
of Technology
Cambridge, MA 02139, USA

L. Liao
Intel Corporation
Santa Clara
CA 95054, USA
ling.liao@intel.com

S. Libertino
CNR – IMM sez. Catania
Stradale Primosole 50
95121 Catania, Italy
sebania@imm.cnr.it

D.R. Lim
Department of Materials Science
and Engineering
Massachusetts Institute
of Technology
Cambridge, MA 02139, USA

A. Liu
Intel Corporation
Santa Clara CA 95054, USA
ansheng.liu@intel.com

J.F. Liu
Department of Materials Science
and Engineering
Massachusetts Institute
of Technology
Cambridge, MA 02139, USA

H.C. Luan
Department of Materials Science
and Engineering
Massachusetts Institute
of Technology
Cambridge, MA 02139, USA

A. Lupu
Institut d'Electronique
Fondamentale
UMR CNRS 8622
Bâtiment 220
91405 Orsay, France

D. Marris
Institut d'Electronique
Fondamentale
UMR CNRS 8622
Bâtiment 220
91405 Orsay, France

J. Michel
Department of Materials Science
and Engineering
Massachusetts Institute
of Technology
Cambridge, MA 02139, USA

M. Moussavi
CEA-LETI 17
Rue des Martyrs
38054 Grenoble, Cedex
09, France
MOUSSAVIME@chartreuse.cea.fr

R. Orobtchouk
Laboratoire de Physique
de la Matière
INSA de Lyon
UMR CNRS 5511
Lyon, France
regis.orobtchouk@uisa-lyon.fr

M. Paniccia
Intel Corporation
Santa Clara CA 95054, USA
mario.paniccia@intel.com

D. Pascal
Institut d'Electronique
Fondamentale
UMR CNRS 8622
Bâtiment 220
91405 Orsay, France

L. Pavesi
Dipartimento di Fisica
via Sommarive 14
38050 Povo (Trento), Italy
lorenzo.pavesi@unit.it

M. Popovic
Department of Electrical Engineering
and Computer Science
Massachusetts Institute
of Technology
Cambridge, MA 02139, USA

G.T. Reed
University of Surrey, UK
G.Reed@surrey.ac.uk

H. Rong
Intel Corporation
Santa Clara CA 95054, USA
haisheng.rong@intel.com

M. Rouvière
Institut d'Electronique
Fondamentale
UMR CNRS 8622
Bâtiment 220
91405 Orsay, France

A. Sciuto
CNR – IMM sez. Catania
Stradale Primosole 50
95121 Catania, Italy

D. Van Thourhout
Ghent University
Departement of Information
Technology
St Pietersnieuwstraat 41
9000 Gent, Belgium
Dries.VanThourhout@ugent.be

L. Vivien
Institut d'Electronique Fondamen-
tale,
UMR CNRS 8622
Bâtiment 220
91405 Orsay, France

K. Wada
Department of Materials Science
and Engineering
Massachusetts Institute
of Technology
Cambridge, MA 02139, USA
and
Department of Materials Engineering
The University of Tokyo
Bunkyo
Tokyo 113-8565, Japan
kwada@material.t.u-tokyo.ac.jp

H. Zimmermann
Vienna University of Technology
A-1040 Vienna, Austria
horst.zimmermann@tuwien.ac.at

1

Advanced Conventional Interconnects: State of the Art, Future Trends, and Limitations

M. Moussavi

Summary. As design rules drop below 90 nm, a variety of challenges emerge such as *RC* delay, electromigration resistance, and heat dissipation exacerbated by increased chip power. The use of copper and thin barrier layers solves resistivity and electromigration problems but not for long due to electron scattering issues' increasing the apparent resistivity. Moreover, reliability issue with respect to an efficient diffusion barrier is a concern. Low *k* dielectrics allowing capacitance reduction have low thermal conductivity and hence poor heat dissipation capability. Integration of copper and low *k* dielectrics is intensively studied worldwide, and this chapter gives an overview of the international state of the art to overcome critical issues of advanced interconnects. Severe limitations of the conventional interconnects in the near future act as a technology push for alternative solutions such as 3D or optical interconnects.

1.1 Introduction

The semiconductor industry has sustained rapid technology development over a long period of time. Since the invention of integrated circuits, the initial material choices (silicon substrate, SiO_2 gate or intermediate dielectrics, and Al wiring) as well as processes such as metal etch or lithography have stood the test of 45 years. With an increase of the integration density and a decrease in the number of the interconnection design rules, the application of physics and materials science laws has become more critical. New challenges from new materials and architectures are growing. According to the International Technology Roadmap for Semiconductors (ITRS) 2003 update (Tables 1.1 and 1.2), 45-nm generation is only 5 years away.

For a resistivity requirement of $2.2 \,\mu\Omega\,cm$ (the sum of barrier and metal resistivities) copper has been chosen to be integrated in a damascene architecture. The damascene architecture is a commonly used term for the process in which the dielectric is deposited first, followed by lithography and etch, and finally the metal (Cu) is deposited and polished. Single Damascene (SD) is used for one-level (line) structure, while Dual Damascene concerns the

Table 1.1. 2003 ITRS MPU interconnect technology requirements (near term)

	2003	2004	2005	2006	2007	2008	2009
Year of prodution	2003	2004	2005	2006	2007	2008	2009
Technology node		hp90			hp65		
DRAM ½ pitch (nm)	100	90	80	70	65	57	50
MPU/ASIC ½ pitch (nm)	120	107	95	85	76	67	60
MPU printed gate length (nm)	65	53	45	40	35	32	28
MPU physical gate length (nm)	45	37	32	28	25	22	20
Number of metal levels	9	10	11	11	11	12	12
Number of optional levels – ground planes/capacitors	4	4	4	4	4	4	4
Total interconnect length (m cm²) – active wiring only, excluding global levels [1]	579	688	907	1002	1117	1401	1559
FIT s/m length cm⁻² × 10⁻³ excluding global levels [2]	8.6	7.3	5.5	5.0	4.5	3.6	3.3
J_{max} (A cm⁻²) – intermediate wire (at 105°C)	3.7×10^5	5.0×10^5	6.8×10^5	7.8×10^5	1.0×10^6	1.4×10^6	2.5×10^6
Metal 1 wiring pitch (nm)*	240	214	190	170	152	134	120
Metal 1 A/R (for Cu)	1.6	1.7	1.7	1.7	1.7	1.8	1.8
Interconnect RC delay (ps) for 1 mm Metal 1 line	191	224	284	355	384	477	595
Line length (mm) where τ = RC delay (Metal 1 wire)	79	65	55	46	41	34	28
Cu thinning at minimum pitch due to crosion (nm), 10% × height, 50% areal density, 500 µm square array	19	18	16	14	13	12	11
Intermediate wiring pitch (nm)	320	275	240	215	195	174	156
Intermediate wiring dual damascene A/R (Cu wire/via)	1.7/1.5	1.7/1.5	1.7/1.5	1.7/1.6	1.8/1.6	1.8/1.6	1.8/1.6
Interconnect RC delay (ps) for 1mm intermediate line	105	139	182	224	229	288	358
Line length (mm) where τ = RC delay (intermediate wire)	107	83	69	58	53	43	37
Cu thinning at minimum intermediate pitch due to erosion (nm), 10% × height, 50% areal density, 500 µm square array	27	23	20	18	18	15	10
Minimum global wiring pitch (nm)	475	410	360	320	290	260	234
Ratio range (global wiring pitches/intermediate wiring pitch)	1.5–5.0	1.5–6.7	1.5–6.7	1.5–6.7	1.5–8.0	1.5–8.0	1.5–8.0
Global wiring dual damascene A/R (Cu wire/via)	2.1/1.9	2.1/1.9	2.2/2.0	2.2/2.0	2.2/2.0	2.3/2.0	2.3/2.0
Interconnect RC delay (ps) for 1 mm global line at minimum pitch	42	55	69	87	92	112	139
Line length (mm) where τ = RC delay (global wire at minimum pitch)	169	132	112	93	83	69	59
Cu thinning at maximum width global wiring due to dishing and erosion (nm), 10% × height, 80% areal density	168	193	176	158	172	160	144
Cu thinning global wiring due to dishing (nm), 100 µm wide feature	30	29	24	21	19	17	15
Conductor effective resistivity (µΩ cm) Cu intermediate wiring	2.2	2.2	2.2	2.2	2.2	2.2	2.2
Barrier/cladding thickness (for Cu intermediate wiring) (nm) [3]	12	10	9	8	7	6	6
Interlevel metal insulator (minimum expected) – effective dielectric constant (κ)	3.3–3.6	3.1–3.6	3.1–3.6	3.1–3.6	2.7–3.0	2.7–3.0	2.7–3.0
Interlevel metal insulator (minimum expected) – bulk dielectric constant (κ)	<3.0	<2.7	<2.7	<2.7	<2.4	<2.4	<2.4

* Refer to Executive Summary Figure 4 for definition of Metal 1 pitch

Manufacturable solutions exist and are being optimized

Manufacturable solutions are known

Interim solutions are known ◆

Manufacturable solutions are NOT known

process of both via and line level simultaneously. In conventional architecture the metal (Al) is deposited first; then the process continues with litho, etch, and dielectric deposition. The interlevel metal–insulator effective dielectric constant requirement in the range of 1.5–2 can be achieved by using porous silica. The following sections describe, step by step, critical issues in the integration of these materials with reference to recent publications. The main issue to address is the power density increase and heat dissipation. This is why conventional interconnects may come to an end and let the field open to alternative solutions such as 3D integration or optical interconnects.

Table 1.2. 2003 ITRS MPU interconnect technology requirements (long term)

year of prodution	2010	2012	2013	2015	2016	2018
Technology node	hp45		hp32		hp22	
DRAM ½ pitch (nm)	45	35	32	25	22	18
MPU/ASIC ½ pitch (nm)	34	42	38	30	27	21
MPU printed gate length (nm)	25	20	18	14	13	10
MPU physical gate length (nm)	18	14	13	10	9	7
Number of metal levels	12	12	12	13	14	14
Number of optional levels – ground planes/capacitors	4	4	4	4	4	4
Total interconnect length (m cm^{-2}) – active wiring only, excluding global levels [1]	1784	2214	2544	3544	4208	5035
FIT s/m length cm^{-2} × 10^{-3} excluding global levels [2]	2.8	2.3	2.0	1.4	1.2	1.0
Jmax (A cm^{-2}) – intermediate wire (at 105°C)	3.0x10^6	3.7x10^6	4.3x10^6	5.1x10^6	5.8x10^6	6.9x10^6
Metal 1 wiring pitch (nm)*	108	84	76	60	54	42
Metal 1 A/R (for Cu)	1.8	1.8	1.9	1.9	2	2
Interconnect RC delay (ps) for 1 mm Metal 1 line	616	963	970	1510	2008	2679
Line length (mm) where τ = RC delay (Metal 1 wire)	25	18	15	11	9	6
Cu thinning at minimum pitch due to crosion (nm), 10% 3 height, 50% areal density, 500 µm square array	10	8	7	6	5	4
Intermediate wiring pitch (nm)	135	110	95	78	65	55
Intermediate wiring dual damascene A/R (Cu wire/via)	1.8/1.6	1.9/1.7	1.9/1.7	1.9/1.7	2.0/1.8	2.0/1.8
Interconnect RC delay (ps) for 1mm intermediate line	380	552	614	908	1203	1552
Line length (mm) where τ = RC delay (intermediate wire)	32	23	19	14	11	8
Cu thinning at minimum intermediate pitch due to erosion (nm), 10% × height, 50% areal density, 500 µm square array	12	10	9	7	7	6
Minimum global wiring pitch (nm)	205	165	140	117	100	83
Ratio range (global wiring pitches/intermediate wiring pitch)	1.5–10	1.5–10	1.5–13	1.5–13	1.5–16	1.5–16
Global wiring dual damascene A/R (Cu wire/via)	2.3/2.1	2.3/2.1	2.4/2.2	2.4/2.2	2.5/2.3	2.5/2.3
Interconnect RC delay (ps) for 1 mm global line at minimum pitch	143	220	248	354	452	618
Line length (mm) where τ = RC delay (global wire at minimum pitch)	52	37	30	23	19	13
Cu thinning at maximum width global wiring due to dishing and erosion (nm), 10% × height, 80% areal density	155	127	148	122	130	130
Cu thinning global wiring due to dishing (nm),100 µm wide feature	14	13	10	9	8	7
Conductor effective resistivity (µΩ cm) Cu intermediate wiring	2.2	2.2	2.2	2.2	2.2	2.2
Barrier/cladding thickness (for Cu intermediate wiring) (nm) [3]	5	4	3.5	3	2.5	2
Interlevel metal insulator – effective dielectric constant (κ)	2.3–2.6	2.3–2.6	2.0–2.4	2.0–2.4	<2.0	<2.0
Interlevel metal insulator (minimum expected) – bulk dielectric constant (κ)	<2.1	<2.1	<1.9	<1.9	<1.7	<1.7

*Refer to Executive summary Figure 4 for definition of Metal 1 pitch

Manufacturable solutions exits,and being optimized
Manufacturable solutions are known
Interim solutions are known
Manufacturable solutions are NOT known

1.2 Interconnect Schemes Analysis

As reported previously [1], simultaneous optimization of RC delay and crosstalk is incompatible and the impact of technological evolutions will involve design solutions to improve crosstalk. For a given generation, RC reduction can be achieved by either relaxing the pitch [2] or changing the material. In this section, we present RC analysis only with material change. The dimensions (metal width and space) for each generation correspond to the minimum requirements of the ITRS. Effective (barrier and metal) resistivities

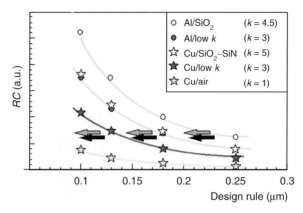

Fig. 1.1. Simulation results on RC evolution with design rules

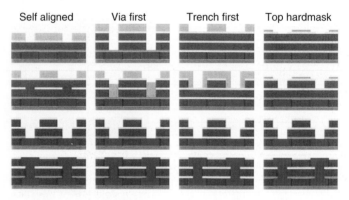

Fig. 1.2. Integration schemes for Dual Damascene interconnects

of $3.5\,\mu\Omega\,cm$ for Al and $2.2\,\mu\Omega\,cm$ for Cu have been chosen. The metal height remains constant in all design rules. The simulation results are presented in Fig. 1.1.

These results show clearly that from 0.25 to 0.1 μm, substitution of Al by Cu and SiO_2 by $k = 3$ and $k = 2$ dielectric materials keeps RC quasiconstant, while an unchanged combination of metal and dielectric drastically increases the parasitics in the following generation.

As no viable etching process is available for Cu, Dual Damascene architecture is adopted. Via and trench filling simultaneously is another advantage. Different strategies for Dual Damascene structure fabrication have been developed and are classified into two categories (self-aligned and not self-aligned structures). Figure 1.2 shows various architectures to achieve Dual Damascene structures with the integration of advanced low k dielectrics:

1. Self-aligned structure
2. Via first at trench level

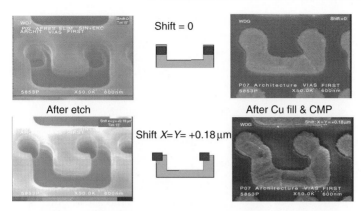

Fig. 1.3. Via first at via level (self-aligned structure) SEM views

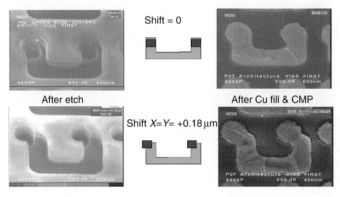

Fig. 1.4. Via first at trench level (not self-aligned structure) SEM views

3. Trench first
4. Top hard mask to avoid any contact between photoresist materials and nitrogen content dielectrics to prevent resist poisoning

In all examples, the absence of shift between lines and vias leads to a well-defined structure. In the case of misalignments, the self-aligned strategy allows reduced via size and as-designed metal space, while the second strategy gives a reduced metal space and as-designed via size as shown in Figs. 1.3 and 1.4.

1.3 Technological Issues

Recent significant progress on 193-nm lithography (immersion) allows the extendibility of DUV techniques to 45 nm. Beam lithography has been used for extremely narrow geometries with available negative photoresists as illustrated in Fig. 1.5.

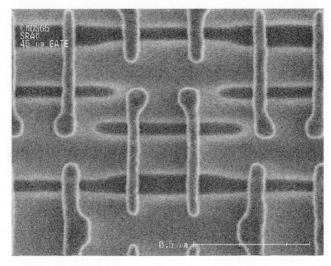

Fig. 1.5. eBeam direct writing with 45-nm target in an SRAM

To verify the results predicted by the simulation model, one has to fabricate a multilevel interconnect structure with new materials. Two approaches, commonly called via first and trench first architectures, have been studied. In both cases similar process challenges exist. Different steps with potential issues are listed:

1. Low k dielectric main issues:
   – Effective dielectric constant due to the use of SiO_2-like materials as a capping layer
   – Adhesion to the under layer, which imposes an adhesion promoter layer
   – Moisture absorption, which increases the dielectric constant
   – Mechanical and thermal stabilities as well as heat dissipation are generally poor due to the porosity of the materials
   – Chemical compatibility with different reactants

Concerning isolation, the material impact on total capacitance and thermal conductivity is summarized in Fig. 1.6. The total capacitance has been measured with intermetal line low k and interlayer SiO_2. Xerogel dielectrics ($k = 2.4$) allow close to 50% reduction of the capacitance but their thermal conductivity is ten times smaller than SiO_2, which may cause reliability issues.

2. Dielectric etching and stripping (after lithography) issues
   – Trench and via profiles
   – Compatibility of strippers with low k materials
   – Cu cleaning (via)

The etching of low k materials can be quite challenging as described by the Bell Labs [3]. Fluorocarbon chemistry with high bias power is commonly

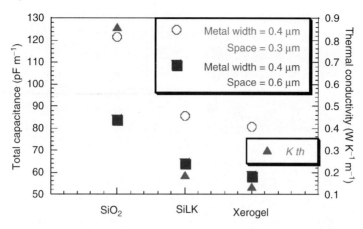

Fig. 1.6. Thermal conductivity and impact of space variation on lateral capacitance for various dielectric materials

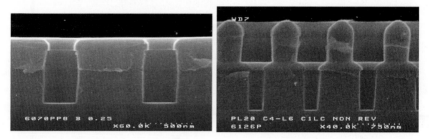

Fig. 1.7. 0.16 µm via etch of low k dielectric showing bowed profile and hard mask undercut

used to disrupt the strong Si–O bonds, whereas in the case of a pure organic dielectric, SiLK material for example, the active etchant in oxygen straight profile is however difficult to achieve (Fig. 1.7). In an HDP tool, the high degree of gas phase dissociation leads to high concentration of atomic oxygen, thereby producing some bowing. Moreover, hard mask undercut after line and via etch of small features is a handicap for void-free metallization (Fig. 1.7). The ultimate solution for dielectric materials is approaching air ($k = 1$). Robust process to perform air gap structures is in progress (Fig. 1.8).

Conventional dilute HF process is efficient for sidewall Cu contamination cleaning. However, Cu loss cannot be avoided. In addition, due to the relative porosity of advanced low k materials, new cleaning chemistries have to be developed for resist stripping at hard mask level [4].

3. Diffusion barrier deposition issues:
   - Conformality
   - Barrier resistivity and performance versus thickness
   - Adhesion to Cu and the underneath dielectric

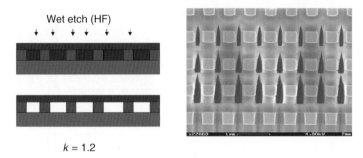

Fig. 1.8. Process for air gap structures

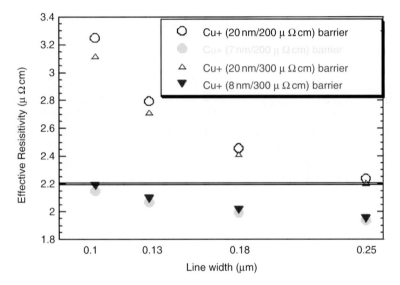

Fig. 1.9. Effective resistivity of metal plus barrier for two different combinations of barrier resistivity and thickness. This plot has been computed using Table 1.1 aspect ratio

New barrier studies suitable for more aggressive features are focused on CVD WxN [5] or TaN [6] and electrolessly deposited Co(P) or Ni(P) [7]. However, the integration capability of these materials for below 0.18-µm geometries has to be demonstrated. For example, the minimum thickness for which barrier performance data of electroless CoW(P) material are available is 50 nm. As shown in Fig. 1.9, effective resistivity of 2.2 µΩ cm can be obtained by either decreasing drastically the thickness of a 200-µΩ cm barrier (down to 7 nm for 0.1 µm metal width) or selecting an 8-nm thick layer of a 30-µΩ cm barrier. For the line level, the thickness impact is much more important than the resistivity value.

4. Cu deposition issues:
   - Conformality (filling capability of new techniques).
   - Adhesion and purity of the films due to CVD precursors or intensive use of additives in electroplating chemistries.
   - Combinations of Cu IMP or CVD seed layer with electroplating techniques have already been reported. The issue of filling small features can be addressed in the case of superconformal or bottom-up Cu filling [8].
5. Cu chemical mechanical polishing (CMP) issues:
   - Process compatibility with low k/barrier/Cu stack in terms of mechanical and chemical stability
   - Copper dishing and oxide erosion

As reported by Sematech [9], polishing xerogel damascene structures is difficult due to the poor mechanical strength of the film, which necessitates the deposition of both conformal oxide liner (at line level and after etch) and capping layer. New approaches combining planar electroplating of copper and electrochemical–mechanical polishing of Cu are being introduced (Fig. 1.10)

Finally, the increase in resistance of Cu as the Cu line shrinks is the major issue that the IC industry will face. This is due to the predominance of electron scattering that will neutralize the bulk favorable resistivity of Cu [10].

1.4 Electrical Results: State of the Art

In this section we give some integration examples, including comparative electrical results in terms of RC reduction with different metal/dielectric combinations.

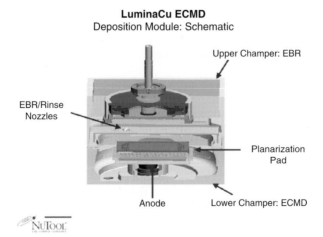

Fig. 1.10. eCMP chamber schematic (courtesy of Nutool, an ASMI company)

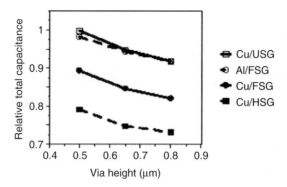

Fig. 1.11. Relative third-metal capacitance as a function of via height (Courtesy of IBM [11])

Our simulation model presented in Sect. 1.2, has been validated by experimental reduction of 40% RC between Cu/SiLK and Cu/SiO$_2$ for 0.18-μm double-level metal interconnect [11].

IBM [12] recently presented a relative third-metal capacitance comparison between various metal/dielectric combinations: Cu/undoped silica glass (USG) ($k = 4.1$), Al/fluorinated silica glass (FSG) ($k = 3.6$ – no barrier layers between either Al line or W vias and the FSG dielectric), Cu/FSG ($k = 3.6$) and Cu/Hitachi silica glass (HSG) ($k = 2.9$). For the minimum pitch of 0.63 μm and constant line resistance, Cu/FSG results in a 10% reduction in the total RC delay relative to Al/FSG case or Cu/HSG compared with Cu/FSG (Fig. 1.11). Using flare material ($k = 2.8$), more than 30% reduction of wiring capacitance was obtained for 0.24/0.24 μm line/space in comparison with TEOS dielectric (Matsushita [13]). Lower permittivity values can be obtained with porous silica. Sematech has already reported Cu/xerogel integration [9] (Figs. 1.12 and 1.13).

The limitations of copper metallization, low k integration as well as examples of alternative materials or combination of materials for barriers and capping layers have been described in a recent article in *Semiconductor International* by Singer [14]. Technology and reliability issues have been reported by Saraswat [15].

Despite new material introduction that is already very challenging, power density increase by design (due to the shrinkage required for the structures and multiple functions to be integrated in a device) is the major issue in advanced circuits ([16] and Fig. 1.14)

1.5 Future Trends

3D integration is motivated by the possibility of:

1. Integration of heterogeneous technologies such as memory and logic together with optical I/O

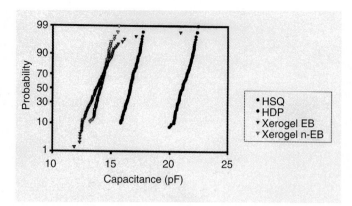

Fig. 1.12. Capacitance comparison among three dielectric materials: xerogels, hydrogen sisesquioxane (HSQ) and high density plasma SiO_2 (HDP) for 0.3 µm line/space comb structures (Courtesy of Sematech [9])

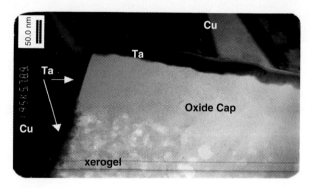

Fig. 1.13. TEM image of Cu filled 0.35 mm trench in xerogel. (Courtesy of Sematech [9])

2. The chip footprint being drastically reduced compared to 2D integration
3. Long horizontal wires being replaced by short vertical wires
4. Interconnect length being reduced and therefore R, L, C that being reduced, thereby contributing to power and delay reduction

Optical interconnects can be the only alternative solution to overcome the issue of speed and power. For a given technology node, optical interconnects are faster than repeated wires beyond a critical length. For a 45-nm node, for example, the critical length is 5 mm, assuming that the total delay in an optical interconnect system is the sum of transmitter delay, receiver delay, and waveguide delay (assumed simplistic of 11.5 ps mm^{-1}). Waveguide delay dominates after 15 mm [17].

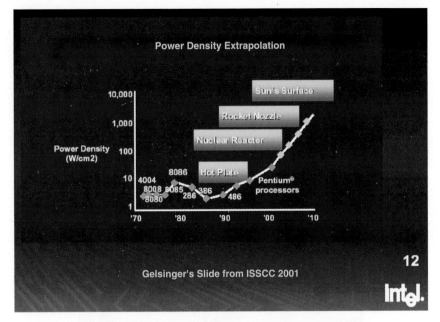

Fig. 1.14. Power extrapolation for different technologies of microprocessors

References

1. G. Lecarval et al., "Advanced Interconnect Scheme Analysis: Real Impact of technological improvements", IEDM 1998, pp. 837–840
2. Takahashi et al., "Interconnect Design Strategy: Structures, Repeaters and Materials toward 0.1 µm ULSIs with a GHz clock operation", IEDM 1998, pp. 833–836
3. H.L. Maynard et al., "Embedded Organic low k structures for sub 0.18 µm CMOS: Integration and Etching issues", IITC 1999, pp. 212–214
4. D. Louis et al., "Post etch cleaning of Dual Damascene System Integrating Cu & SiLK", IITC 1999, pp. 103–105
5. A. Vijayendran and M. Vanek, "Copper Barrier Properties of Ultra Thin PECVD WxN", IITC 1999, pp. 122–124
6. A. Paranjpe et al., "CVD TaN Barrier for Copper Metallization and DRAM Bottom Electrode", IITC 1999, pp. 119–121
7. E.J. O'Sullivan et al., "Electrolessly deposited diffusion barriers for microelectronics", IBM J. Res. Dev., Vol. 42, No. 5, 1998
8. P.C. Andricacos et al., "Damascene copper electroplating for chip interconnections", IBM J. Res. Dev., Vol. 42, No. 5, 1998
9. E. Todd Ryan et al., "Material Property Characterization and Integration Issues for Mesoporous Silica", IITC 1999, pp. 187–199
10. Y. Zhou, I. Bray, and I. McCarthy, "Model Calculations of electron scattering from copper", J. Phys. B: At. Mol. Opt. Phys., Vol. 32, 1999, 1033–1039
11. O. Demolliens et al., "Copper–silk integration in a 0.18 µm double level metal interconnects", IITC 1999, pp. 198–199

12. S. Crowder et al., "A 0.18 μm high-performance logic technology", 1999 VLSI Symposium Technology Digest, pp. 105–106
13. N. Aoi et al., "A novel clustered hard mask technology for Dual Damascene multilevel interconnects with self-aligned via formation using an organic low k dielectric", 1999 VLSI Symposium Technology Digest, T4 B-1
14. P. Singer, "Copper challenges for the 45 nm node", Semicond. Int., 1 May 2004
15. P. Kapur, J. McVittie, and K. Saraswat, "Technology and reliability constrained future copper interconnects", IEEE Trans. Electron. Devices, Vol. 49, No. 4, April 2002, 590–597
16. http://www.intel.com/research/silicon/micron.htm
17. K. Saraswat, EE311/Interconnect course, Stanford University, http://www.stanford.edu/class/ee311/welcome.htm

2

Optical Gain in Silicon and the Quest for a Silicon Injection Laser[*]

L. Pavesi

Summary. In this chapter, a review of the reported approaches to achieve an injection silicon laser is presented. After an initial discussion of the basics on light amplification and gain in semiconductor, we consider the limitations of silicon, in particular its band structure. A short introduction to the first optically pumped silicon laser is given. Then the various approaches to get an injection silicon laser are presented and evaluated: bulk silicon, silicon nanocrystals, and Er-coupled silicon nanocrystals.

2.1 Basics on Light Amplification and Gain

A laser is based on three main components (Fig. 2.1): an active material that is able to generate and amplify light by stimulated emission of photons, an optical cavity that provides the optical feedback to sustain the laser action, and a pumping mechanism that is able to excite the active material such that population inversion can be achieved [1]. In an injection diode laser the pumping mechanism is provided by carrier injection via a p–n junction and the optical feedback is provided by a Fabry–Peròt cavity. The use of electrical injection makes the device particularly interesting for integration with microelectronics.

The light generation property of an active material is quantified by the internal quantum efficiency, η_{int}. η_{int} gives the ratio between the number of photons generated to the number of electron–hole pairs that recombine. This number is given by the ratio between the electron–hole (e–h) radiative recombination probability over the total e–h recombination probability, i.e., by the fraction of all the excited e–h pairs that recombine radiatively. It easy to demonstrate that $\eta_{int} = \tau_{nr}/(\tau_{nr} + \tau_{r})$, where τ_{nr} and τ_{r} are the nonradiative

---

[*] This chapter is an up-to-date version of the article published in *Materials Today* 8 (January 2005) 18

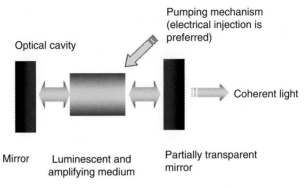

Optical cavity

Pumping mechanism
(electrical injection is
preferred)

Coherent light

Mirror   Luminescent and
amplifying medium

Partially transparent
mirror

Fig. 2.1. Basic components of a laser

and radiative lifetimes, respectively. Thus in order to have a high η_{int} either the radiative lifetime should be short (as in direct banggap semiconductors) or the nonradiative lifetime should be long (as in color center systems).

The property of amplifying light is given by the gain spectrum of the material. For a bulk semiconductor it is related to the joint density of states $\rho(\hbar\omega)$, the Fermi inversion factor $f_{\text{g}}(\hbar\omega)$, and the radiative lifetime:

$$g\left(\hbar\omega\right) d\Phi\left(\hbar\omega\right) = dr_{\text{stim}}\left(\hbar\omega\right) - dr_{\text{abs}}\left(\hbar\omega\right)$$
$$= \frac{\lambda^2}{8\pi\tau_r}\rho\left(\hbar\omega\right)f_{\text{g}}\left(\hbar\omega\right)\Phi\left(\hbar\omega\right)dz, \tag{2.1}$$

where g is the gain coefficient, $d\Phi$ the change in the photon flux, dr_{stim} or dr_{abs} the rate of stimulated emission or absorption at a given photon energy $\hbar\omega$, respectively, $f_{\text{g}}(\hbar\omega, E_{\text{F}}^{\text{e}}, E_{\text{F}}^{\text{h}}, T) = \left[f_{\text{e}}(\hbar\omega, E_{\text{F}}^{\text{e}}, T) - \left(1 - f_{\text{h}}(\hbar\omega, E_{\text{F}}^{\text{h}}, T)\right)\right]$, f_{e} and f_{h} are the thermal occupation functions for electrons and holes, and Φ the photon flux density. E_{F}^{e} and E_{F}^{h} are the quasi-Fermi levels for electrons and holes, respectively. When no external pumping is present, the Fermi inversion factor reduces to the simple Fermi statistics for an empty conduction band and a filled valence band ($f_{\text{g}} < 0$) and the gain coefficient reduces to the absorption coefficient α. When an external pump excites a large density of free carriers, the splitting of the quasi-Fermi levels increases and when $E_{\text{F}}^{\text{e}} - E_{\text{F}}^{\text{h}} > \hbar\omega$ the condition of population inversion is verified and $f_{\text{g}} > 0$. This means that (2.1) is positive and hence the system shows positive net optical gain ($g > 0$). Note that in (2.1) a critical role is played by the radiative lifetime: the shorter the lifetime the stronger the gain.

For an atomic system, the expression of the gain coefficient reduces to

$$g\left(\hbar\omega\right) = \sigma_{\text{em}}\left(\hbar\omega\right)N_2 - \sigma_{\text{abs}}\left(\hbar\omega\right)N_1, \tag{2.2}$$

where σ_{em} is the emission cross-section, σ_{abs} the absorption cross-section, N_2 and N_1 the density of active centers in the excited and ground states, respectively. If $\sigma_{em} = \sigma_{abs}$, the condition to have positive optical gain is that $N_2 > N_1$, i.e., the condition of population inversion. The emission cross-section is related to the radiative lifetime by the Einstein relation

$$\sigma_{em}(\nu) = \frac{\lambda^2}{8\pi\tau_{rad}}\aleph(\nu), \qquad (2.3)$$

where ν is the frequency, $\aleph(\nu)$ the lineshape function associated to the optical transition. Also in (2.3) the radiative lifetime enters. If a piece of active material of length L is used to amplify light, one achieves light amplification whenever the material gain g is positive *and* larger than the propagation losses α_p of the light through the material, i.e., $g > 0$ and $g > \alpha_p$. If the system is forged as a waveguide of length L, and we call I_T and I_0 the intensity of the transmitted and of the incident beams, the amplification factor of the light is then

$$G = I_T/I_0 = \exp\left[(\Gamma g - \alpha_p)L\right] > 1, \qquad (2.4)$$

where Γ is the optical confinement factor of the optical mode in the active region. In a laser, optical feedback is usually provided by a Fabry–Peròt cavity so that the round trip gain (the overall gain experienced by a photon traveling back and forth across the cavity) can be larger than 1. This condition is expressed by the relation $G^2 R_1 R_2 > 1$, where R_1 and R_2 are the back and front mirror reflectivities. For an injection laser it is simple to demonstrate that the injection current density at threshold J_{th} is equal to

$$J_{th} = \frac{\alpha_r + \alpha}{\alpha}\frac{el}{\eta_i\tau_r}\Delta n_T, \qquad (2.5)$$

where α_r are the resonator losses experienced by light traveling in the cavity, which also includes the mirror losses $\alpha_m = (1/2L)\ln(1/R_1 R_2)$, and ℓ the active region thickness and Δn_T the transparency excess free carrier density. Δn_T is related to the material properties, while the other parameters can be optimized by proper cavity design to reduce the threshold current. Note that in (2.4) the internal quantum efficiency is present.

2.2 Limitation of Silicon for Light Amplification

Among the various semiconductor materials that have been used to form lasers, the absence of silicon is striking. Let us review why Si has not been used as a laser material until now [2–4]. Si is an indirect bandgap semiconductor. As a consequence, the probability for a radiative recombination is low, which in turn means that the e–h radiative lifetime is long, of the order of some milliseconds. An e–h pair has to wait on average a few milliseconds to recombine radiatively. During this time both the electron and the hole move around and cover a volume of the order of $10\,\mu m^3$. If they encounter a defect or a trapping center, the carriers might recombine nonradiatively. Typical nonradiative recombination lifetimes in Si are of the order of some nanoseconds. Thus, in electronic-grade silicon the internal quantum efficiency η_{int} is about 10^{-6}. This is the reason why silicon is a poor luminescent material: the efficient nonradiative recombinations that deplete the excited carriers rapidly. Many strategies have been researched over the years to overcome this silicon limitation, some of which are based on the spatial confinement of the carriers, some on the introduction of impurities, some on the use of quantum confinement, and others on the use of Si–Ge alloys or superlattices [2].

In addition, two other phenomena limit the use of Si for optical amplification (see Fig. 2.2). The first is a nonradiative three-particle recombination mechanism where an excited electron (hole) recombines with a hole (electron) by releasing the excess energy to another electron (hole). This is called nonradiative Auger recombination mechanism (Fig. 2.2a). This recombination mechanism is active as soon as more than one carrier is excited. The probability of an Auger recombination is directly proportional to the number of excited carriers and inversely proportional to the bandgap energy [5]. For our discussion this is a very relevant mechanism because the more excited is the semiconductor, the more effective is the Auger recombination. The probability for an Auger recombination in a bulk material is proportional to Δn^3, we can thus write a nonradiative recombination lifetime due to Auger as $\tau_A = 1/C\Delta n^2$, where C is a constant which depends on the doping of the material. For silicon $C \sim 10^{-30}\,\mathrm{cm^6\,s^{-1}}$ [5]. For $\Delta n \sim 10^{19}\,\mathrm{cm^3}$, $\tau_A = 10\,\mathrm{ns}$. The Auger recombination is the dominant recombination mechanism for high-carrier injection rate in Si.

The second phenomenon is related to free carrier absorption (see Fig. 2.2b). Excited carriers might absorb photons and thus deplete the inverted population and, at the same time, increase the optical losses suffered by the signal beam. The free carrier absorption coefficient can be empirically related to the Si free carrier density n_{fc} and to the light wavelength λ as $\alpha_n \sim 10^{-18} n_{fc}\lambda^2$ at 300 K [6]. For $n_{fc} = 10^{19}\,\mathrm{cm^{-3}}$ and $\lambda = 1.55\,\mathrm{\mu m}$, $\alpha_n = 24\,\mathrm{cm^{-1}}$. For heavily

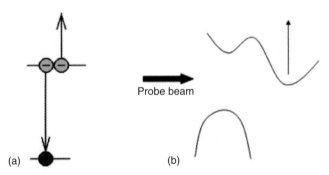

Fig. 2.2. (a) Auger process. (b) Free carrier absorption process. The *wavy lines* represent the band structure of silicon

doped Si this is the main limitation to lasing, while for intrinsic Si this contribution can be exceedingly small, unless n_{fc} is very high as in a laser. In a confined system, such as Si nanocrystal, this recombination mechanism is due to confined carriers and, hence, is called confined carrier absorption.

2.3 Various Approaches to a Silicon Laser

In the early 2000s a series of articles appeared that questioned the common belief that silicon cannot be used to form a laser [7–10]. In October 2004, the first report on a silicon laser appeared [11], while in February 2005 the first continuous wave (CW) Raman laser integrated in silicon was reported [12].

2.3.1 Silicon Raman Laser

It is based on the stimulated Raman scattering (SRS) mechanism, which appears in a silicon waveguide when a very high-power pump laser is coupled in [11]. SRS is a well-known effect in fiber optics, indeed it is used to realize all optical fiber amplifiers. Raman scattering is a nonlinear optical process that can be seen as an inelastic scattering of photons by lattice vibrations (phonons) (Fig. 2.3a). Conservation of the energy requires that if a phonon is excited (Stokes' process)

$$\hbar\omega_{in} = \hbar\omega_{out} + \hbar\omega_{ph}, \tag{2.6}$$

where a photon of energy $\hbar\omega_{in}$ has been scattered to an energy $\hbar\omega_{out}$ by creating a phonon whose energy $\hbar\omega_{ph}$ is almost equal to 15.6 THz in Si. When a high-intensity pump beam is launched in an optical fiber together with a weak signal beam with an energy difference, that is equal to a phonon energy, the weak signal beam stimulates Raman scattering processes that involve the strong pump beam. In this way optical power is transferred from the pump beam to the signal beam, yielding an overall amplification of the signal beam. It appears that this SRS is much more efficient in silicon waveguides than in silica optical fiber: the Raman gain cross-section in silicon is five orders of magnitude larger [12]. In fact Raman amplifiers have been demonstrated in silicon waveguides with waveguide length of a few cm while in optical fiber amplifiers some km long fibers are needed to have the same gain.

Unfortunately, when an intense light beam travels through silicon it may happen that two photons add up their energies so that silicon is no longer transparent, i.e., two-photon absorption occurs, and electrons and holes are created (panel b in the figure). A cloud of free carriers forms in the waveguide channel and the optical losses of the waveguide increase dramatically due to free carrier absorption, which weakens both the pump and signal beam: in this situation no SRS is possible and the gain coefficient becomes negative (Fig. 2.3b).

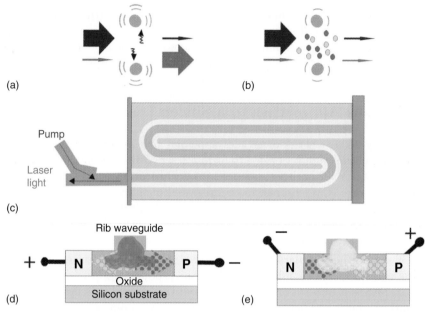

Fig. 2.3. (**a**) SRS effect: blue pump light, red signal light, black wavy lines phonons, gray circles Si atoms; (**b**) free carriers induced absorption of pump and signal light: yellow and light blue disks are free carriers; (**c**) a top view of the silicon Raman structure: the pump light is channeled into the laser by an optical fiber (brown lines), the laser light exits the laser via the same optical fiber, the active waveguide is S-shaped to increase the effective length and reduce the device footprint, reflective coating (orange) is used to increase the facet reflectivity; (**d**) cross-section of the active part of the device: light gray refers to the n-hyphen and p-type doped regions of the device, white is the oxide cladding layer, dark gray is the active region formed by a rib waveguide in silicon; the red region corresponds to the optical laser mode. When the device is reverse biased electrons (yellow disks) and holes (light blue) are swept away from the optical mode region. (**e**) The same device forward biased: free carriers are injected into the active region and the optical mode (pink) is strongly attenuated

To avoid free carrier absorption, a pulsed pump beam with pulses shorter than the typical lifetime of the free carriers was used. Then by using an 8-m long single mode optical fiber, the exit of a 3-cm long waveguide was looped back into the entry: the signal photons recycled into the waveguide. Thus the signal got multiple amplifications and, eventually, the laser threshold was overcome and laser light exited from the waveguide. The first pulsed silicon laser with a fiber loop cavity was fabricated [11].

Shortly after, in January 2005 Intel's team demonstrated the first all-silicon-pulsed Raman laser [12]. It was based on an S-shaped silicon rib waveguide resting on a silicon on insulator (SOI) wafer (Fig. 2.3c). The S-shape was used to have a 4.8-cm long waveguide while maintaining a small footprint as real estate on a silicon chip is extremely expensive. The exact emission wavelength is determined by the pump wavelength and the cavity details. By changing the cavity length it is possible to get various wavelengths out of a single pump wavelength within a very narrow interval.

In February 2005, the same team reported a continuous wave all-silicon Raman laser [12]: the free carriers generated by two-photon absorption in the waveguide channel are driven away by using an electric field generated by a reverse biased p–i–n diode (Fig. 2.3d). The two electrodes collect the free carrier generated by the two-photon absorption allowing continuous wave and stable laser emission. Interestingly, the microelectronic properties of silicon (a current can flow through) are applied to drive its nonlinear optical properties.

The same idea was used to chop a pulsed Raman laser on and off by injection of free carriers in the silicon waveguide with a forward bias p–i–n diode formed across the silicon waveguide [11]: free carriers absorb light and switch off the laser (Fig. 2.3e). Despite the interest of direct modulation to code-in data stream electrically, the modulation frequency is still too low.

2.3.2 Bulk Silicon Light-Emitting Diodes

The common belief that bulk silicon cannot be a light-emitting material has been severely questioned in a series of recent works. The most interesting one was published in 2001 [9]. An Australian group noticed that world-record solar cells are characterized by extremely long carrier recombination lifetimes of the order of some milliseconds. That is, the recombination lifetime is of the order of the radiative lifetime, hence η_{int} is of the order of 1. Then, if the solar cell is biased in the forward regime instead of the usual reverse regime, the solar cell could behave as a very efficient light-emitting diode (LED).

Figure 2.4 shows a schematic of the device and a room temperature emission spectrum. To increase the light extraction efficiency, the LED surface was texturized so that most of the internally generated light impinged on the external surface of the cell with an incident angle lower than the critical angle for total internal refraction. Thus the light extraction efficiency was increased from a few percentages, typical of a flat surface to almost 100% for the texturized LED. Finally, to reduce free carrier absorption to a minimum, the electrodes, i.e., the heavily doped regions, were confined in very thin and small lines. By using these three practices, a plug-in efficiency (ratio of the optical power emitted from the LED to the electrical driving power) larger than 1% at 200 K was achieved. Most interesting is that the turn-on voltage of the device was the same as the forward bias of the solar cell, i.e., less than 1 V.

The same research group also published a theoretical paper [14] that questioned one common belief that indirect bandgap materials could not show optical gain because of parasitic absorption processes due to free carrier [15].

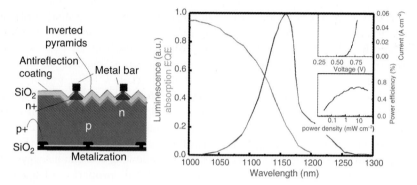

Fig. 2.4. Summary of the results of the Australian group on a bulk silicon LED. *Left*: Sketch of the LED geometry. *Right*: Luminescence spectrum (red), absorption spectrum (green), power efficiency versus injected electrical power density (blue), and $I-V$ characteristics (*inset*) at room temperature. Data have been redrawn from [9, 13]

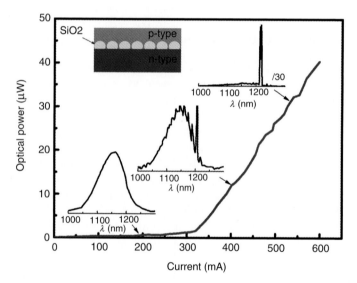

Fig. 2.5. Optical power versus injected current for an LED containing SiO_2 nanoclusters in the junction region (*inset*). Also shown are a few electroluminescence spectra for different injection rate (*arrows*). The data have been redrawn from [16]

Indeed they demonstrated that optical gain is theoretically possible and pointed out that the most suitable energy region is the sub-bandgap region where processes involving phonons could help in achieving gain.

These theoretical arguments have been partially confirmed in a recent study where stimulated emission has been observed (see Fig. 2.5) [16]. As the limit to efficient light generation in Si is the short nonradiative lifetime, the idea was to avoid carrier diffusion and to spatially localize free carriers in a small device region where nonradiative recombination centers can be easily saturated. To localize carriers, in a study ion implantation has been used to induce dislocation loops at the junction of a p–n diode [10]. The dislocation loops cause local strain fields, which in turn increase locally the energy gap causing a potential barrier for carrier diffusion. LEDs based on this idea were realized.

Other researchers realized carrier localization by spin-on doping of small silica nanoparticles at the junction of a p–n diode (Fig. 2.5) [16]. The current–voltage $I-V$ characteristic of the diode shows rectifying behavior with a clear threshold in the light–current $L-I$ characteristic. A change from a broad emission spectrum characteristic of band-to-band emission below threshold to sharp peaks due to stimulated emission above threshold is also observed. Stimulated emission is observed for a two-phonon indirect transition as it was theoretically predicted. Furthermore, when the injection current significantly exceeds the threshold, a single peak dominates. All these results are very encouraging since the proposed systems have excellent electrical qualities as they are p–n junctions. The main problem with the bulk Si approach is related to the presence of a large enough gain to overcome possible free carrier losses, which to date is still unclear. However, in light of the present state of the art, a laser made with bulk silicon seems within easy reach.

2.3.3 Optical Gain in Si Nanocrystals

Another interesting approach to form light emitters and amplifiers in Si is to use small Si nanoclusters (Si-nc) dispersed in a dielectric matrix, most frequently SiO_2 [2]. With this approach one maximizes carrier confinement, improves the radiative probability by quantum confinement, shifts the emission wavelength to the visible and controls it by Si-nc dimension, decreases the confined carrier absorption due to the decreased emission wavelength, and increases the light extraction efficiency by reducing the dielectric mismatch between the source materials and the air. Various techniques are used to form Si-nc whose size can be tailored in a few nm range (Fig. 2.6).

Starting with a silicon-rich oxide, which can be formed by deposition, sputtering, ion implantation, cluster evaporation, etc., a partial phase separation is induced by thermal annealing. The duration of the thermal treatment, the annealing temperature, the starting excess Si content all determine the final size of the cluster, the dispersion in size, which can be significant, and the Si-nc crystalline nature. The size dispersion is usually claimed as the source of the broad emission lineshape that at room temperature is typical of the Si-nc emission spectra. However, both size-selected deposition [17] and single Si-nc luminescence experiments [18] demonstrate that most of the luminescence broadening is intrinsic in nature. The active role of the interface region in determining the optical properties of Si-nc has been highlighted

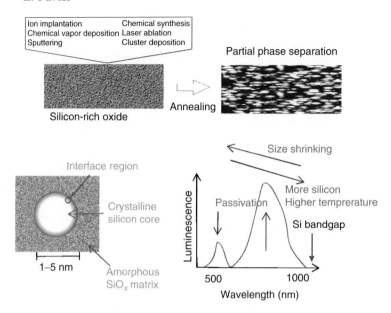

Fig. 2.6. Si nanocrystals formation, structure, and luminescence spectrum

in a joint theoretical and experimental article [19]. The origin of the luminescence in Si-nc is still unclear; many authors believe that it comes from confined exciton recombination in the Si-nc [20] while others support a defect-assisted recombination mechanism where luminescence is due to recombination of carriers trapped at radiative recombination centers, which form at the interface between Si-nc and the dielectric [21] or even in the dielectric [22]. One candidate for these centers is the silanone bond, which is formed by a double Si–O bond [23]. The most probable nature of the luminescence in Si-nc is a mechanism that involves both recombination paths: excitons at about 800 nm and trapped carriers on radiative interface state at about 700 nm. Indeed passivation experiments show that the intensity and lineshape of the emission can be influenced by exposition to hydrogen gas or by further oxidation [24].

A number of papers reported observation of optical gain in these systems [7,25–32]. Gain was observed in many different experiments in Si-nc formed by many different techniques. Figure 2.7 reports a summary of the most relevant data taken on Si-nc formed by plasma enhanced chemical vapor deposition (PECVD) [25–27]. Two techniques are reported here: the variable stripe length method (VSL), which is sketched in the inset of Fig. 2.7 and is based on the one-dimensional amplifier model [26]; and the pump–probe technique, which is based on the probe amplification in the presence of a high-energy and high-intensity pump beam [27]. In the VSL method by varying the pumped region extent (whose length is z) one measures the amplified spontaneous emission

(I_{ASE}) signal coming out from an edge of a waveguide whose core is rich in Si-nc:

$$I_{\text{ASE}}(z) = \frac{J_{\text{sp}}(\Omega)}{g_{\text{mod}}}\left(e^{g_{\text{mod}}z} - 1\right),$$

where $J_{\text{sp}}(\Omega)$ is the spontaneous emission intensity emitted within the solid angle Ω and g_{mod} is the *net modal gain* of the material, defined as $g_{\text{mod}} = \Gamma g_{\text{m}} - \alpha$.

Data reported in Fig. 2.7 show that the ASE intensity increases sublinearly with the pumping length when the pumping power is lower than a threshold. For pumping power higher than threshold, the ASE signal increases more than exponentially. This is a consequence of the pump-induced switching from absorption ($g_{\text{mod}} < 0$) to gain ($g_{\text{mod}} > 0$).

In addition if time-resolved measurements are performed (Fig. 2.7, right, top panel) [27], the ASE decay lineshape shows two time regimes: a fast decay within the first ns and a slow time decay with typical time constant of few μs. It is well known that Si-nc have time decay constant of some μs, so the appearance of a ns-time decay is at first surprising. What is important is the fact that

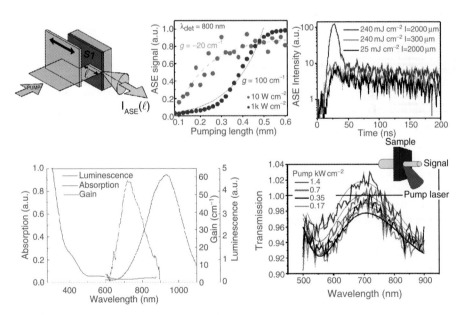

Fig. 2.7. Summary of various experimental proofs of gain in Si-nc. *Top left* panel, geometry used to measure the ASE; *top center panel*, ASE versus the pumping length for two pumping powers; *top right*, ASE time decay for the various pumping conditions indicated in the *inset* (*l* is the pumping length); *bottom left*, luminescence, absorption, and gain spectra at room temperature for a Si-nc-rich waveguide; *bottom right*, transmission spectra for various pumping powers (the *inset* shows the experimental geometry used). Data have been redrawn from [25–27]

the fast decay appears only if both the pumping power and excitation volume are large. If one decreases the excitation volume at high power or the pumping power at large excitation volume the fast decay disappears. This can be understood if the fast decay is due to stimulated emissions. In fact, at high pumping rates three competitive paths open: stimulated emission, Auger recombination, and confined carrier absorption. All these could be the cause of the fast decay. In particular, the Auger lifetime and the confined carrier absorption lifetime can be modeled in a Si-nc by $\tau_A = (1/C_A)N_{ex}$ and $\tau_{CC} = 1/2C_{CC}N_{ex}$, where C_A and C_{CC} are coefficients and N_{ex} is the density of excited recombination centers. N_{ex} is directly proportional to the pumping power and not to the pumping volume. Thus when decreasing the pumping length the ASE lineshape should be unchanged. On the other hand, by a simple rate equation modeling [33], the stimulated emission lifetime turns out to be $\tau_{se} = \frac{4}{3}\pi R_{NS}^3(1/\xi\sigma_g c n_{ph})$, where R_{NS} is the average radius of the Si-nc, ξ their packing density, σ_g the gain cross-section, and n_{ph} the photon flux density. Note that τ_{se} depends on the material properties (R_{NS},ξ,σ_g) and also on the photon flux density n_{ph} that exists in the waveguide. n_{ph} depends, in turn, on the waveguide losses, the Si-nc quantum efficiency and the pumping rates. In addition, if the sample shows gain, by increasing the excitation volume, n_{ph} exponentially increases, i.e., τ_{se} decreases. τ_{se} shortens when either the pumping length or the pumping power increases as both increase n_{ph}. It is important to note that the calculations show that the Auger lifetime in Si-nc is in the interval 0.1–10 ns [34], which means that Auger is a strong competitive process that should always be considered. In some Si-nc systems, due to either material problems or poor waveguide properties or both, Auger and confined carrier absorption might prevail and no optical gain could be observed.

Figure 2.7 (bottom left) shows a summary of the wavelength dependence of the luminescence, absorption, and gain spectra in a sample with 4-nm Si-nc [25]. It is seen that the gain spectrum is on the high energy side of the emission band and that absorption is negligible in the region of gain and luminescence. These facts suggest a four-level model to explain the gain where the levels can be associated to both different Si-nc populations and to a radiative state associated to a Si=O double bond for which optical excitation causes a large lattice relaxation of the Si=O bond [35] as in the silanone molecule (Fig. 2.8).

Pump–probe measurements were attempted with contradictory results worldwide [27, 36]. Our group was able to show probe amplification under pumping conditions (see Fig. 2.7 bottom right panel) [27], while another group reported pump-induced absorption probably associated with confined carrier absorption [36]. Literature results show that the confined carrier absorption cross-section σ_{fc} in Si-nc is at least one order of magnitude reduced with respect to bulk Si [37]: $\sigma_{fc} \approx 10^{-18}\,\mathrm{cm}^2$ at 1.55 µm in P-doped Si-nc. This cross-section should be further reduced to 700 nm due to the λ^2 dependence of the confined carrier absorption. Transmission measurements of a probe beam through a Si-nc slab deposited on a quartz substrate show the typical interference fringes due to multiple reflection at the slab interfaces (Fig. 2.7).

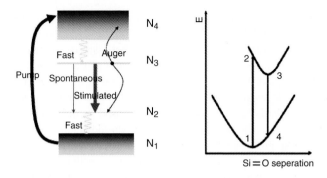

Fig. 2.8. The four-level model introduced to explain gain in Si-nc

When the pump power is raised the transmission is increased and, at the maximum power used, net probe amplification with respect to the input probe intensity in air is observed in a narrow wavelength interval. Note that the probe amplification spectrum overlaps the fast luminescence spectrum measured by time-resolved technique. Based on these results, design of optical cavity for a Si-nc laser has been published [38]. In addition, very favorable results have been published with respect to Si-nc-based LED where turn-on voltage as low as few volts can be demonstrated by using thin Si-nc active layers [39]. Electroluminescence in these LEDs was due to impact excitation of electron–hole pairs in the Si-nc. Another recent work [40] reports on a field effect transistor (FET) structure where the gate dielectric is rich with Si-nc. In this way by changing the sign of the gate bias, separate injection of electrons and holes in the Si-nc is achieved. Luminescence is observed only when both electrons and holes are injected into the Si-nc. By using this pulsing bias technique, bipolar injection is achieved, which should lead to high efficiency in the emission of the LED. In addition, extremely high external quantum efficiency in Si-nc formed in Si_3N_4 or in oxynitride has been reported [41]. Channel optical waveguide with a core layer rich of Si-nc shows optical losses of only a few dB/cm mainly due to direct Si-nc absorption and Mie scattering due to the index mismatch among the Si-nc and the dielectric [42]. All these different experiments have still to be merged in a laser cavity structure to demonstrate a Si-nc-based laser.

2.3.4 Light Amplification in Er-Coupled Si-nc

The radiative transitions in the internal 4f shell of erbium ions (Er^{3+}) are exploited in the erbium-doped fiber amplifier (EDFA) [43]: an all-optical amplifier, which has revolutionized the optical communication technology. During the 1990s, several experimental efforts have been made in order to develop an efficient and reliable light source by using Er^{3+} in Si [2]. The idea was to excite the Er^{3+}, which emits 1.535 μm photons, by an energy transfer from the electrically injected e–h pairs in a p–n Si diode. The most successful results have been the demonstration of room temperature emission with an external quantum efficiency of 0.1% in a MHz modulated Er^{3+}-doped Si LED [44]. The main problem associated to Er^{3+} in Si is the back transfer of energy from the Er^{3+} ions to the Si host, which causes a lowering of the emission efficiency of the diode [45]. This is due to a resonant level that appears in the Si bandgap due to the Er^{3+} doping and which couples with the Er^{3+} levels. In order to reduce this back-transfer process, it was proposed to enlarge the bandgap of the Er^{3+} host so that the resonance between the defect level and the internal Er^{3+} levels is lost [46]. Si-nc in a SiO_2 dielectric were thus proposed as the host [47]. Indeed it turns out that Si-nc are very efficient sensitizers of the Er^{3+} luminescence with typical transfer efficiency as high as 70% and with a typical transfer time of $1\,\mu s$ [48]. In addition, the Er^{3+} are dispersed in SiO_2, where they found their most favorable chemical environment. Quite interestingly the transfer efficiency gets maximized when the Si-nc are not completely crystallized but still in the form of Si-nc [49]. Some reports claim even that the Er^{3+} can be excited through defects in the matrix [50]. Still under debate is the number of Er ions that can be excited by a single Si-nc: a few or many ions.

Figure 2.9 summarizes the various mechanisms and defines the related cross-sections for this system. Excitation of Er^{3+} occurs via an energy transfer from photoexcited e–h pairs that are excited in the Si-nc: the overall efficiency of light generation at $1.535\,\mu m$ through direct absorption in the Si-nc is described by an effective Er^{3+} excitation cross-section σ_{exc}. On the other hand, the direct absorption of the Er^{3+} ions, without the mediation of the Si-nc, and the emission from the Er ions are described by an absorption σ_{abs} and an emission σ_{em} cross-sections, respectively. The typical radiative lifetime of Er^{3+} is of $7\,ms$, which is similar to the one of Er^{3+} in pure SiO_2 [51]. Figure 2.10 (left) reports the luminescence and the absorption spectra measured in an Er^{3+}-coupled Si-nc ridge waveguide at room temperature [52].

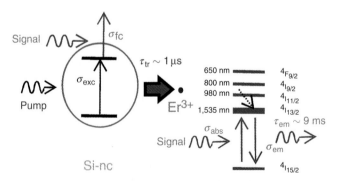

Fig. 2.9. Diagram of the excitation process of (Er^{3+}) ions via a Si-nc, with the main related cross-sections. On the *left*, the main internal energy levels of the (Er^{3+}) are shown.

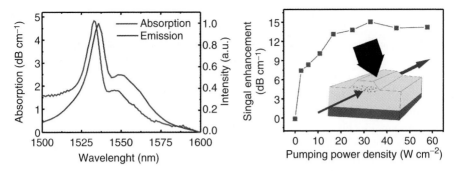

Fig. 2.10. (*left*) Absorption and luminescence spectra of an Er^{3+}-coupled Si-nc waveguide. Data after [52]. (*right*) Signal enhancement at 1.535 μm in an Er^{3+}-coupled Si-nc waveguide versus the pumping power density by using top pumping as shown in the *inset*. Data after [56]

Table 2.1. Summary of the various cross-sections related to Er^{3+} in various materials

	Er in SiO_2 (cm^2)	Er in Si (cm^2)	Er in Si-nc (cm^2)	reference for Er in Si-nc
effective excitation cross-section of luminescence at a pumping energy of 488 nm	$1–8 \times 10^{-21}$	3×10^{-15}	$1.1–0.7 \times 10^{-16}$	[53,54]
effective excitation cross-section of electroluminescence		4×10^{-14}	1×10^{-14} by impact ionization	[55]
emission cross-section at 1.535 μm	6×10^{-21}		2×10^{-19}	[56]
absorption cross-section at 1.535 μm	4×10^{-21}	2×10^{-20}	5×10^{-21}	[52]

The best reported results are shown and taken in the references listed in the last column

Table 2.1 summarizes the best results for the various cross-sections reported in the literature. It is important to note the five order of magnitude increase in σ_{exc} and the fact that this value is also conserved when electrical injection is used to excite the Si-nc [55]. In addition, it is striking that σ_{em} gets enhanced by one order of magnitude with respect to the value for Er^{3+} in pure SiO_2 [57]. A cautious note should be drawn here as measurements for σ_{abs} do not confirm this enhancement [52]. If one places the Er^{3+} ions in a Si-nc ridge waveguide (see inset of Fig. 2.10, right), one can perform experiments on signal amplification at 1.535 μm with the aim to demonstrate an Er-doped waveguide amplifier (EDWA). The main advantage of an EDWA with respect to an EDFA is the reduced size, the decreased pump power to achieve the same gain, and the wide spectrum range to optically pump the system. A few groups have performed such an experiment [52, 56, 57]. The most successful result was reported in [56], see Fig. 2.10, right. In this work, a very low Si-nc concentration has been used and an internal gain of $7\,\mathrm{dB\,cm^{-1}}$ has been deduced. A successful experiment of pumping the EDWA with LED was also reported [58]. In other experiments, with a large Si-nc concentration, no or weak signal enhancement was observed [52, 57]. The reason is attributed to the presence of a strong confined carrier absorption, which introduces a loss mechanism at the signal wavelength and prevents the sensitizing action of the Si-nc. Indeed, the energy transfer is in competition with confined carrier absorption at the signal wavelength (see Fig. 2.9). A confined carrier cross-section of $10^{-18}\,\mathrm{cm^2}$ is usually assumed [48]. Propagation losses, saturation of Er^{3+} excitation, up-conversion of the pumped light, and confined carrier absorption make difficult the proper design of EDWA where optical amplification can be observed. If the results of [56] are confirmed, it seems that the right direction is to use a small Si-nc concentration, short annealing times, and a factor 100 more Er^{3+} than Si-nc. Having got internal gain, electrically injected LED [55, 59], and optical cavities [60], a laser that uses the Er^{3+}-coupled Si-nc system as active material seems feasible. In this respect, it is worth noticing that toroidal microcavities formed in silica doped with Er^{3+} have demonstrated optically pumped lasing at room temperature [61].

2.4 Conclusions

After 4 years of the first observation of optical gain in Si-nc, the first silicon laser has been demonstrated, though not using Si-nc, but Raman effects. This shows that the perspectives to achieve an injection laser in Si are nowadays more solid than ever. We are quite optimistic that a laser will be realized in the near future by using one of the various approaches presented here.

Acknowledgments

It is a pleasure to thank the hard work of my coworkers and students. The support of EC through the SEMINANO and SINERGIA projects is also acknowledged.

References

1. O. Svelto and D.C. Hanna, *Principles of Lasers* (Plenum, New York, 1998)
2. S. Ossicini, L. Pavesi, and F. Priolo, *Light Emitting Silicon for Microphotonics*, Springer Tracts in Modern Physics, Vol. 194 (Springer Berlin Heidelberg New York, 2003)
3. *Towards the First Silicon Laser*, edited by L. Pavesi, S. Gaponenko, and L. Dal Negro, NATO Science Series (Kluwer, Dordrecht 2003)
4. *Silicon Photonics*, edited by L. Pavesi and D. Lockwood, Topics in Applied Physics, Vol. 94 (Springer Berlin Heidelberg New York, 2004)

5. P. Jonsson, H. Bleichner, M. Isberg, and E. Nordlander, J. Appl. Phys. **81**, 2256 (1997)
6. D.K. Schroder, R.N. Thomos, and J.C. Swartz, IEEE Trans. Electron. Dev. **ED-25**, 254 (1978)
7. L. Pavesi, L. Dal Negro, C. Mazzoleni, G. Franzò, and F. Priolo, Nature **408**, 440 (2000)
8. G. Dehlinger, L. Diehl, U. Gennser, H. Sigg, J. Faist, K. Ensslin, and D. Grützmacher, Science **290**, 2277 (2000)
9. M.A. Green, J. Zhao, A. Wang, P.J. Reece, and M. Gal, Nature **412**, 805 (2001)
10. W.L. Ng, M.A. Lourenço, R.M. Gwilliam, S. Ledain, G. Shao, and K.P. Homewood, Nature **410**, 192 (2001)
11. O. Boyraz and B. Jalali, Optics Express **12**, 5269 (2004); Optics Express **11**, 1731, 59 (2003); Optics Express **13**, 796 (2005)
12. A. Liu, H. Rong, M. Paniccia, O. Cohen, and D. Hak, Optics Express **12**, 4261 (2004); Nature **433**, 292 (2005); Nature **433**, 625 (2005)
13. J. Zhao, M.A. Green, and A. Wang, J. Appl. Phys. **92**, 2977 (2002)
14. T. Trupke, M.A. Green, and P. Wurfel, J. Appl. Phys. **93**, 9058 (2003)
15. W.P. Dumke, Phys. Rev. **127**, 1559 (1962)
16. M.J. Chen, J.L. Yen, J.Y. Li, J.F. Chang, S.C. Tsai, and C.S. Tsai, Appl. Phys. Lett. **84**, 2163 (2004)
17. M. Zacharias, J. Heitmann, R. Scholz, U. Kahler, M. Schmidt, and J. Bläsing, Appl. Phys. Lett. **80**, 661 (2002)
18. J. Valenta, R. Juhasz, and J. Linnros, Appl. Phys. Lett. **80**, 1070 (2002)
19. N. Daldosso, M. Luppi, S. Ossicini, E. Degoli, R. Magri, G. Dalba, P. Fornasini, R. Grisenti, F. Rocca, L. Pavesi, S. Boninelli, F. Priolo, C. Bongiorno, and F. Iacona, Phys. Rev. B **68**, 085327 (2003)
20. J. Heitmann, F. Muller, L. Yi, M. Zacharias, D. Kovalev, and F. Eichhorn, Phys. Rev. B **69**, 195309 (2004)
21. L. Khriachtchev, M. Räsänen, S. Novikov, O. Kilpelä, and J. Sinkkonen, J. Appl. Phys. **86**, 5601 (1999)
22. L. Khriachtchev, M. Räsänen, S. Novikov, and L. Pavesi, Appl. Phys. Lett. **85**, 1511 (2004)
23. Y.J. Chabal, K. Raghavachari, X. Zhang, and E. Garfunkel, Phys. Rev. B **66**, 161315 (2002)
24. J.S. Biteen, N.S. Lewis, and H.A. Atwater, and A. Polman, Appl. Phys. Lett. **84**, 5389 (2004)
25. L. Dal Negro, M. Cazzanelli, N. Daldosso, Z. Gaburro, L. Pavesi, F. Priolo, D. Pacifici, G. Franzò, and F. Icona, Physica E **16**, 297 (2003)
26. L. Dal Negro, M. Cazzanelli, L. Pavesi, S. Ossicini, D. Pacifici, G. Franzò, F. Priolo, and F. Iacona, Appl. Phys. Lett. **82**, 4636 (2003)
27. L. Dal Negro, M. Cazzanelli, B. Danese, L. Pavesi, F. Iacona, G. Franzò, and F. Priolo, J. Appl. Phys. **96**, 5467 (2004)
28. L. Khriachtchev, M. Rasanen, S. Novikov, and J. Sinkkonen, Appl. Phys. Lett. **79**, 1249 (2001)
29. J. Ruan, P.M. Fauchet, L. Dal Negro, M. Cazzanelli, and L. Pavesi, Appl. Phys. Lett. **83**, 5479 (2003)
30. K. Luterova, K. Dohnalova, V. Svrcek, I. Pelant, J.-P. Likforman, O. Cregut, P. Gilliot, and B. Honerlage, Appl. Phys. Lett. **84**, 3280 (2004)
31. M.H. Nayfeh, S. Rao, and N. Barry, Appl. Phys. Lett. **80**, 121 (2002)

32. M. Cazzanelli, D. Kovalev, L.D. Negro, Z. Gaburro, and L. Pavesi, Phys. Rev. Lett. **93**, 207402 (2004)
33. V.I. Klimov, A.A. Mikhailovsky, S. Xu, A. Malko, J.A. Hollingsworth et al., Science **290**, 314 (2000)
34. C. Delerue, M. Lannoo, and G. Allan, Phys. Rev. Lett. **75**, 2228 (1995)
35. F. Zhou and J.D. Head, J. Phys. Chem. B **104**, 981 (2000); A.B. Filonov, S. Ossicini, F. Bassani, F. Arnaud D'Avitaya, Phys. Rev. B **65**, 195317 (2002)
36. R.G. Elliman, M.J. Lederer, N. Smith, and B. Luther-Davies, Nucl. Instrum. Methods B **206**, 427 (2003)
37. A. Mimura, M. Fujii, S. Hayashi, D. Kovalev, and F. Koch, Phys. Rev. B **62**, 12625 (2000)
38. S.L. Jaiswal et al., Appl. Phys. A **77**, 57 (2003)
39. G. Franzò, A. Irrera, E.C. Moreira, M. Miritello, F. Iacona, D. Sanfilippo, G. Di Stefano, P.G. Fallica, and F. Priolo, Appl. Phys. A **74**, 1 (2002)
40. R.J. Walters, R.I. Bourianof, and H. Atwater, Nat. Mater. **4**, 143 (2005)
41. K.S. Cho et al., Appl. Phys. Lett. **86**, 071909 (2005)
42. P. Pellegrino, B. Garrido, C. Garcia, J. Arbiol, J.R. Morante, M. Melchiorri, N. Daldosso, L. Pavesi, E. Schedi, and G. Sarrabayrouse, J. Appl. Phys. **97**, 074312 (2005)
43. E. Desurvire, *Erbium-Doped Fiber Amplifiers: Principles and Applications* (Wiley, New York, 1994)
44. G. Franzò, S. Coffa, F. Priolo, and C. Spinella, J. Appl. Phys. **81**, 2784 (1997)
45. F. Priolo, G. Franzò, S. Coffa, and A. Carnera, Phys. Rev. B **57**, 4443 (1998)
46. A.J. Kenyon, P.F. Trwoga, M. Federighi, and C.W. Pitt, J. Phys. Condens. Matter **6**, L319 (1994)
47. M. Fujii, M. Yoshida, Y. Kanzawa, S. Hayashi, and K. Yamamoto, Appl. Phys. Lett. **71**, 1198 (1997)
48. D. Pacifici, G. Franzò, F. Priolo, F. Iacona, and L. Dal Negro, Phys. Rev. B **67**, 245301 (2003)
49. G. Franzò et al., Appl. Phys. Lett. **82**, 3871 (2003)
50. D. Kuritsyn, A. Kozanecki, H. Przybylin'ska, and W. Jantsch, Appl. Phys. Lett. **83**, 4160 (2003)
51. M. Wojdak, M. Klik, M. Forcales, O. B. Gusev, T. Gregorkiewicz, D. Pacifici, G. Franzò, F. Priolo, and F. Iacona, Phys. Rev. B, **69**, 233315-1 (2004)
52. N. Daldosso et al., submitted to Appl. Phys. Lett.
53. F. Priolo, G. Franzò, D. Pacifici, V. Vinciguerra, F. Iacona, and A. Irrera, J. Appl. Phys. **89**, 264 (2001)
54. A.J. Kenyon, C.E. Chryssou, C.W. Pitt, T. Shimizu-Iwayama, D.E. Hole, N. Sharma, and C.J. Humphreys, J. Appl. Phys. **91**, 367 (2002)
55. F. Iacona, D. Pacifici, A. Irrera, M. Miritello, G. Franzó, F. Priolo, D. Sanfilippo, G. Di Stefano, and P.G. Fallica, Appl. Phys. Lett. **81**, 3242 (2002)
56. H.-S. Han, S.-Y. Seo, J.H. Shin, and N. Park, Appl. Phys. Lett. **81**, 3720 (2002)
57. P.G. Kik, and A. Polman, J. Appl. Phys. **91**, 534 (2002)
58. J. Lee, J. Shin, and N. Park, *"Optical gain in Si-nanocrystal sensitized, Er-doped silica waveguide using top-pumping 470nm LED"*, Proceedings of Optical Fiber Communication Conference, Los Angeles, CA, 2004
59. M.E. Castagna, S. Coffa, M. Monaco, A. Muscara, L. Caristia, S. Lorenti, and A. Messina, Mater. Sci. Eng. B **105**, 83 (2003)
60. F. Iacona, G. Franzò, E.C. Moreira, and F. Priolo, J. Appl. Phys. **89**, 8354 (2001)
61. A. Polman, B. Min, J. Kalkman. T.J. Kippenberg, and K.J. Vahala, Appl. Phys. Lett. **84**, 1037 (2004)

3

Silicon Raman Laser, Amplifier, and Wavelength Converter

O. Boyraz

Summary. Exploiting the strong optical confinement and the large Raman gain in the material, the first optical amplifiers and lasers in silicon have been demonstrated. Wavelength conversion, between the technologically important wavelength bands of 1,300 and 1,500 nm, has been demonstrated through Raman-enabled four-wave mixing. Carrier lifetime is the single most important parameter affecting the performance of silicon Raman devices. The desired reduction in lifetime is attained by reducing the lateral dimensions of the optical waveguide and actively removing the carriers with a reverse-biased diode. An integrated diode also offers the ability to electrically modulate the optical gain. The rigid Raman spectrum of silicon can be engineered by using germanium–silicon alloys and superlattices.

3.1 Introduction

Silicon is the bread and butter of the electronics industry. Its combination of technological sophistication and economics of scale is unparalled in the history of the Industrial Age. Ideally, photonic devices should also be manufactured in silicon. However, with the exception of one or two devices that are in low-volume production [1], the silicon photonics industry is virtually nonexistent. Technologically, this can be attributed to the premature perception that physical properties of silicon do not lend themselves to important functions such as light emission, amplification, and wavelength conversion. Yet, photonic devices are essential building blocks of fiber optic networks that form the backbone of the Internet. Being able to tap into silicon's vast manufacturing base will reduce the cost of photonic devices, which in turn will accelerate penetration of optics into the access portion of the Internet. Silicon photonics can thus be viewed as a key to broadband access for the masses. Additionally, the technology can solve important problems in today's computing systems as well as spawn new industries of its own. For example, as the trend to reducing device dimensions continues, a significant bottleneck has appeared at the electronics interconnect level, where a large gap exists between individual device speeds and the speed of interconnects that link them [2,3]. Optical interconnects can potentially solve this important problem.

The compatibility of silicon photonics with silicon IC manufacturing, and separately with silicon micro-electromachining system(MEMS) technology, has generated significant attention to this field. In 2004, the US Defense Advanced Research Project Agency (DARPA) launched the Electronic and Photonic Integrated Circuit (EPIC) program, the first of its kind dedicated to silicon photonics. The four-year program funds research and development at universities and industry with the ultimate goal of producing production-capable silicon optoelectronic circuits. The first International Conference on Group IV Photonics was also launched in 2004. The year was also witness to the demonstration of the first silicon laser [4]. The rapid pace of progress is continuing and the first quarter of 2005 has already seen the demonstration of direct electrical modulation of the Raman laser [5] and report of the first continuous wave (CW) silicon Raman laser [6].

Compared to other integrated optics platforms, a distinguishing property of silicon is the tight optical confinement made possible by the large index mismatch between silicon and SiO_2. While a myriad of high-performance passive devices were demonstrated in the 1990s [7], creation of active devices proved to be much more difficult. Unfavorable physical properties such as the near absence of Pockels effect caused by the symmetric crystal structure and the lack of efficient optical transitions, due to the indirect band structure, were the culprits.

Raman scattering was proposed and demonstrated in 2002 as a means to bypass these limitations and create optical amplifiers and lasers in silicon [8]. The approach was motivated by the fact that the stimulated Raman gain co-efficient in silicon is 10^3–10^4 times larger than that in fiber. The modal area in a silicon waveguide is roughly 100 times smaller than that in fiber, resulting in a proportional increase in optical intensity. The combination makes it possible to realize chip-scale Raman devices that normally require kilometers of fiber to operate. The initial demonstration of spontaneous Raman emission from silicon waveguides in 2002 was followed by the first demonstration of both stimulated Raman scattering [9] and parametric Raman wavelength conversion [10] in 2003. Other merits of Raman effect include the fact that it occurs in pure silicon and hence does not require rare-earth dopants (such as erbium), and that the spectrum is widely tunable through the pump laser wavelength.

3.2 Raman Amplification in Silicon

Classical electrodynamics provides a simple and intuitive macroscopic description of the Raman scattering process. In the spontaneous scattering, thermal vibrations of lattice at frequency ω_v (15.6 THz in silicon) produce a sinusoidal modulation of the susceptibility. The incident pump field induces an electric polarization that is given by the product of the susceptibility and the incident field. The beating of the incident field oscillation (ω_p) with oscillation of the susceptibility (ω_v) produces induced polarizations at the sum frequency, $\omega_p + \omega_v$, and at the difference frequency $\omega_p - \omega_v$. The radiation produced by these two polarization components is referred to as anti-Stokes and Stokes waves, respectively. Quantum statistics dictates that the ratio of Stokes power to anti-Stokes power is given by $(1 + N)/N$, where $N = [\exp(\hbar\omega_v/kT) - 1]^{-1}$ is the Bose occupancy factor and has a value of ~ 0.1 for silicon at room temperature.

The same model can be extended to describe stimulated Raman scattering [11]. Here, we assume that the pump field and the Stokes field are present, with a frequency difference equal to the atomic vibrational frequency. The latter can be due to spontaneous emission, or in the case of a Raman amplifier, it is the input signal that is to be amplified. The two fields (pump and Stokes) create a force that stimulates atomic vibrations, even in the absence of a dipole moment. This can be understood as follows. If E is the total field comprising pump and Stokes, and χ is the susceptibility, the energy stored in the field, $V = ((1 + \chi)/2)EE^*$, will have a component oscillating at $\omega_p - \omega_s = \omega_v$. Through the modulation of the susceptibility with displacement, Q, this will produce a force

$$F \propto \frac{\partial V}{\partial Q} \approx \frac{\partial \chi}{\partial Q} E_p E_s \exp(-\omega_v t).$$

This driving force will enhance atomic oscillations, which in turn will increase the amplitude of the Stokes field, E_s. This positive feedback phenomenon is called stimulated Raman scattering and results in amplification of the Stokes field.

While providing an intuitively appealing description of Raman scattering, the macroscopic model described earlier does not account for detailed processes responsible for Raman scattering in silicon. The microscopic picture reveals that the direct coupling of light with atomic vibrations, described by the interaction Hamiltonian involving photons and phonons, is very weak. This is generally true in semiconductors due to the large atomic mass that appears (squared) in the denominator of the cross-section. In silicon, the lack of lattice polarization further underscores this fact. Electrons mediate the Raman scattering process in silicon. Microscopically, the scattering proceeds in three steps [12]. In step 1, the incident photon excites the semiconductor into an intermediate step by creating an electron–hole pair. In step 2, the pair is scattered into another state by emitting a phonon via the electron-phonon interaction Hamiltonian. In step 3, the electron–hole pair in the intermediate step recombines radiatively with emission of a scattered photon. While electrons mediate the process, they remain unchanged after the process. Furthermore, transitions involving electrons are virtual and hence need not conserve energy although momentum must be conserved.

Raman describes the scattering process involving the optical phonon branches of atomic vibrations (as opposed to Brillouin who describes scattering involving acoustic phonons). In first-order scattering, only one phonon is involved and momentum conservation implies that only zone-center phonons can participate. Higher order Raman scattering involves multiple phonons, which can be from any point in the Brillouin zone as long as their total momentum equals the (negligible) photon momentum. In silicon, the zone-center optical phonon is triply degenerate with a frequency of 15.6 THz. Figure 3.1 shows the typical Raman spectrum of silicon with the pump-Stokes separation of 15.6 THz highlighted. The first-order resonance, which is of primary importance here, has a FWHM of approximately 100 GHz [13]. This imposes a maximum information bandwidth of approximately 100 GHz that can be amplified. The Raman linewidth becomes broader when a broadband pump is used.

Crystal symmetry imposes a selection rule that dictates which scattering geometries are allowed. The spontaneous scattering efficiency, S, is given by

$$S = S_0 \sum_{k=1,2,3} |\hat{e}_s \cdot R_k \cdot \hat{e}_p|^2 . \tag{3.1}$$

Unit vectors $\hat{e}_p$ and $\hat{e}_s$ denote the polarization of the pump and Stokes electromagnetic fields. S_0 contains intrinsic microscopic property of silicon including derivatives of the polarizability, and the absolute amplitude of the displacement of the zone-center optical phonons. The sum runs over the three Raman matrices, each corresponding to the phonon displacement along one of the three principal axes of the crystal [14]:

$$\vec{R}_1 = \begin{bmatrix} 0 & 1 & 0 \\ 1 & 0 & 0 \\ 0 & 0 & 0 \end{bmatrix}, \quad \vec{R}_2 = \begin{bmatrix} 0 & 0 & 1 \\ 0 & 0 & 0 \\ 1 & 0 & 0 \end{bmatrix}, \quad \vec{R}_3 = \begin{bmatrix} 0 & 0 & 0 \\ 0 & 0 & 1 \\ 0 & 1 & 0 \end{bmatrix}. \tag{3.2}$$

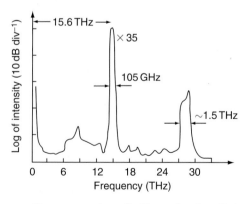

Fig. 3.1. Spontaneous Raman spectra of silicon showing first- and second-order Stokes emission. Frequency is plotted relative to pump frequency. Anti-Stokes spectrum is not shown

Table 3.1. Spontaneous scattering efficiency for various wavevector and polarization directions of pump and Stokes fields

$\hat{k}_p$ and $\hat{k}_s$	$\hat{e}_p$	$\hat{e}_s$	relative efficiency
$[1\bar{1}0]$	$[110]$	$[110]$	S_0
$[1\bar{1}0]$	$[110]$	$[001]$	S_0
$[11\bar{2}]$	$[111]$	$[111]$	$4S_0/3$
$[111]$	$[1\bar{1}0]$	$[11\bar{2}]$	$2S_0/3$
$[111]$	$[11\bar{2}]$	$[11\bar{2}]$	S_0
$[0\bar{1}1]$	$[011]$	$[100]$	S_0
$[0\bar{1}1]$	$[011]$	$[011]$	S_0

Table 3.1 shows the relative spontaneous Raman intensities obtained for different scattering configurations in silicon. The vectors $\hat{k}_p$ and $\hat{k}_s$ denote the pump and Stokes propagation directions, respectively.

The Raman gain coefficient, g_R, can be obtained from the spontaneous efficiency, S, using the Einstein relation [15]:

$$g_R = \frac{8\pi c^2 \omega_p}{\hbar \omega_s^4 n^2(\omega_s)(N+1)\Delta\omega} S. \tag{3.3}$$

Substituting the appropriate values, the gain coefficient is obtained as ~ 76 cm GW^{-1} [8]. This uses $S = 8.4 \times 10^{-7}$ cm^{-1} Sr^{-1}, which was obtained by extrapolating the values measured (1.1–1.55 μm wavelength using λ^{-4} relation). This is in the same order of magnitude but several times larger than the values extracted from Raman gain measurements (~ 20 cm GW^{-1}) at 1.55 μm, nonetheless, when compared to silica (0.93×10^{-2} cm GW^{-1}), the Raman gain in silicon is 10^3–10^4 times larger. Such a large difference has its origin in the much narrower linewidth of the Raman spectrum in *crystalline* silicon, compared to the *amorphous* fiber.

It is customary to describe the induced polarization for the case of stimulated Raman scattering through the nonlinear susceptibility, χ^R_{ijmn}, defined by the expression:

$$P_i^{NL}(\omega_s) = \varepsilon_0 \chi^R_{ijmn} E_j(\omega_p) E_m(-\omega_p) E_n(\omega_s). \tag{3.4}$$

On the other hand, the atomic displacement can be obtained using a classical harmonic oscillator model [11] with the driving force described earlier. By comparing the induced polarization suggested by the displacement with the definition given here, one arrives at the following expression for the induced Raman susceptibility:

$$\chi^R_{ijmn} = 2\Gamma\omega_v \frac{2ncg_R}{\omega_s(\mu_0/\varepsilon_0)^{1/2}} \frac{\sum_{k=1,2,3} (R_{ij})_k (R_{mn})_k}{(\omega_v^2 - (\omega_p - \omega_s)^2 - 2i\Gamma(\omega_p - \omega_s))}, \tag{3.5}$$

where Γ is the damping term in the harmonic oscillator equation (related to phonon dephasing time) and n is the refractive index. Crystal symmetry consideration, described by the Raman tensor, R, leads to a total of 12 equal nonvanishing components that have of the form:

$$1221 = 1212 = 2112 = 2121 = 1331 = 1313 = 3113 = 3131 = 2332$$
$$= 2323 = 3223 = 3232. \tag{3.6}$$

The induced susceptibility is related to the Raman gain coefficient as [14]:

$$\chi^{R}_{1221}(\omega_p - \omega_s = \omega_v) = \frac{i}{(\mu_0/\varepsilon_0)^{1/2}} \frac{2ncg_R}{\omega_s} = 11.2 \times 10^{-14} i [\text{cm}^2 \, \text{V}^{-2}]. \tag{3.7}$$

Another nonlinear optical effect in semiconductors is the two-photon absorption (TPA). This is a deleterious effect that results in pump depletion and generation of free carriers. These carriers give rise to a broadband absorption spectrum through the free carrier plasma effect. TPA has been shown to be negligible from the point of view of pump depletion [9]. This is plausible since the TPA coefficient in silicon, β, is relatively small, $\sim 0.5 \, \text{cm} \, \text{GW}^{-1}$. On the other hand, absorption by TPA-generated free carriers is a broadband process that competes with Raman gain. The effect has been identified as a limiting factor in all-optical switching in III–V semiconductor waveguides [16–20]. It has also been discussed as a potential limit to achievable Raman gain in GaP waveguides [21], although a Raman gain of 24 dB was demonstrated in these waveguides [22]. More recently, TPA-induced free carrier absorption (FCA) has been measured in silicon waveguides in the context of Raman process [23, 24] and in transmission of ultrashort pulses in silicon waveguides [25].

The magnitude of TPA-induced FCA depends on free carrier concentration through the relation: $\alpha^{\text{FCA}} = 1.45 \times 10^{-17} (\lambda/1.55)^2 \Delta N$, where λ is the wavelength, in micrometers, and ΔN is the density of electron–hole pairs [26, 27]. The latter is related to the pump intensity, I_p, by

$$\Delta N = \beta I_p^2 \tau_{\text{eff}}/(2h\nu), \tag{3.8}$$

where $h\nu$ is the pump photon energy and τ_{eff} is the effective recombination lifetime for free carriers. This equation neglects the contribution to free carrier generation due to pump-signal TPA and hence is valid in the regime where Stokes intensity (I_s) $I_s \ll I_p$.

The fundamental parameter that governs the TPA-induced loss and hence the success of Raman based devices is the recombination lifetime, τ_{eff}. It is well known that the recombination lifetime in silicon-on-insulator (SOI) is much shorter than that in a bulk silicon sample with comparable doping concentration. This lifetime reduction is due to the presence of interface states at the boundary between the top silicon and the buried oxide layer. This effect depends on the method used for the preparation of SOI wafer and the film thickness, with measured and expected values ranging between 10 and 20 ns [28–30]. In SOI waveguides the lifetime is further reduced to a few nanoseconds, or even below in the case of submicron waveguides, due to the recombination at the etched waveguide facets and in the case of rib waveguides due to diffusion into the slab regions [30]. The lifetime can be further reduced by application of a reverse bias p–n junction [23, 24, 30] or by introduction of midgap states through high energy irradiation, and gold or platinum doping. Modest amount of CW gain has been observed in deep submicron waveguides [31] where the impact of surface and interface recombination plays a critical role in reducing the lifetime. CW gain has also been demonstrated by sweeping the free carriers using a reverse bias p–n junction [32]. This approach is further discussed in the context of Raman laser later.

The plot of the net Raman gain as a function of CW pump intensity for a waveguide of length, $L = 1\,\text{cm}$ and propagation loss of $1\,\text{dB}\,\text{cm}^{-1}$, is shown in Fig. 3.2 [24] for different free carrier lifetime values. The plot shows that more than $5\,\text{dB}$ of gain can be obtained with a pump intensity of about $100\,\text{MW}\,\text{cm}^{-2}$. Gain increases with intensity while the loss rises as intensity squared and dominates when lifetime is long. The pump is assumed to be a monochromatic source. The finite linewidth of the pump laser will result in a lower gain than what is predicted in Fig. 3.2. It is clear that to create a successful amplifier, an effective lifetime of $\leq 1\,\text{ns}$ is required.

3.3 Raman Wavelength Conversion

As mentioned earlier, Raman scattering spectrum also contains an anti-Stokes wave that is upshifted from the pump by the 15.6 THz phonon frequency. The gain coefficient for the anti-Stoke wave will have a negative sign, indicating

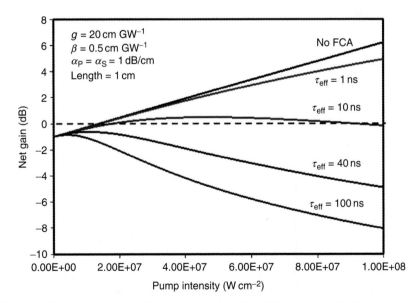

Fig. 3.2. Impact of carrier lifetime on achievable CW Raman gain. Gain increases with intensity while loss rises as intensity squared and dominates when lifetime is long

that an incident anti-Stoke wave will be attenuated. However, anti-Stoke signal can be generated through four wave mixing (FWM) induced through the Raman susceptibility, in much the same way that conventional FWM takes place via the electronic third-order nonlinear susceptibility (responsible for the Kerr effect). In the Raman process, energy conservation dictates that $\omega_{as} = 2\omega_p - \omega_s$, and momentum conservation results in the so-called phase-matching condition, with the total phase mismatch defined as:

$$\Delta\beta = 2\beta_p - \beta_s - \beta_{aS} , \qquad (3.9)$$

where β is the wave vector for a given wavelength and the corresponding mode of polarization μ (1 for TE_0 and 2 for TM_0). As $\Delta\beta$ approaches 0, pump, Stokes, and anti-Stokes waves experience coherent interaction. This phenomenon is referred to as coherent anti-Stokes Raman scattering (CARS) in the spectroscopy literature and was first observed in silicon in 2003 [10]. As shown in Fig. 3.3, the process of creation of anti-Stokes photon is accompanied by creation and annihilation of a zone-center phonon.

The conversion efficiency is highly sensitive to the phase mismatch and, in general, the efficiency has a $sinc^2$ dependence on phase mismatch. In silicon, phase mismatch is dominated by material dispersion as waveguide dispersion is relatively negligible unless submicron modal dimensions. In such devices, waveguide dispersion provides a means to compensate for material dispersion. Other means of phase matching include the use of waveguide birefringence and/or strain. At phase matching, neglecting nonlinear loss mechanisms, the evolution of Stokes and anti-Stokes fields, $E(z)$, along the waveguide length, z, is given by

$$E_S(z) = E_S(0) + (E_S(0) + E_{aS}^*(0))g_R I_p z/2 ,$$

$$E_{aS}^*(z) = E_{aS}^*(0) - (E_S(0) + E_{aS}^*(0))g_R I_p z/2 . \qquad (3.10)$$

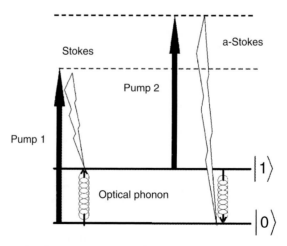

Fig. 3.3. Energy-level representation of Raman wavelength conversion process. $|1\rangle$ and $|0\rangle$ are vibrational states and arrows represent virtual transitions. A phonon is created and annihilated leaving the phonon population unchanged

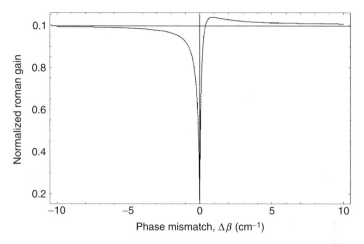

Fig. 3.4. The variation of normalized Raman gain as a function of phase mismatch. At zero phase mismatch, Raman gain is suppressed in favor of parametric Stokes to anti-Stokes conversion

Equation 3.10 predicts a linear increase in the fields with distance, which holds true for small propagation lengths. Once the Stokes and anti-Stokes fields become equal in amplitude, no further change takes place leading to a saturation effect. The characteristic length is very long, therefore, this regime is not expected to occur in chip-scale devices.

Figure 3.4 shows the normalized Raman gain as a function of phase mismatch. At large values of $|\Delta\beta|$, stimulated Raman amplification is predominant and leads to amplification of Stokes signal with an effective gain given by g_R. Under phase-matching condition, Raman gain is suppressed and the parametric coupling of pump and Stokes to anti-Stokes dominates. In this region, Stokes, anti-Stokes, and pump fields are strongly coupled and parametric conversion dominates. For small, positive values of $\Delta\beta$, the normalized gain slightly exceeds unity due to modulation instability. This effect has also been predicted and observed in optical fibers [33].

3.4 Experimental Demonstrations

Experimental results shown in Fig. 3.5a highlight the competition between Raman gain and TPA-induced. Here a CW pump emitting at 1,427 nm was used along with a tunable CW signal laser. The plot shows amplification of the signal laser as it is tuned across the Raman resonance. The data clearly shows the competition between the resonant Raman amplification and the broadband pump-induced absorption. It is clear that further increase in pump intensity is futile as the intensity-squared increase of TPA-induced loss dominates over Raman gain, which increases linearly with pump intensity. Also shown, in Fig. 3.5b, is the measured spontaneous emission spectra

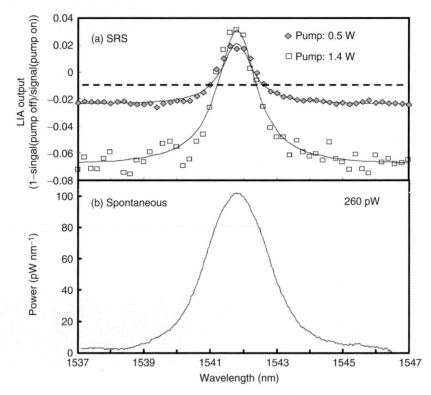

Fig. 3.5. (**a**) Measured spectral characteristic of stimulated Raman scattering (SRS) in an SOI waveguide, under CW pumping. Two different input pump powers are shown, to illustrate the effect of TPA-induced free carrier absorption. (**b**) Measured spontaneous emission spectrum

from the same waveguides. The observed FWHM is approximately twice the 100 GHz intrinsic Raman linewidth, a feature caused by the finite linewidth of the pump laser.

One method for avoiding free carrier accumulation and concomitant loss is pulsed pumping [34–38]. As long as the pump-pulse period is longer than the free carrier lifetime the free carrier accumulation can be mostly eliminated. Figure 3.6 shows the measured change in CW signal beam (tuned to the peak of Stokes resonance) due to the pump pulse. The pump source was a mode-locked fiber laser with 25 MHz repetition rate and ∼1 ps output pulse width. Since phonon response time in silicon is more than 3 ps, the pulse width is broadened by using a spool of standard single mode optical fiber. Maximum pump on–off gain of 20 dB has been obtained. Taking into account the losses in the waveguide, we obtain a net waveguide gain of 13 dB [38]. This gain includes waveguide propagation losses but not the fiber–waveguide coupling

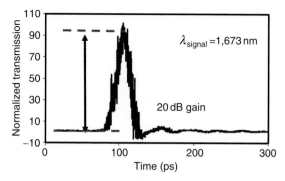

Fig. 3.6. Time resolved Raman amplification with the signal laser at 1673 nm. A pump on–off gain of 20 dB is obtained. Pump pulse wavelength is 1,540 nm

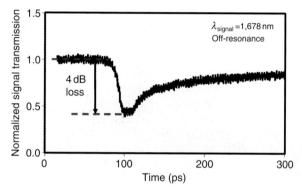

Fig. 3.7. Time resolved signal loss with the signal laser at 1,678 nm, i.e., outside the Raman resonance. Maximum loss of 4 dB and carrier lifetime of 4 ns is obtained

loss. In separate devices that have adiabatic mode tapers, we have shown net fiber-to-fiber gain of 11 dB under similar pulse pumping scheme [34].

The net free carrier loss and the free carrier lifetime can be measured by performing the same measurement with the signal laser tuned away from the Raman peak. This is shown in Fig. 3.7. Maximum loss due to the combined FCA loss is measured to be 4 dB. Thus the intrinsic Raman gain in the silicon waveguide is 24 dB. By extrapolating the exponential decay of the carriers we estimate a free carrier lifetime of ~4 ns.

Variation of optical gain as a function of peak pump power coupled to the waveguide is shown in Fig. 3.8. Optical gain is found to saturate around 37 W of peak pump power. This can be attributed to the pulse breakup and excessive spectral broadening of pump laser in the fiber pigtail preceding the waveguide [34].

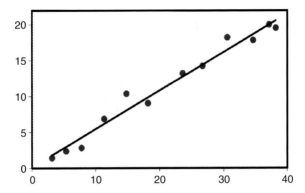

Fig. 3.8. On–off optical gain as a function of peak pump pulse power. Maximum gain of 20 dB is obtained

Stokes/a-Stokes conversion

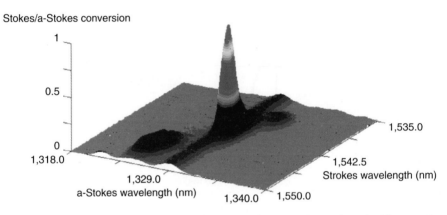

Fig. 3.9. Measured anti-Stokes spectra vs. Stokes signal wavelength. The z-axis represents the conversion efficiency, normalized to unity. CW pump laser wavelength is 1,428 nm

Figure 3.9 shows experimental verification of Raman wavelength conversion in silicon. The pump laser (at 1,428 nm) is coupled to the TE_0 mode, and the signal laser is coupled to the TM_0 mode. The Stokes signal laser is scanned in the range from 1,530 to 1,560 nm. Figure 3.9 shows a-Stokes spectra measured as a function of the Stokes signal wavelength. The CW pump power in the waveguide was 0.7 W. There is a clear peak at 1,328.8 nm of anti-Stokes emission when the Stokes laser is tuned to Stokes wavelength of 1,542.3 nm. The nature of the satellite peaks may be due to the $sinc^2$ dependence of the conversion efficiency with phase mismatch.

Figure 3.10a shows converted a-Stokes signal spectrum [39]. The FWHM for wavelength conversion, which is approximately 250 GHz, is determined solely by the pump laser linewidth. Figure 3.10b shows the conversion of

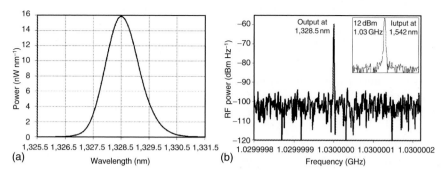

Fig. 3.10. (a) Anti-Stokes spectrum showing the converted signal (b) RF spectrum at 1,328.5 nm showing conversion of a 1.03 GHz RF modulation from Stokes to anti-Stokes. The *inset* shows the RF spectrum of the input signal at 1,542.3 nm

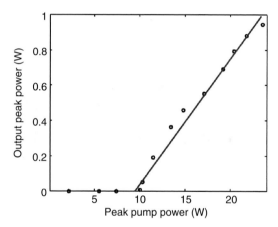

Fig. 3.11. Measured laser output power with respect to peak pump power. Lasing threshold is measured to be at 9 W peak power level. The slope efficiency obtained by dividing the output peak pulse power by that of the input is 13%

1.03 GHz RF modulation from 1542 to 1328.5 nm [39]. The input RF signal power applied to the Stokes wavelength is shown in the inset. The measured electrical signal to noise ratio (SNR) is 34 dBe. We note that from the application point of view, the 1,320 and 1,550 nm bands are the two most important bands in optical communication. The measured conversion efficiency was approximately 10^{-5}. As mentioned previously, a number of design approaches are available for phase matching a silicon waveguide and for realizing high conversion efficiency [40].

Figure 3.11 shows the measure input–output curve for the first silicon Raman laser [4]. The laser, demonstrated in 2004, operated in the pulse mode and consisted of a 1.7 cm long silicon waveguide gain medium and an external cavity formed via a fiber loop. The laser has a threshold of 9 W peak pulse power (corresponding to a few mW of average power) and was able to produce output pulses with over 2.5 W peak power at 25 MHz repetition rate. The demonstration was a major milestone as it clearly showed that silicon can indeed lase. The strong lasing characteristics and high conversion efficiency (~13%) of the prototype laser showed that silicon Raman lasers must be considered as a practical and compact alternative to fiber Raman lasers.

In 2005, Intel Corporation demonstrated the first CW silicon Raman laser [6]. The device was a 5 cm long silicon waveguide with a cavity formed by HR coating the chip facets. CW operation was achieved by using a reverse bias p–n junction to sweep out the TPA generated free carriers, an approach that was previously proposed in 2004 [23, 24, 30]. Figure 3.12 shows the laser input-output behavior at reverse bias voltages of 5 and 25 V. The laser produced a maximum output power of ~9 mW at 25 V reverse bias with 600 mW of CW pump power inside the waveguide. The CW operation is an important step in the development of silicon Raman lasers. One drawback of the reverse bias carrier sweep out approach is the electrical power dissipated on the chip. With a reverse current of approximately 50 mA expected for this device, the laser dissipates an on-chip electrical power of more than 1 W. From this perspective, methods that can drastically reduce carrier lifetime, and hence mitigate the need for active carrier removal, are desired. As a compromise, reduction of the required sweep out voltage will reduce the electrical power dissipation.

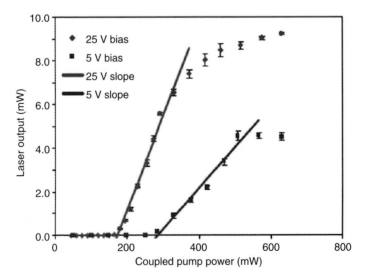

Fig. 3.12. Threshold characteristics of CW silicon Raman laser demonstrated in a 5 cm silicon waveguide and with using reverse biased p–i–n diode for carrier sweep out [6]

The ability to integrate a p–n junction along with the gain medium offers the exciting possibility of intracavity gain switching [5]. By injecting free carriers into the gain medium, cavity loss can be modulated leading the direct electrical modulation of the laser output. Using this technique, direct laser switching with 30 dB extinction ratio has been demonstrated [5]. This is a unique feature of silicon Raman lasers that is not possible in fiber Raman lasers. It allows the laser to be interfaced with on-chip electronic circuitry in all-silicon optoelectronic integrated circuits. This idea, first demonstrate by UCLA, has been extended by Intel in a demonstration of an electrically switched Raman amplifier, with the device representing a loss-less modulator [41].

3.5 GeSi Raman Devices

The introduction of germanium in the overall scheme of nonlinear Raman processes in silicon offers new avenues for tailoring the device characteristics. In particular [42]:

1. The strain caused by the difference in the lattice constants of Si and Ge along with the composition effect provide mechanisms for tuning the Stokes shift associated with the dominant Si–Si ($500\,cm^{-1}$) vibrational mode [43]. In addition, the presence of Si–Ge modes ($400\,cm^{-1}$) and Ge–Ge modes ($300\,cm^{-1}$) provides flexibility in pump and signal wavelengths.
2. Spectral broadening can be achieved via graded Ge composition.
3. The strain resulting in birefringence [44, 45] can provide an additional degree of freedom for phase matching in the wavelength conversion process. Stress also results in broadening of the gain spectrum via splitting of the degenerate optical phonon modes.
4. When grown on an SOI substrate, the use of double cladding in the vertical direction can improve fiber–waveguide coupling efficiency.
5. Higher carrier mobility, and hence diffusion constant, in SiGe reduces the effective lifetime in the waveguide. This reduces the losses associated with the free carriers that are generated by two-photon absorption (TPA). However, this benefit will be countered by the higher TPA coefficient in GeSi.

Recently, the first GeSi optical amplifier and laser has been demonstrated [42]. A pulsed gain of 14 dB and lasing with sharp threshold characteristics were observed for $Ge_{0.08}Si_{0.92}$ rib waveguides. The Stokes spectrum, shown in Fig. 3.13, exhibits a 37 GHz red shift, which is in qualitative agreement with a model that takes into account the effect of composition and strain on the optical phonon frequency [46]. These results suggest that the spectrum of Raman scattering can be engineered using the GeSi material system. As a result, GeSi Raman devices represent an exciting topic for future research and development.

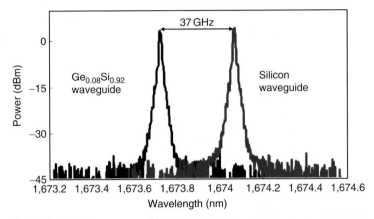

Fig. 3.13. Stimulated Raman spectra of GeSi waveguides compared to that of silicon waveguide. A 37 GHz red shift in the Stokes wavelength is observed. Pump wavelength was 1,539 nm [42]

3.6 Conclusions

This chapter has outlined light generation, amplification, and wavelength conversion in silicon and GeSi using stimulated Raman processes. These effects are routinely observed in optical fiber; however, several kilometers of length are required to do so. What has enabled us to achieve these processes on millimeter-length scales on a chip are two fundamental differences between an optical fiber and a silicon microstructure. The first is the difference in atomic structures. Vibrational modes in silica broaden into bands; hence the Raman gain has a very large bandwidth and a low peak value, requiring long interaction lengths for the effect to be observed. In contrast, silicon is a single crystal and supports only three degenerate optical vibration modes. The result is a much narrower bandwidth gain but a much higher peak gain (nearly $10^3 - 10^4$ times higher). Second is the difference in modal areas. The large index–contrast in silicon/SiO$_2$ waveguides results in a mode area that is approximately 100 times smaller (assuming 0.8 µm^2 waveguides) than in a standard single mode fiber (mode area = 80 µm^2). The proportionally higher power density in the silicon waveguide lowers the threshold for nonlinear optical processes.

The intrinsic Raman bandwidth in silicon is 105 GHz and is broadened in the experiments by the pump laser linewidth (typically ~2 nm). The resulting bandwidth is sufficient for amplifying several WDM channels. The bandwidth can be broadened by broadening the pump linewidth even further (although this reduces the peak gain) or by using multiple pump wavelengths. The Raman phenomenon is fully tunable; the tuning range is only limited by the available pump wavelengths. This is an advantage over the nanocrystal approach for light generation and amplification. In addition, the Raman effect can also perform wavelength conversion. Another important advantage of this approach is that it does not require rare-earth dopants or nanostructures. Hence, it is truly compatible with silicon IC manufacturing. A limitation of the Raman approach is it is optically pumped. However, it has been shown that silicon Raman lasers can be electronically switched or modulated using intracavity gain–loss modulation [6]. Therefore, these lasers or similar amplifiers can be interfaced with on-chip electronic circuitry. Raman amplification and lasing in GeSi waveguides has recently been demonstrated. The GeSi material system provides an opportunity to engineer the, otherwise rigid, Raman spectrum of silicon. Owing to the enhance carrier mobility, GeSi is pursued by the CMOS IC industry for future high-speed circuits. This provides another impetus for investigating GeSi Raman devices.

Going forward, low-loss waveguides with small cross-sections are required for realizing high-performance devices. Surface roughness produces strong scattering and high propagation loss due to the high index contrast between the silicon waveguide core and the cladding (air or SiO_2). As such, the losses of silicon waveguides tend to increase with in cross-section. Fortunately, new waveguide-fabrication processes that are in development promise low-loss waveguides with submicron cross-sections [47,48]. Naturally, one must be able to efficiently couple light into these structures. Several novel approaches for high-efficiency coupling into submicron waveguides have also been developed [48].

References

1. See for example http://www.kotura.com
2. M. Paniccia, M. Morse, M. Salib, "Integrated photonics", chap. 2 in L. Pavesi, D.J. Lockwood (Eds.), Silicon Photonics, Topics in Appl. Phys. 94, 51–88 (2004)
3. Z. Gaburro, "Optical interconnect", chap. 4 in L. Pavesi, D.J. Lockwood (Eds.), Silicon Photonics, Topics in Appl. Phys. 94, 121–176 (2004)
4. O. Boyraz, B. Jalali, "Demonstration of a silicon Raman laser", Opt. Exp. 12, 5269–5273 (2004)
5. O. Boyraz, B. Jalali, "Demonstration of directly modulated silicon Raman laser", Opt. Exp. 13, 796–800 (2005)
6. H. Rong, R. Jones, A. Liu, O. Cohen, D. Hak, A. Fang, M. Pannicia, "A continuous-wave Raman silicon laser", Nature 433, 725–728 (2005)
7. B. Jalali, S. Yegnanarayanan, T. Yoon, T. Yoshimoto, I. Rendina, F. Coppinger, "Advances in silicon-on-insulator optoelectronics", IEEE J. Sel. Top. Quant. Electron. 4, 938–947 (1998)
8. R. Claps, D. Dimitropoulos, Y. Han, B. Jalali, "Observation of Raman emission in silicon waveguides at 1.54 m", Opt. Exp. 10, 1305–1313 (2002)
9. R. Claps, D. Dimitropoulos, V. Raghunathan, Y. Han, B. Jalali, "Observation of stimulated Raman scattering in Silicon waveguides", Opt. Exp. 11, 1731–1739 (2003)
10. R. Claps, V. Raghunathan, D. Dimitropoulos, B. Jalali, "Anti-stokes Raman conversion in silicon waveguides", Opt. Exp. 11, 2862–2872 (2003)
11. A. Yariv, Quantum Electronics, 3rd edn., Wiley, New York (1988), ISBN 0-471-60997-8

12. P.Y. Yu, M. Cardona, Fundamentals of Semiconductors, 3rd rev. edn., Springer Berlin Heidelberg New York (2005)
13. P.A. Temple, C.E. Hathaway, "Multiphonon Raman Spectrum of Silicon", Phys. Rev. A 7, 3685 (1973)
14. D. Dimitropoulos, B. Houshmand, R. Claps, B. Jalali, "Coupled-mode theory of Raman effect in silicon-on-insulator waveguides", Opt. Lett. 28, 1954–1956 (2003)
15. J.M. Ralston, R.K. Chang, "Spontaneous-Raman-scattering efficiency and stimulated scattering in silicon", Phys. Rev. B 2, 1858–1862 (1970)
16. J.H. Yee, H.H.M. Chau, "Two-Photon indirect transition in GaP crystal", Opt. Commun. 10, 56–58 (1974)
17. K.W. DeLong, G.I. Stegeman, "Two-photon absorption as a limitation to all-optical waveguide switching in semiconductors", Appl. Phys. Lett. 57(20) 2063–2064 (1990)
18. A. Villeneuve, C.C. Yang, G.I. Stegeman, C.N. Ironside, G. Scelsi, R.M. Osgood, "Nonlinear absorption in a GaAs waveguide just above half the band gap", IEEE J. Quant. Electron. 30, 1172–1175 (1994)
19. A.M. Darwish, E.P. Ippen, H.Q. Lee, J.P. Donnelly, S.H. Groves, "Optimization of four-wave mixing conversion efficiency in the presence of nonlinear loss," Appl. Phys. Lett. 69, 737–739 (1996)
20. Y.-H. Kao, T.J. Xia, M.N. Islam, "Limitations on ultrafast optical switching in a semiconductor laser amplifier operating at transparency current", J. Appl. Phys. 86, 4740–4747 (1999)
21. K. Suto, T. Kimura, T. Saito, J. Nishizawa, "Raman amplification in GaP-AlxGa1-xP waveguides for light frequency discrimination", IEE Proc. Optoelectron. 145, 105–108 (1998)
22. S. Saito, K. Suto, T. Kimura, J.I. Nishizawa, "80-ps and 4-ns pulse-pumped gains in a GaP–AlGaP semiconductor Raman amplifier", IEEE Photon. Technol. lett. 16, 395–397 (2004)
23. T.K. Liang, H.K. Tsang, "Role of free carriers from two-photon absorption in Raman amplification in silicon-on-insulator waveguides," Appl. Phys. Lett. 84(15) 2745–2747 (2004)
24. R. Claps, V. Raghunathan, D. Dimitropoulos, B. Jalali, "Influence of nonlinear absorption on Raman amplification in Silicon waveguides," Opt. Exp. 12, 2774–2780 (2004)
25. A.R. Cowan, G.W. Rieger, J.F. Young, "Nonlinear transmission of 1.5 m pulses through single-mode silicon-on-insulator waveguide structures," Opt. Exp. 12, 1611–1621 (2004)
26. R.A. Soref, B.R. Bennett, "Electrooptical effects in Silicon," IEEE J. Quant. Electron. QE-23, 123–129 (1987)
27. R.J. Bozeat, S. Day, F. Hopper, F.P. Payne, S.W. Roberts, M. Asghari, "Silicon based waveguides," in L. Pavesi, D.J. Lockwood (Eds.) Silicon Photonics, chap. 8, 269–294 (2004)
28. M.A. Mendicino, "Comparison of properties of available SOI materials," Properties of Crystalline Silicon, by Robert Hull 18.1 992–1001 (1998)
29. J.L. Freeouf, S.T. Liu, IEEE Int. SOI conf. proc. Tucson, AZ, USA, 3–5 October, 74–75 (1995)
30. D. Dimitropoulos, R. Jhaveri, R. Claps, J.C.S. Woo, B. Jalali, "Lifetime of photogenerated carriers in silicon-on-insulator rib waveguides", Appl. Phys. Lett. 86, 071115 (2005)

31. R. Espinola, J. Dadap, R. Osgood, S.J. McNab, Y.A. Vlasov, "Raman ampli-
    fication in ultrasmall silicon-on-insulator wire waveguides", Opt. Exp. 12(16),
    3713–3718 (2004)
32. R. Jones, H. Rong, A. Liu, A.W. Fang, M.J. Paniccia, D. Hak, O. Cohen, "Net
    continuous wave optical gain in a low loss silicon-on-insulator waveguide by
    stimulated Raman scattering", Opt. Exp. 13, 519–525 (2005)
33. E. Golovchenko, P.V. Mamyshev, A.N. Pilipetskii, E.M. Dianov, "Mutual in-
    fluence of the parametric effects and stimulated Raman scattering in optical
    fibers," IEEE J. Quant. Electron. 26, 1815–1820 (1990)
34. O. Boyraz, B. Jalali, "Demonstration of 11dB fiber-to-fiber gain in a silicon
    Raman amplifier," IEICE Electron. Exp. 1(14), 429–434 (2004)
35. T.K. Liang, H.K. Tsang, "Efficient Raman amplification in silicon-on-insulator
    waveguides," Appl. Phys. Lett. 85, 3343–3345 (2004)
36. Q. Xu, V.R. Almeida, M. Lipson, "Time-resolved study of Raman gain in highly
    confined silicon-on-insulator waveguides," Opt. Exp. 12, 4437–4442 (2004)
37. A. Liu, H. Rong, M. Paniccia, O. Cohen, D. Hak, "Net optical gain in a low
    loss silicon-on-insulator waveguide by stimulated Raman scattering", Opt. Exp.
    12(18), 4261–4268 (2004)
38. V. Raghunathan, O. Boyraz, B. Jalali, "20 dB on-off Raman amplification in
    silicon waveguides", CLEO 2005, Baltimore, MD, CMU1 May (2005)
39. V. Raghunathan, R. Claps, D. Dimitropoulos, B. Jalali, "Wavelength conversion
    in Silicon waveguides using Raman-induced four wave mixing," Appl. Phys. Lett.
    85(1), 34–36 (2004)
40. D. Dimitropoulos, V. Raghunathan, R. Claps, B. Jalali, "Phase-matching and
    nonlinear optical processes in Silicon waveguides," Opt. Exp. 12, 2774 (2003)
41. R. Jones, A. Liu, H. Rong, M. Paniccia, O. Cohen, D. Hak, "Lossless optical
    modulation in a silicon waveguide using stimulated Raman scattering," Opt.
    Exp. 13(5), 1716–1723 (2005)
42. R. Claps, V. Raghunathan, O. Boyraz, P. Koonath, D. Dimitropoulos, B. Jalali,
    "Raman amplification and lasing in SiGe waveguides,." Opt. Exp. 13, 2459–2466
    (2005)
43. M. Robillard, P.E. Jessop, D.M. Bruce, S. Janz, R.L. Williams, S. Mailhot,
    H. Lafontaine, S.J. Kovacic, J.J. Ojha, "Strain-induced birefringence in Si1-
    xGex optical waveguides," J. Vac. Sci. Technol. B 16(4), 1773–1776 (1998)
44. D.-X. Xu, P. Cheben, D. Dalacu, A. Delage, S. Janz, B. Lamontagne,
    M.-J. Picard, W.N. Ye, "Eliminating the birefringence in silicon-on-insulator
    ridge waveguides by use of ridge cladding stress," Opt. Lett. 29(20), 2384–2386
    (2004)
45. R. People, J.C. Bean, "Calculation of critical layer thickness versus lattice mis-
    match for GexSi1-x/Si strained-layer heterostructures", Appl. Phys. Lett. 47(3),
    322–324 (1985)
46. J.C. Tsang, P.M. Mooney, F. Dacol, J.O. Chou, "Measurements of alloy compo-
    sition and strain in thin Ge_xSi_{1-x} layers", J. Appl. Phys. 75, 8098–8108 (1994)
47. P. Koonath, K. Kishima, T. Indukuri, B. Jalali, "SIMOX sculpting of 3-D nano-
    optical structures," LEOS Annual Meeting, Tucson, AZ October (2003)
48. T. Tsuchizawa, K. Yamada, H. Fukuda, T. Watanabe, J. Takahashi, M. Taka-
    hashi, T. Shoji, E. Tamechika, S. Itabashi, H. Morita, "Microphotonics devices
    based on Silicon microfabrication technology", IEEE J. Sel. Top. Quant. Elec-
    tron. 11(1), 232–240 (2005)

4

Electro-Optical Modulators in Silicon

S. Libertino and A. Sciuto

Summary. The aim of this study is to provide an overview of the physics principles ruling optical modulation in Si and the efforts made in the last 25 years to fabricate Si-based light modulators. In the first part the fundamental parameters to describe a modulator are defined, and both the mechanisms for light modulation and the kind of device that can be fabricated are discussed. The second part is devoted to provide an overview of the devices proposed so far in literature. Of course, just a few examples are presented. The different modulator working principles are described and the structures limits evidenced. Finally, a bipolar mode field effect transistor will be taken as an example to describe how to make and, more importantly, also characterise, an electro-optical modulation in a nonconventional way.

4.1 Introduction

The strongest driving force in studying optoelectronic devices arises from the need for better and more performing devices required by the modern technology society. In fact, the progressive shrinking of the feature sizes in microelectronic devices, the increasing demand of higher clock frequencies, and the simultaneous growth in integrated circuit (IC) complexity push the "traditional" microelectronic devices towards their physical limits. As an example, the International Technology Roadmap for Semiconductors (ITRS) [1] shows that by 2010, high-performance ICs will count up to two billion transistors per chip and work with clock frequencies of the order of 10 GHz. Electrical interconnects under these conditions pose severe limitations in speed, power consumption, crosstalk, and voltages needed to bias the devices.

One of the most promising solutions, also foreseen by the ITRS is to replace electrical interconnections with an optical interconnect layer. Such a solution could provide an enormous increase in bandwidth, thereby eliminating any electromagnetic noise problem. Moreover, a decrease in the power consumption is expected, which is associated to an increase in the overall device speed. Of course, to obtain such results a price must be paid. All the devices fabrication steps have to be compatible with future IC technology and the additional cost incurred must be contained. As a result of these efforts, new disciplines connected with optics have appeared: electro-optics, optoelectronics, waveguide technology, etc. Thus, classical optics, initially dealing with lenses, mirrors, filters, etc., has been forced to describe a new family of much more complex devices such as lasers, semiconductor detectors, light modulators, to name a few. The operation of these devices must be described in terms of optics as well as electronics, giving birth to photonics. Photons can control the flux of electrons, in the case of detectors, for example, and electrons themselves can determine the properties of light propagation, as in the case of semiconductor lasers or electro-optic modulators. Today, photonic and optoelectronic devices clearly dominate long-distance communications through optical fibres [2]; many sensor devices are based on optoelectronic devices [3] and they are starting to penetrate the information processing technology field [4]. In fact, all-optical computation and communication systems are beginning to be considered as the future of information processing.

To accomplish the goal of replacing or coexisting with electronic devices, a high level of integration is required. It can be achieved only by using monolithic integration, where all the optical elements including light sources, light control, electronics, and detectors are incorporated in a single substrate. The material of choice for such an integration is Si nowadays. In fact, besides the difficulties arising from the indirect bandgap (an issue for light generation) and the weak electro-optical performances [5], Si has many advantages over "traditional" optoelectronic materials. Most of the microelectronic industry relies on Si as a semiconductor of choice for the fabrication of microelectronic devices. This makes Si the best-known semiconductor material with a long tradition; silicon substrates have a low price and a good optical (and electrical) quality. Last, but not least, there is a wide experience developed in the microelectronic industry in Si processing. On the other hand, there is no unique competitor to Si in the optical domain. In fact, compound semiconductors, such as indium phosphide (InP) or gallium arsenide (GaAs), are widely used to make light sources [6], thanks to their direct bandgap. On the other hand, electro-optical crystals, such as lithium niobate ($LiNbO_3$), are used for light modulation [7], via the Pockels effect (see later). None of these materials succeeded as the material of choice for optical devices, and their manufacturing costs are quite high compared to Si.

Light modulation is one of the main issues in Si-based technology, since Si, as described in Sect. 4.2, does not show a good electro-optical effect. Before entering into the physics part, it is interesting to mention some of the special features of systems based on integrated photonic technology. In particular, an electro-optical modulator must satisfy some requirements: (a) *Low voltage control*. Goal achievable also thanks to the small device dimensions, e.g. the waveguide width. In fact, the width of the channel waveguides allows one to drastically reduce the distance between the control electrodes. This implies that the voltage required to obtain certain electric field amplitude can be considerably reduced. For example, while the typical voltage for electro-optic control in conventional optical systems is of the order of several kV, in integrated optic devices the voltage required is only few volts. (b) *Fast operation.* The small size of the control electrodes in an electro-optic integrated photonic device implies low capacitance, allowing fast switching speed and high modulation bandwidth. (c)*High optical power density.* (d) *Compact and low weight.* The use of a single substrate with an area of several millimetres squared for integrating different photonic components makes the optical chip very compact and light weight. and (e) *Low cost.* The development of integration techniques makes mass production possible via lithographic techniques and mask replication; also, the planar technology reduces the quantity of material necessary to fabricate the photonic devices. These aspects are the basis of a low-cost device that would thus have an easy introduction into the market.

4.2 Introduction to Optical Modulation

To understand the general approach to optical modulation it is important to remember that the typical behaviour in approaching something new is to handle it as something we know. Optical signals and devices are an exception to this general approach. Coherent optical waves can be considered as quasi-sinusoidal carrier waves, characterised by their amplitude, polarisation direction, frequency, and phase. In complete analogy to radiofrequency waves, information can, in principle, be imprinted on optical waves by four modulation techniques: optical amplitude modulation (AM, and intensity modulation), optical frequency modulation (FM), optical phase modulation (PM), optical polarisation modulation. Regardless of the chosen modulation method, the optical signal modulation is achieved through a change in the optical field, generally due to an electric signal.

In the electrical modulation it is possible to distinguish between the *carrier* signal and the *modulating* signal. The information is carried by inducing variations in the carrier signal due to the modulating signal. Once the carrier signal characteristic is known (through a standard), it is possible to recover the information bit from its changes [8]. On the other hand, in optical modulation, the final goal so far pursued is to change the signal amplitude by using either direct AM or AM due to a phase shift induced in an interferometric structure. In fact, in this case the carrier signal frequency (e.g. 200 THz for a wavelength of 1.55 μm, the one most used in telecommunications) is too high to be detected, hence it would not be possible to recover the information coded therein through either PM or FM.

Optical modulation can, in principle, be achieved using either internal or external modulation of the light source. In the first case the source (either a laser or an LED) current can be modulated to modify the light signal intensity (AM). This approach is straightforward, inexpensive, and compact since two functions are achieved with the same device. Direct modulation transmitters have impedance matched and lasers directly connected to the modulating signal. This technique is very cost effective and can provide modulations up to $2.5\,\mathrm{Gb\,s}^{-1}$ [9]. The main drawback is the limited extinction ratio (see later for definition), since we do not want to turn off a laser at the 0-bit. This will cause an impact on the distance per bit-rate product: the output frequency shift with the drive signal. The last effect is due to two main causes: the carrier-induced wavelength variation, the chirp, caused by variations in electron densities in the lasing area; and the temperature variation due to carrier modulation. When the laser current is quickly changed, the different electron densities affect the refractive index and, therefore, produce a wavelength variation. This effect is highly undesirable in, e.g. long-haul systems because of the chromatic dispersion present in such systems.

The chirp effect is solved by going to external modulation. In this case, the laser is biased to a constant value, its specific operating point. Once generated, the optical beam is passed through an optical intensity modulator. The main advantages of external modulation are higher speed, data rates greater than $40\,\mathrm{Gb\,s}^{-1}$ have been reported [10]; low chirp, even if strongly reduced the chirp is still present; large extinction ratio, the optical signal can be fully absorbed in the modulator region; and low modulation distortion. Also in this case some drawbacks must be faced: it is an additional component, which means additional costs; additional loss, named insertion loss, in the range 5–7 dB, is usually obtained including an external modulator in the optical path. Nevertheless, external modulators provide the best signal quality for both PM and AM.

The generation of a phase-modulated signal requires an external modulator capable of changing the optical phase in response to an applied voltage. The most common phase modulator is based on the electro-optic effect: an electric field applied to an electro-optic material, such as $LiNbO_3$, induces a change in its refractive index [7]. If the electric field is applied through a channel waveguide, the change in the refractive index induces a change in the propagation constant of the structure, and therefore the light travelling through that region undergoes a certain phase shift.

The other approach to optical modulation is using pure amplitude modulators [11]. They work as very fast switches. One of the simplest ways to perform this task is to build an integrated Mach-Zehnder interferometer (MZ or MZI, see later) on an electro-optic substrate. Another approach to fabricate amplitude modulators is to make wave-vanishing modulators. A more detailed description of the different modulation approaches is provided in Sect. 4.3.

The next important step in the introduction to optical modulation is the definition of the parameters that identify a device as a modulator. The performances of an optical modulator can be characterised by four parameters [12]:

1. The contrast ratio (CR). It is defined as the On/Off ratio of the power or transmission coefficients in amplitude modulators. Directly from the CR another parameter can be identified, the modulation depth (M), which is defined as:

$$M = \left(1 - \frac{P_{\mathrm{ON}}}{P_{\mathrm{OFF}}}\right) \times 100, \tag{4.1}$$

where P_{ON} and P_{OFF} are the output power when the device is in the On and Off state, respectively. It is indicated in percentage and a device can be considered a modulator if M is above 20–25% [13].

2. The voltage V_π required to obtain a π phase shift for phase modulators or the voltage V_d to obtain a switch from the Off to the On state in amplitude modulators. In phase modulators it is fundamental to change the signal phase without modifying its amplitude, as it will be clearer when the typical device structure, the MZI, is described. The phase change $\Delta\phi$ is obtained, as mentioned earlier, by changing the material refractive index (Δn) within the device length L according to the following formula:

$$\Delta\phi = \frac{2\pi\Delta n L}{\lambda_0}, \tag{4.2}$$

where λ_0 is the optical signal wavelength. The π variation is obtained as a trade-off between the maximum voltage, required to obtain the refractive index variation, and the maximum device length allowed in the device structure. Once V_π is defined, the optimum device length (L_π) is obtained as:

$$L_\pi = \frac{\lambda_0}{2\Delta n}. \tag{4.3}$$

3. The bandwidth Δf.
4. *The insertion loss.* It is defined as the total loss in the device. It is obtained by adding the coupling loss that can be reduced to below 1 dB and the transmission loss in the device region. Typical insertion loss is of 5–7 dB. Of course, the insertion loss increases with the device length and there is a trade-off with the CR.

Finally, there is a trade-off among these parameters. As an example the CR is proportional to the device length whereas the bandwidth is inversely proportional to L. Therefore, a fifth parameter can be defined, the figure of merit (F), for an overall characterisation of the device:

$$F = \frac{\Delta f \lambda}{V_\pi} \quad [\text{GHz } \mu\text{m/V}]. \tag{4.4}$$

4.3 Mechanisms for Light Modulation

Light can be modulated using different physical effects, depending on the material used to fabricate the modulator. In this section a brief overview of the different physical phenomena is provided.

Certain transparent materials change their optical properties when subjected to an external perturbation, as an acoustic wave, an electric field, or an increase in temperature. Not all the materials are affected in the same way, as it depends on their crystal structure.

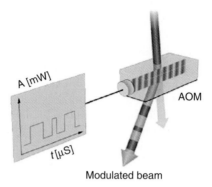

A [mW]

AOM

t[μS]

Modulated beam

Fig. 4.1. The acousto-optical effect from [14]

4.3.1 Acousto-Optical Effect

The refractive index of an optical material can be modified by the presence of sound, hence sound can control light, and this effect is known as acousto-optical effect. Such an effect is a branch of the photoelastic or optoelastic effects [5]. An acousto-optical crystal (see Fig. 4.1) is a material that modifies its refractive index when an acoustic wave travels within its structure. Sound waves in the crystal act as in a gas and travel through compressions and rarefactions of the medium. In those regions where the medium is compressed, the density is higher and the refractive index larger; where the medium is rarefied, its density and refractive index are smaller. In particular, in solids it produces vibration of the molecules around their equilibrium positions, which alter the optical polarisability and consequently of the refractive index. The light passing through the crystal is diffracted or transmitted, depending on the acoustic drive signal applied to the crystal. Regions of different density are created when the sound travels through the crystal. A schematic is shown in Fig. 4.1, from [14]. The light beam passing through the crystal is diffracted according to the pattern (lines in the crystal) defined by the acoustic wave. The diffraction pattern is a function of time, following the drive signal (on the left). When the fixed frequency drive signal is turned on light is diffracted, and when the signal is off light passes through the crystal without diffraction. The light speed is many orders of magnitude larger than the acoustic waves, hence at every moment light views the acoustic wave as a standing wave. The crystal then acts as a grating for the incoming light diffracting the transmitted light. The angle of the diffracted light depends on the wavelength of the density variations, i.e. the drive signal frequency. By changing the frequency the diffraction angle can be controlled. Silicon does not exhibit acousto-optical effect.

4.3.2 Electro-Optical Effects

Certain materials change their refractive index when subjected to a static (d.c.) or a low-frequency electric field. This is caused by forces that distort the positions, orientations, or shapes of the molecules forming the material. The origin of this effect is the atomic response to the electric field. In particular, the application of a d.c. or slowly varying electric field is able to displace atoms and electrons away from their equilibrium positions, thus modifying the polarisability of the medium. Since optical waves have wavelengths three orders of magnitude bigger than the atomic dimensions, the optical response is the average of the response of many atomic orbitals [5, 15].

The dependence of the refractive index on the applied field can be either linear or quadratic. In the first case the effect is known as linear electro-optical effect or Pockels effect, in the second case it is known as quadratic electro-optical effect or Kerr effect. The refractive index of an electro-optical medium is a function of the applied electric field E. Since this function varies only slightly with E, it can be expanded in a Taylor's series about $E = 0$, where the second- and higher-order terms can be neglected as they are many orders of magnitude lower than n:

$$n(E) = n - \frac{1}{2}rn^3 E - \frac{1}{2}\xi n^3 E^2 + \cdots, \tag{4.5}$$

where r and ζ are Pockels and Kerr coefficients, respectively.

When the third term provides a contribution much lower than the second the medium is a Pockels medium:

$$n(E) = n - \frac{1}{2}rn^3 E. \tag{4.6}$$

Typical Pockels coefficient values are in the range 10^{-12} to $10^{-10}\,\mathrm{m\,V^{-1}}$ and typical Pockels media are $NH_4H_2PO_4$ (ADP), KH_2PO_4 (KDP), $LiNbO_3$, $LiTaO_3$, and CdTe [15]. In general n is dependent upon the direction of the applied field with respect to the crystal axes, hence it is necessary to univocally define the field direction with respect to the crystal axis. The devices are labelled as longitudinal or transverse according to the electric field direction with respect to the light direction. The most commonly used material is the lithium niobate ($LiNbO_3$) having $r \sim 3.1 \times 10^{-11}\,\mathrm{m\,V^{-1}}$. This effect is not present in silicon due to its centro-symmetric crystal structure.

In centro-symmetric crystals the coefficient r is 0, the second term of (4.4) vanishes and the refractive index dependence on the applied electric field is:

$$n(E) = n - \frac{1}{2}\xi n^3 E^2. \tag{4.7}$$

In this case the material is known as a Kerr medium. Typical Kerr coefficient values are in the range 10^{-18} to $10^{-14}\,\mathrm{m\,V^{-1}}$ in crystals and 10^{-22} to $10^{-19}\,\mathrm{m\,V^{-1}}$ in liquids [5].

A very careful study on the Kerr effect in Si has been performed by Soref and Bennet [16]. In particular, they calculated the refractive-index change (Δn) of crystalline Si produced by an applied electric field (E) at room temperature for wavelengths ranging 1.0–2.0 μm. They estimated the strength of the Kerr effect in crystalline Si using the anharmonic oscillator model of Moss [17] and assuming oscillation frequency ω much lower than the oscillator resonance frequency ω_0 ($\omega \ll \omega_0$). It gives the result:

$$\Delta n = -3e^2 \left(n^2 - 1\right) \frac{E^2}{2nM^2\varpi_0^4 x^2}, \tag{4.8}$$

where e is the electronic charge, n the unperturbed refractive index, M the effective mass, and x the average oscillator displacement. It is interesting to note that the n perturbation is independent of optical wavelength.

Assuming $n = 3.50$ at $\lambda = 1.3\,\mu m$, M as the electron mass, a resonance frequency ω_0 of $2\pi \times 10^{15}\,\text{rad s}^{-1}$, and an x value of $10^{-9}\,m$, (4.7) provides the refractive index variation Δn as a function of the applied electric field E. The results of the calculation are summarised as a solid line in Fig. 4.2 from [16]. The calculation provides a Δn of 10^{-4} when the electric field reaches $10^6\,\text{V cm}^{-1}$. Unfortunately, such electric field value is above breakdown in lightly doped silicon.

Another electro-optical effect in Si is the Franz–Keldysh effect. It is known that the photoabsorption coefficient (α) rises when the band edge energy is reached. If a semiconductor has a direct bandgap α rises very steeply, while, as in the case of silicon, indirect bandgap semiconductors have a smaller α increase. In any case, the absorption threshold is slightly shifted towards lower energies if a strong electric field is applied to the semiconductor. This effect

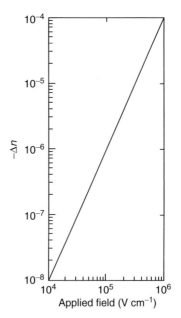

Fig. 4.2. Kerr effect in c-Si vs. E as determined from anharmonic oscillator mode, see [16]

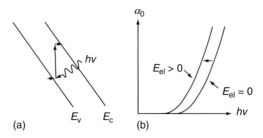

Fig. 4.3. Franz–Keldysh effect. (**a**) A strong electric field tilts the band structure. When a valence electron absorbs a photon tunnels into the conduction band. (**b**) A red shift is observed. Figure from [15]

is due to the distortion of the energy bands of Si. Upon the application of an electric field the electrons can tunnel across the bandgap, thus a lower energy than in band-to-band absorption is required. As explained in Fig. 4.3 [15], the tunnelling process is strongly favoured if a photon, having energy close to bandgap, is absorbed. Such effect is more evident at wavelengths close to the bandgap. The Franz–Keldysh effect is also called photon-assisted tunnelling. It actually involves both electroabsorption ($\Delta\alpha$ vs. E effects) and electrorefraction (Δn vs. E effects), but the last effect is much lower. The electroabsorption spectrum at the indirect edge has been measured in detail by Wendland and Chester [18]. Also in this case, Soref and Bennet [16] performed a numerical calculation to investigate the electrorefraction and the results are summarised in Fig. 4.4 $\Delta n(h\nu)$ was calculated over the range from $h\nu = 0.77$–$1.23\,\mathrm{eV}$. This range includes a transparent region and a $0.11\,\mathrm{eV}$ excursion above the nominal gap. Then, Δn was expressed as a function of optical wavelength from 1.00 to 1.60 μm.

It is found that Δn reaches a maximum at 1.07 μm, a wavelength slightly below the nominal λ_g. Then, as λ increases Δn decreases very rapidly. At 1.07 μm the Δn is 1.3×10^{-5} for and external field of $100\,\mathrm{kV\,cm^{-1}}$. The change in the index as a function of the applied field is plotted in Fig. 4.4 at the optimum 1.07 μm wavelength and at the nearby 1.09 μm wavelength. The figure clearly shows that the Frank–Keldysh effect decreases very rapidly increasing the wavelength, to become close to 0 at 1.55 μm, the wavelength of interest in optical telecommunications. This effect is preferably used in direct bandgap semiconductors, where the absorption threshold is well defined.

4.3.3 Thermo-Optical Effects in Silicon

In order to explain the thermo-optical effect, the refractive index can be written as a complex number, $n + \mathrm{i}k$, where n (conventional refractive index) is the real part and the optical extinction coefficient k is the imaginary part. This last coefficient is directly related to the optical absorption coefficient of the material (α) by the relation $k = \alpha\lambda/4\pi$.

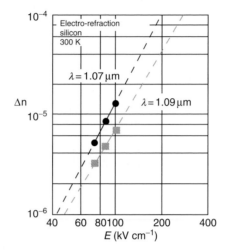

Fig. 4.4. Field dependence of electrorefraction at 1.07 μm (*circle*) and 1.09 μm (*square*) for Si, see [16]. *Dashed lines* are extrapolations

The thermo–optical effect lies on the fact that the refractive index is not a constant with respect to temperature. Thermally induced modulations of the complex refractive index originate from three major effects, namely the changes in the distribution functions of carriers and phonons, the temperature-induced shrinkage of the bandgap, and the thermal crystal expansion. As the temperature rises, the latter mechanism decreases the optical density and, consequently, the refractive index. Since an increase is experimentally observed, the two former effects are obviously decisive. The variation of the refractive index with the temperature at a constant pressure is called the thermo-optical coefficient.

To evaluate the thermo-optical effect the n variation as a function of temperature $(\partial n/\partial T)$, the thermo-optical coefficient, must be observed. Its unit is per degree centigrade or Kelvin, and it ranges between 10^{-3} and 10^{-6} [K^{-1}]. The thermo-optical coefficient of most of the semiconductors commonly used in optoelectronics has been measured by Bertolotti et al. [19] using the prism technique, in a wide wavelength range, from 0.5 to 12 μm, for temperatures in the range 15–35°C. The results for silicon and germanium are shown in Fig. 4.5. They found a value of 1.86×10^{-4} K^{-1} at 1.55 μm for the thermo-optical coefficient in Si.

Typically, the Si thermo-optical coefficient is three times greater than in classical thermo-optical materials and eight times greater than in silica-based materials [20]. Studies [21] performed as a function of the temperature in a wider range, 300–580 K, and reported in Fig. 4.6 show a linear dependence of the thermo-optical coefficient in the observed temperature range with a variation from the 1.86×10^{-4} K^{-1} value measured at room temperature up to 2.4×10^{-4} K^{-1} at the maximum temperature measured. The data indicate no dependence on sample doping or crystal orientation.

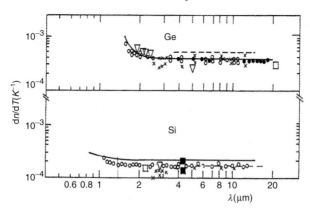

Fig. 4.5. Temperature coefficients of the refractive indexes of Ge and Si. Calculated $\mathrm{d}n/\mathrm{d}T$ dependence [19] (*solid lines*)

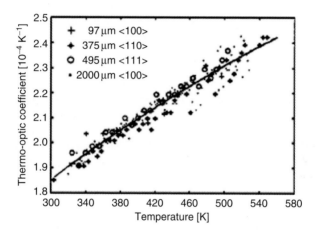

Fig. 4.6. Measured thermo-optical coefficient in various silicon samples. Interpolation of data as a second-order polynomial is shown as a *solid line* [21]

The linear dependence was fitted by a polynomial showing a weak quadratic dependence. In particular,

$$\frac{\partial n}{\partial T} = 9.48 \times 10^{-5} + 3.47 \times 10^{-7}T - 1.49 \times 10^{-10}T^2 \quad (\mathrm{K}^{-1}). \quad (4.9)$$

Finally, the Si room temperature value is double of the $\mathrm{LiNbO_3}$ coefficient at the same wavelength and 15-fold higher than the fibre silica coefficient [22].

As mentioned earlier, not only is the real part of the complex refractive index affected by optical properties modifications, e.g. due to temperature changes, but also the absorption coefficient varies. Nevertheless, for the thermo-optic effect, the absorption variation ($\mathrm{d}\alpha$) as a function of the temperature ($\mathrm{d}T$) for a lightly doped (1×10^{16} cm^{-3}) Si sample at a wavelength of 1.55 μm is of $d\alpha/dT \approx 10^{-4} \mathrm{cm}^{-1}\mathrm{K}^{-1}$, hence the absorption coefficient variation is low [23, 24].

4.3.4 Free Carrier Absorption

The most important physical phenomenon affecting the silicon refractive index is the free carrier absorption. In fact, the optical properties of silicon are strongly modified by injection of charge carriers into an undoped sample $(+\Delta N)$ or by removal of free carriers from a doped sample $(-\Delta N)$. In particular, a photon can be absorbed by free carriers causing intra-band transitions, the inter-band contribution being negligible. On account of the high density of states at the band edges, the optical excitation of an electron from the valence band to the conduction band exhibits only a very weak dependence on the carrier concentration [16]. Therefore, a change of the complex refractive index due to the injection of carriers originates from an increase of the free carrier absorption. Optically, it does not make much difference whether carriers come from impurity ionisation or from injection, thus they will be assumed as equivalent in the rest of the discussion. Three carrier effects are important (1) traditional free carrier absorption, (2) Burstein–Moss bandfilling that shifts the absorption spectrum to shorter wavelengths, and (3) Coulombic interaction of carriers with impurities, an effect that shifts the spectrum to longer wavelengths. They act simultaneously. What is experimentally observed is a red shift of the absorption spectrum. Hence, Coulombic effects are stronger than bandfilling in crystalline Si [16,25,26]). To theoretically describe the free carriers-optical effect, the Drude model [27], which regards the charge carriers as harmonic oscillators with vanishing binding energies, has been used. Solving the equation of motion in the frequency domain the electrical polarisations due to the displacement of the electrons and the holes are obtained. The complex refractive index is then calculated by expanding the square root of the dielectric constant. The calculation provides the variation of both the real part of the refractive index Δn and the absorption coefficient $\Delta \alpha$ as a function of the variation of the electrons and holes concentrations, ΔN_e and ΔN_h, respectively,

$$\Delta n = -\frac{e^2 \lambda^2}{8\pi^2 c^2 \varepsilon_0 n} \left(\frac{\Delta N_e}{m_{ce}^*} + \frac{\Delta N_h}{m_{ch}^*} \right), \tag{4.10}$$

$$\Delta \alpha = \frac{e^3 \lambda^2}{4\pi^2 c^3 \varepsilon_0 n} \left(\frac{\Delta N_e}{\left(m_{ce}^*\right)^2 \mu_e} + \frac{\Delta N_h}{\left(m_{ch}^*\right)^2 \mu_h} \right), \tag{4.11}$$

where n is the refractive index of unperturbed crystalline Si, e the electron charge, λ the light wavelength, ε_0 the permittivity of free space, c the speed of light, m_{ce}^* and m_{ch}^* are the conductivity effective mass of electrons and holes, respectively, while μ_e and μ_h are the electrons and holes mobility, respectively. Substituting the material parameters of silicon at the two wavelengths most important in telecommunication (1.3 and 1.55 μm) they become:

$$\Delta n = \Delta n_e + \Delta n_h = -\left(8.35 \times 10^{-22}\Delta N_e + 5.70 \times 10^{-22}\Delta N_h\right)$$
$$\Delta \alpha = \Delta \alpha_e + \Delta \alpha_h = 2.51 \times 10^{-19}\Delta N_e + 3.91 \times 10^{-19}\Delta N_h$$, (4.12)

$$\Delta n = \Delta n_e + \Delta n_h = -\left(1.2 \times 10^{-23}\Delta N_e + 8.0 \times 10^{-24}\Delta N_h\right)$$
$$\Delta \alpha = \Delta \alpha_e + \Delta \alpha_h = 3.6 \times 10^{-19}\Delta N_e + 5.1 \times 10^{-19}\Delta N_h$$, (4.13)

respectively, where $\Delta n_e(\Delta \alpha_e)$ and $\Delta n_h(\Delta \alpha_h)$ are the $n(\alpha)$ changes due to electrons and holes contributions, respectively. However, an accurate description of the plasma-optical effect needs to consider all scattering processes that are assisted by phonons or impurities since free carrier absorption requires the interaction with a third particle for momentum conservation. Although these effects are implicitly included in the mobility, they are not treated rigorously by the Drude model. In order to obtain a theory closer to the experimental observations, Soref and Bennet [16] collected experimental data about the absorption spectra of heavily doped samples [28,29] and calculated the real part of the refractive index from the Kramers–Kronig relation [30]. In fact, since n and k are related by the Kramers–Kronig dispersion relations, the same relations hold for their variations, Δn and Δk. The results of their calculations are at 1.3 μm:

$$\Delta n = \Delta n_e + \Delta n_h = -\left[6.2 \times 10^{-22}\Delta N_e + 6.0 \times 10^{-18}\left(\Delta N_h\right)^{0.8}\right], \quad (4.14)$$

$$\Delta \alpha = \Delta \alpha_e + \Delta \alpha_h = 6.0 \times 10^{-18}\Delta N_e + 4.0 \times 10^{-18}\Delta N_h, \quad (4.15)$$

and at 1.55 μm:

$$\Delta n = \Delta n_e + \Delta n_h = -\left[8.8 \times 10^{-22}\Delta N_e + 8.5 \times 10^{-18}\left(\Delta N_h\right)^{0.8}\right], \quad (4.16)$$

$$\Delta \alpha = \Delta \alpha_e + \Delta \alpha_h = 8.5 \times 10^{-18}\Delta N_e + 6.0 \times 10^{-18}\Delta N_h. \quad (4.17)$$

The most important difference with respect to the Drude model shown here is a Δn dependence on $\Delta N_h^{0.8}$. Actually, they also report a Δn dependence on $\Delta N_e^{1.05}$. The last dependence is quite close to the Drude theory value and is generally not considered. More recently, Huang et al. [30] reported a concentration dependence proportional to $\Delta N_e^{1.07}$ and $\Delta N_e^{1.03}$ at a wavelength of 1.3 μm.

The real part of the complex refractive index and the absorption coefficient, as derived from (4.14) and (4.15), are plotted in Fig. 4.7a, b, respectively.

4.4 Device Structures

Once the physical principles for light modulation in silicon have been reviewed, a brief review of the devices proposed in literature can be done. Before such a review, special attention must be dedicated to a device structure widely used

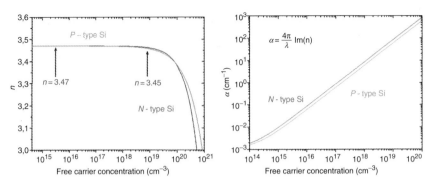

Fig. 4.7. (**a**) Real part of the refractive index and (**b**) absorption coefficient for a wavelength of $1.55\,\mu$ m, as derived from (4.16) and (4.17), respectively

to design light modulators. As mentioned in earlier sections, optical modulation is achieved using either pure amplitude modulators or phase modulators integrated in an interferometric structure. The most widely used structure is the MZI; hence the first part of this section is devoted to the study of such a device structure.

If light coming from a coherent source is split into two beams, and each follows a slightly different optical path, when recombining them, an interference pattern is obtained due to the phase change between both beams. It provides information concerning the difference between the optical paths. This operation principle lies at the heart of MZIs. A schematic of an MZ is shown in Fig. 4.8 (from [31]). The MZI starts with a channel single-mode waveguide and then splits into two symmetric branches by means of a Y-branch (input coupler). After some distance, the two branches become parallel. The MZI continues with a symmetric reverse Y-branch and ends in a straight waveguide. If the MZI is exactly symmetric and if the optical path on both branches is exactly equal, the input light splits at the first Y-junction into the two parallel branches and then recombines constructively into the final straight waveguide. On the contrary, if in one of the interferometer's arms the light suffers a phase shift of $180°$, at the end of the second Y-branch the light coming from the two branches will recombine out of phase and will give rise to destructive interference, with no light at the output. The light is then leaked away in the substrate. In practice, the phase shift in one arm is carried out via one of the above-mentioned effects, by applying a voltage across the waveguide. By adequately designing the active area (proper length, see earlier), the electrode geometry, and the applied voltage, a total phase shift of $180°$ can be obtained for a specific wavelength. Once the MZ working principle is known, it becomes clear that the light absorption in the "active" interferometer arm must be as low as possible, in order to maximise the device modulation depth. This device is designed in order to have $M = 100\%$. It can be proven that the best performances are obtained when the splitting is exactly the same at both branches.

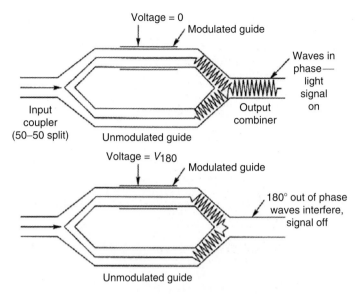

Fig. 4.8. Structure and operation principle of an MZI. In the upper scheme the light is transmitted, in the lower scheme a V_π is applied to the "active" device arm, a $180°$ phase shift is obtained with respect to the "reference" arm and the signal output is 0

Although this division can be easily achieved in classical optics (using a beam splitter), this is not so easy in integrated optics. So far, there are two main problems with integrated MZI. First of all the precision and the size of the Y-junction edge depend on the technology used. Moreover, the angle between waveguides rarely has a value above $1°$. Thence, waveguides slowly separate, and total device lengths of 1 cm have been reported for Si-based high-speed modulators [32]. Standard technology is unable to produce this kind of edges accurately and normally they become rounded when the distance between the waveguides is the same than the minimum dimension obtainable with the technology used. This rounding is uncontrollable. The second problem with MZI arises from its modal behaviour. The simple description done before perfectly describes the MZI operation if there is a single mode light signal on each branch, that is, all waveguides in the MZI are single mode. If it is not so, the intensity pattern at the output will not be defined by simple two-beam interference, since more field contributions must be considered. Hence, if waveguides have to be single mode in the cross-section, their ribs have to be as small as possible. However, as the rib decreases, the confinement is worse and losses at the Y-junctions dramatically increase. Hence, it is necessary to reach an agreement between the confinement factor and the modal properties of the MZI waveguides.

Finally, an MZI can be designed in a "push–pull" configuration, as shown in Fig. 4.9. The phase shift is obtained by applying a phase shift of $\Delta\phi/2$ to

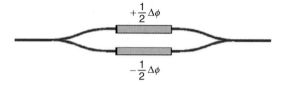

Fig. 4.9. MZI in a push–pull configuration. The phase shift is obtained by biasing both arms

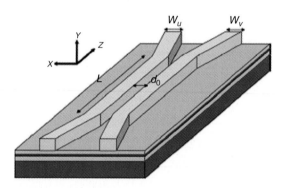

Fig. 4.10. A directional coupler. Two single-mode waveguides (u and v) having width W_u and W_v are parallel over a length L with a distance d_0 between them

both arms. The bias sign is opposite in order to get a final $\Delta\phi$ shift of 180°. This configuration can be used, e.g. to diminish the peak voltage.

Another interesting device structure is the directional coupler. If it is based on the evanescent field, it may be one of the most unused light properties in confined media. Any guided mode has an evanescent tail that propagates off from the core. If a second waveguide is closely placed so as to overlap the evanescent field of the modes propagating by both waveguides, a coupling between them will be produced. The distance that the waveguides have to be parallel so as to transfer all the power (L) depends on the working wavelength and morphology. If light is injected to one of the waveguides, a periodic interchange of power between both waveguides will be obtained. Moreover, if the waveguides are identical, there will be a complete power transfer. A schematic of such a structure is presented in Fig. 4.10.

Assuming weakly interacting fields and single-mode waveguides, the coupled-mode theory [33] can be applied, by obtaining the expressions for the field evolution along the z direction. A coupling constant can be defined, depending on the evanescent field overlap and being a function of the mode confinement inside the waveguide (which, in turn, is a function of the waveguide width and the rib etch), the separation between waveguides and the working wavelength [34]. There exists a periodical power exchange as a function of the propagation distance. As an application example, directional couplers with identical waveguides can be used for the design of optical switches. If light is injected only in one waveguide and the parallel distance is the coupling length, all the power injected from waveguide u is transferred to waveguide v. If a change in the propagation constants is introduced so as to reduce the coupling length to half, power will be no longer at v, but will return to u. This change in the optical constants can be done using the electro-optic effect [35]. If a given oscillating voltage is applied, an oscillating value of the propagation constant will be obtained, causing the output signal modulation. It should be noted, however, that this operation principle can only be applied to those compounds, such as $LiNbO_3$ or III–V compounds that have extremely high electro-optical coefficient. Unfortunately, this is not the case for silicon. Hence, directional couplers based on silicon technology cannot be applied for electro-optical devices.

4.5 Devices Proposed in Literature

Many devices have been presented in literature, based on the physical principles described earlier. Approaches to modulation using either the thermo-optical or the carrier injection (or accumulation), modulating the refractive index in a one arm of an MZI, or total internal-reflection-based structures, cross-switches, Y switches, and Fabry–Perot (FP) resonators will be reviewed. The absence of mechanical elements in these devices makes them more reliable than micro-electromechanical systems (MEMS). A list of the devices proposed in literature to date is shown in Table 4.1.

Most of the listed devices present common features: long interaction distance, high injection current densities and power consumption in order to obtain a useful modulation depth. Such length and power requirements impose difficulties in integrating these devices on-chip. There is, therefore, an urgent need for structures that can be implemented in a micron-size region offering low current density, low power consumption, and high modulation depth. In fact, according to Cadien [36] from Intel, a Si-based electro-optical modulator must satisfy very stringent requirements in order to go into the market. The requirements for a modulator to be monolithically integrated in telecommunication systems are summarised in Table 4.2.

4.5.1 MEMS Devices

As mentioned in Sect. 4.2 light modulation is achieved, using different approaches, by AM of the beam. The majority of light modulators can be considered as shutters, i.e. when activated they cut off a beam of light or disperse it. This can be done mechanically, e.g. by introducing an opaque object on the beam path, as shown in Fig. 4.11.

In general, a MEMS modulator is formed by an input waveguide, an actuated region, and an output waveguide. The actuator may act on a waveguide, as shown in Fig. 4.12, e.g. bending it with a piezoelectric element. When the

Table 4.1. Si electro-optic modulators in literature

year	author	el. struct.	optical struct.	λ (μm)	M (%)	J (kA cm^{-2})	τ_s (ns)	ν_w (MHz)	length (μm)	P/F	ref.
1987	Lorenzo	p–i–n	coupler	1.3	50	1.26			2,000	F	[37]
1989	Hemenway	p–i–n	EA	1.3	10	100	18	200	vertical	F	[38]
1991	Treyz	p–i–n	EA	1.3	75	3.0	50	≈40	500	F	[39]
1991	Treyz	p–i–n	MZ	1.3	65	1.6	50	≈40	500	F	[40]
1991	Xiao	p–i–n	FP	1.3	10	6.0	25	≈80	vertical	F	[41]
1993	Huang	impact	EA	1.3			1	1000	2,000	P	[42]
1994	Liu	p–i–n	EA	1.3	90	9.0	200	≈10	800	F	[43]
1994	Wang	μMech	Bragg	1.5	40				300	P	[44]
1994	Liu	p–i–n	coupler	1.3	90	12.5	100	≈20	200	F	[45]
1995	Zhao	p–i–n	MZ	1.3	98		200	≈10	816	F	[46]
1995	Liu	p–i–n	FP		60				100	P	[47]
1996	Zhao	p–i–n	coupler	1.3	97	1.027	210	≈10	1,103	F	[48]
1996	Marxer	μMech	FP	1.3			180	≈10	55	F	[49]
1997	Cutolo	p–i–n	Bragg	1.5	50		24	≈80	3,200	P	[50]
1997	Cutolo	BMFET	EA	1.5	20	2.3	6	≈300	1,000	P	[51]
1997	Zhao	p–i–n	FP		88	8.8	110	≈20	190	F	[52]
1998	Breglio	BMFET	Bragg	>1.2	20		5.6	≈ 300	1,000	P	[53]
2000	Hewitt	p–i–n	EA	>1.2			39	≈40	500	P	[54]
2000	Irace	BMFET	coupler	>1.2			8	200	5900	P	[55]
2001	Coppola	p–i–n	Bragg	1.5	94		9.3	≈200	3,200	P	[56]
2001	Coppola	BMFET	coupler	>1.2			8	≈200	5,000	P	[57]
2003	Sciuto	BMFET	EA	>1.2	90s				100	F	[58]
2003	Barrios	p–i–n	FP	1.55	80	0.116	21		20	P	[20, 59]
2003	Barrios	p–i–n	FP	1.55	80	0.61	1.3		10	P	[20, 60]
2004	Liu	MOS	MZ	>1.2				1 GHz		F	[32]
2004	Png	p–i–n	MZ	>1.2				5 GHz		P	[61]
2005	Paniccia	MOS	MZ	>1.2				2.5 GHz		F	[62]

BMFET: bipolar mode field-effect transistor; *EA*: free carrier absorption; *J*: current density; τ_s: switching time; P: proposed, F: fabricated

Table 4.2. Requirements for a modulator to be monolithically integrated in telecommunication systems

requirements	metric
high bandwidth	>20 Gb s^{-1}
low insertion loss	<3 dB
low capacitance	< 10 fF
low voltage operation	<1 V
small area	<100 μm^2
compatible CMOS processing	meets EHS and micro-contamination spec.
operational stability	>100°C, >7 years
withstand thermal processing conditions during chip making and packaging	>250°C, >1 h

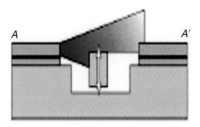

Fig. 4.11. Example of MEMS modulator. An actuated shutter is moved in/out of the optical path

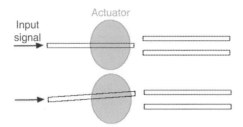

Fig. 4.12. MEMS operation principle. The actuated element is the end of the input waveguide that faces one of the two output waveguides, depending on the actuator state

waveguide is bend, it faces the output guide. Other examples of actuators are galvanostatic mirrors [63], gates shifting thanks to springs [64], and Bragg reflectors whose grating dimensions vary using actuated springs [65].

One of the main drawbacks with MEMS is their slow response, with switching speeds of about 1–100 ms. Hence, they are used more as switches than as modulators.

4.5.2 Thermo-Optical Modulators

A variation of the refractive index can be achieved using the thermo-optical effect. The device may be placed as an active arm of an MZI. Of course the waveguide warming must be limited to one arm of the MZI; hence, an particular care must be taken in designing the device structure. The first prototype was proposed by Treyz [66] using silicon-on-insulator (SOI) technology. Such a device was heated by means of electrical power dissipation in a resistive layer, covering one of, or both, the MZI arms. The modulator bandwidth was few tens of kilohertz, the speed limited by the Si thermal dissipation. An improved version of such device was proposed by Cocorullo et al. [67,68]. A schematic of their device is shown in Fig. 4.13. They used both silicon-on-silicon (SOS) and SOI substrates, the last preferred in order to reduce the insertion loss (well below 1 dB). However, the thermal conductivity of SiO_2 is about two orders of magnitude lower than that of Si ($K_{SiO_2} = 0.013 \, \mathrm{W \, K \, cm^{-1}}$). The

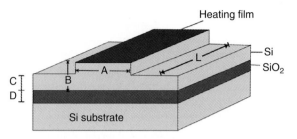

Fig. 4.13. The thermo-optical modulator proposed by Cocorullo et al. [67]

SiO$_2$ layer represents a bottleneck to the heat dissipation towards the heat sink at the wafer bottom. This consideration has its pros and cons. Due to the better isolation with respect to SOS, lower power is dissipated to achieve the same temperature variation. On the other hand, the draining of the stored heat may be slower; hence longer switching times could be obtained. Their calculations show that this is not the case. Their device has $A = B = 3\,\mu m$ (see figure), C = 1.5 μm for an active region length of 1,000 μm. The power pulse is 2 W, to achieve a temperature increase of 10°C, and the maximum operation frequency achieved is 2.3 MHz with a measured M (see (4.1) of 55%. Their calculation indicates M up to 70%.

Smaller temperature ranges for thermo-optical device operation have been proposed. In particular, a temperature rise of 7°C can be achieved on a 500 μm-long device by applying 10-mW power as shown in [69]. The thermo-optical effect, however, is rather slow and can only be used for below 10-MHz modulation frequencies.

4.5.3 Free Carrier Absorption-Based Modulators

One of the first free carrier absorption-based modulators in Si was proposed by Soref and Bennett in 1986 [70]. A device schematic (see [61]) is shown in Fig. 4.14. It is a p$^+$–n–n$^+$ silicon junction, designed in a single-mode rib waveguide. Their device is polarisation independent, in a first approximation, and their calculations indicate a device length below 1 mm to obtain a π radian phase shift, an MZI structure. The corresponding loss is less than 1 dB at 1.3 μm. The buried SiO$_2$ layer allows achieving low loss keeping the possibility to have a back contact. The device structure is vertical.

Few years later, Treyz et al. [39] reported the Si waveguide intensity modulators operating in the range of 1.3–1.55 μm shown in Fig. 4.15. The device is a vertical p–i–n structure integrated in a Si rib waveguide; the intrinsic region has a thickness of 7.7 μm with a p-type doping concentration of ~5×10^{15} cm^{-3}. The p$^+$ layer has a doping concentration of ~5×10^{19} cm^{-3} and a thickness of 0.5 μm. By applying a forward bias to the device, free carriers are injected into the p-type Si (guiding) region of the device. The maximum M value achieved was 76% for a high current density of $3.4 \times 10^3\,\mathrm{A\,cm^{-2}}$.

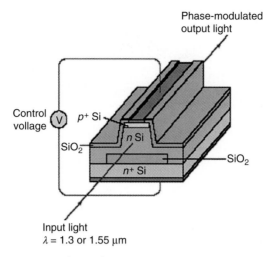

Phase-modulated
output light

Control
voltage

Input light
λ = 1.3 or 1.55 μm

Fig. 4.14. Proposed SOI p$^+$–n–n$^+$ channel-waveguide electro-optical phase modulator [61, 70]

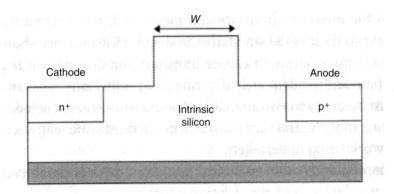

Fig. 4.15. Schematic diagram of the device proposed in [61]. Rib height is 4.6 μm with widths of 13 and 24 μm. Diode lengths are 500 and 1000 μm

The measured response time (10–90% of the transient) is less than 50 ns. However, the performances of this device are strongly limited by the high propagation loss of the waveguide, above 20 dB cm^{-1}, and the devices were multimode.

Since then, many p–i–n structures have been proposed either as full devices [39, 43, 54, 71], or as a part of an MZI [40, 41, 46, 47, 52, 61, 72], to convert PM into AM. The p–i–n structure was modified from vertical (see Fig. 4.15) to horizontal [73, 74]. The main advantage of horizontal p–i–n structures is the possibility to use SOI wafers, thus strongly reducing the insertion loss, since all the electrical contacts can be taken on the sample surface. As an example, in Fig. 4.16 some of the proposed structures are reported. The horizontal p–i–n structures may be asymmetrical, as in Fig. 4.16a, having the

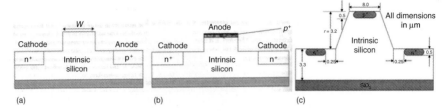

Fig. 4.16. Horizontal p–i–n structures (from [73]): (**a**) the pockets are on the rib sides, (**b**) symmetrical structure, the p$^+$ region is on top of the rib and the n$^+$ regions on both sides, (**c**) symmetrical p–i–n proposed by Tang [74] having non-vertical rib walls

n$^+$ and p$^+$ doped region one at each side of the rib. The main drawback of this structure is that the upper rib region may not be filled by free carriers when the device is in the on state (forward bias), hence the light may pass through the device partially unabsorbed. To overpass this limit a symmetrical structure was proposed. In this case, see Fig. 4.16b, one of the doping pockets (p$^+$) is on top of the rib, while two symmetrical n$^+$ doped regions are placed on both sides of the rib. In this case, when the modulator is switched on, the free carriers completely fill the optical path region. Studies performed on such structures [54,61,73] show that a three-terminal device requires less drive current (2.8 vs. 8 mA) and are faster (29 vs. 39 ns) than a two-terminal device for an equivalent injection concentration. This is because three-terminal devices offer more efficient carrier injection. The drawback is the higher insertion loss due to doping contact at the rib top. Finally, the rib walls can be designed with an angle below 90°, see Fig. 4.16c, in order to improve the efficiency of the free carrier collection. Such a device was proposed in 1995 by Tang and co-workers [74]. It is integrated in a single-mode waveguide and fabricated on SOI wafers. The device length is 500 µm and it was integrated in one arm of an MZI. The maximum operation current is 7 mA and exhibits passive loss of 0.7 dB cm^{-1}, while the loss in the active device is 1.3 dB. The maximum operation frequency reported is 6 MHz. In a very recent paper [72], a series of p–i–n devices having reduced dimensions were modelled. They are phase modulators working with drive currents as low as 0.5 mA. The simulated devices exhibit bandwidths ranging from 70 MHz up to above 1 GHz, having transient rise times of 0.3 ns and fall times of 0.12 ns. The single-mode waveguide dimensions are approximately 1 µm in height and 0.5–0.75 µm in width. Finally, Iodice et al. [71] simulated an integrated waveguide-vanishing-based modulator in SOI technology. It can be used either as a phase modulator or as a pure AM device. Their active device is 3×3 µm and the lateral confinement is guaranteed by two highly doped n$^+$ (8×10^{19} As cm^{-3}) and p$^+$ (2×10^{19} B cm^{-3}) regions with a depth of 1 µm. This p–i–n structure has the main advantage of being planar. Moreover, since the two doped regions face each other, the free carrier injection efficiency is the highest possible. The main drawback of this structure is the high insertion loss due to the lateral confinement. In fact, the device propagation loss is of ~18 dB cm^{-1}, in a single-mode waveguide. The modulator length is 1 mm. The p–i–n diode allows injecting free carriers into the rib volume between the doped regions. The resulting optical behaviour is the vanishing of the confinement in the rib region and the light absorption/dispersion. The biasing voltage is 1.0 V, with an electrical power consumption of 2 mW. According to their simulations a modulation depth close to 100% is reached with a rise time of ~10 ns. Fall time can be faster if a proper reverse polarisation is applied to the device.

An interesting wave-vanishing structure was earlier proposed by Pirnat et al. [75]. A structure cross-section schematic is shown in Fig. 4.17b, while the full device structure perspective is shown in Fig. 4.17a.

The device waveguide consists of an n-type Si core layer (3.5×10^{17} cm^{-3}) on a p-type Si (6×10^{17} cm^{-3}) substrate. When the device is forward biased, the refractive index of the core decreases and the fundamental guided mode is displaced downwards into the substrate. However, the mode is not extinguished because the substrate is bounded by a p$^+$ contact. A spatial filter at the output converts the mode displacement into optical intensity modulation.

Another interesting device is the one proposed by Huang in 1993 [30]. It is based on an impact avalanche transit time-type diode to generate carriers by avalanche multiplication. Such a device, according to the authors, should be free from the minority-carrier lifetime limitation. Their device uses two regions: the first is a small, high-field avalanche region where impact ionisation occurs and carriers are generated; the second is a long, low-field drift region where generated carriers cause the refractive index to be modulated. Modulation speed is improved since the drifting carriers are majority carriers. They simulate switching time below 1 ns, hence there is a modulation bandwidth in the gigahertz range. However, the predicted current density is high, as is the insertion loss.

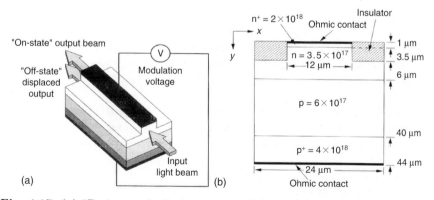

Fig. 4.17. (a) 3D view mode-displacement modulator. (b) Cross-sectional view of device geometry with dimensions and dopant concentrations. Device proposed in [75]

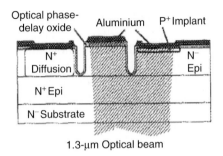

Fig. 4.18. Cross-section of the modulator proposed in [38]

A completely different structure was proposed by Hemenway et al. [38]. They propose an all-silicon reflection-type optical amplitude modulator. Free-carrier dispersion effect is used to spatially modulate the phase of an optical beam. Subsequent PM to AM conversion occurs by mode selection in the coupled single-mode fibre. A schematic of the device is shown in Fig. 4.18. Inside the device half of the beam is phase delayed relative to the other half. The phase delay consists of two parts: a static delay of $\lambda/4$ and a separate, modulated, delay due to free carrier injection. The modulation depth is $\sim$35%, at a driving peak current of 26 mA. A bandwidth of 200 MHz for lower injection currents is reported.

Quite a different approach to light modulation was followed by Wang et al. [44]. They propose an electrically induced Bragg reflector integrated in a SOI waveguide, shown in Fig. 4.19. Interdigitated metal fingers are fabricated on top of a silicon rib waveguide, forming Schottky junctions with the heavily doped silicon film. The Bragg reflector is formed when the metal fingers are alternatively forward and reverse biased. The resulting longitudinal periodic refractive index structure, whose variation Δn is about 5.0×10^{-3}, acts as a Bragg mirror. The simulated device length was 500 μm, and M was 70%, while the device speed, limited by the RC time constant, is in the range of 50 GHz.

A similar device was proposed more recently by Cutolo and co-workers [50]; a schematic of the device structure is shown in Fig. 4.20. Their Bragg reflector was placed in the middle of a lateral p–i–n diode. The Bragg grating has a period of 227 nm, a depth of 45 nm, for a waveguide width of 3 μm, while the interaction length is 3.2 mm. AM was achieved by changing, through injection of free carriers, the Bragg resonance wavelength.

Numerical simulations, made using MEDICI, indicate a modulation depth of 50% obtained with an electrical power consumption of 4.0 mW. A response time of 12 ns, corresponding to a maximum frequency of 42 MHz, and insertion loss of 1.0 dB are reported for this device. The device's performances are similar to those observed for MZI-based modulators in terms of driving currents and response times, but the power consumption is about three times higher

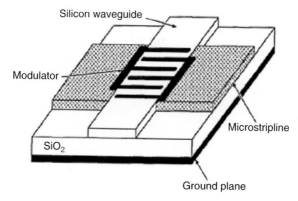

Fig. 4.19. The Bragg modulator proposed in [44]

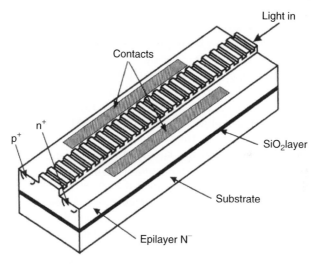

Fig. 4.20. The Bragg optical modulator proposed in [50]

to achieve the same modulation depth. Recent calculations on the same device [76] indicate 1.4-GHz operating bandwidth reducing the rib width. This last device layout has a waveguide height and rib width of 1 μm, etch depth of 100 nm, and an interaction length of 3 mm. The authors attribute this improvement in device performances to both the reduction in dimension and the application of a pre-bias to the modulator to an "off" level (0.6 V) just below the turn-on voltage (0.8 V), which allowed faster movement of the injected carriers.

Finally, a more recent work [59] proposed a modulator based on an FP microcavity formed by two distributed Bragg reflectors (DBRs) embedded in a SOI rib waveguide. Both DBRs have the same number of Si/SiO₂ periods. The transmission of this structure is highly sensitive to small index changes in

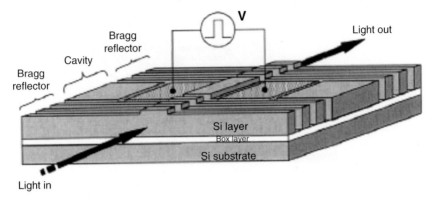

Fig. 4.21. FP microcavity with high-reflectivity Bragg reflectors in an SOI waveguide [59]

the cavity, making it adequate for intensity modulation applications in a short length. Moreover, the device requires a low concentration of injected carriers to switch to a non-resonant position. A schematic of the proposed device is shown in Fig. 4.21.

For illustration purposes, the trenches down to the buried oxide (BOX) layer are drawn as empty, while in the simulated device they are assumed to be completely filled with SiO_2. Heavily doped p^+ and n^+ regions are defined in the cavity region, at both sides of the rib and contacted by metal electrodes. The rib width and Si thickness are 1.5 μm, and the etch depth is 0.45 μm. The 20 μm-long device is predicted to require a d.c power of the order of 25 μW at an operating wavelength of 1.55 μm to achieve 31 MHz operating bandwidth with transmittance of 86%, M of 80%, and no significant thermo-optical effect.

Another approach to Si electro-optical modulation was proposed for the first time by Cutolo and co-workers [51]. Their device is a three-terminal BMFET integrated, in their design, in a single-mode low-loss SOI waveguide. Since the device needs a back contact to work, the BOX layer is removed from the region under the active device. A schematic is shown in Fig. 4.22.

The device dimensions are $6 \times 6\,\mu m^2$ for the rib width and the core height, while a etch depth of 1.5 μm is assumed in the passive waveguide region. In the active region, the lateral confinement is ensured by the gate-doped region in a planar configuration. This approach is very similar to the one used more recently by Iodice and co-workers [71] for their p–i–n device. A more detailed description of the operation principle of this device is provided in Sect. 4.6 According to the authors, this three-terminal structure is more efficient than the p–i–n structures previously proposed since the third terminal allows a faster removal of the free carriers from the optical path. Their simulations indicate an AM of 20%, when 126-mW power is applied to the device. A switching time of 5.6 ns is also defined. Finally, when the device is operated as a phase modulator (in an arm of an MZI), it exhibited a very high figure of

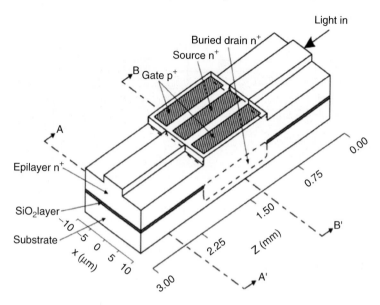

Fig. 4.22. BMFET proposed in [51]

merit, $215° \, V^{-1} \, mm^{-1}$ when a driving power of 43 mW is used. The switching time is lower than before, being below 3.5 ns, thus indicating a maximum operating frequency above 100 MHz. The main drawback of this device is the propagation loss, which is $11.8 \, dB \, cm^{-1}$ in the passive device. Experimental measurements, carried out by Breglio et al. [53], provide a modulation depth of 20%, with a driving power of 500 mW. The much higher power value could be due to contact resistance in the fabricated devices. Finally, Sciuto et al. [58] provide an experimental static M of 75% with a driving power of 160 mW for a similar device. The transient behavior of this device is reported in Sect. 4.6.

The most promising results, a milestone for the production of a fast electro-optical modulator, have been reported by Liu and co-workers [32] and their device is shown in Fig. 4.23. they report the fabrication of a Si-based optical modulator having a maximum operation frequency above 1 GHz. It is a MOS capacitor integrated in a single-mode SOI waveguide. This device operates by the free carrier effect and bears a close resemblance to a CMOS transistor. Its structure consists of n-type crystalline Si with an upper rib of p-type polysilicon. The n-type and p-type regions are separated by a thin insulating oxide layer. Upon application of a positive voltage to the polysilicon, charge carriers accumulate at the oxide interface by changing the refractive index distribution in the device, hence there is a phase shift in the optical wave propagating through the device. The measured extinction ratio is above 16 dB, by applying a voltage of ~7.7 V. The insertion loss is ~15.3 dB in the on state, including ~4.3 dB per interface coupling loss to and from the waveguide, and

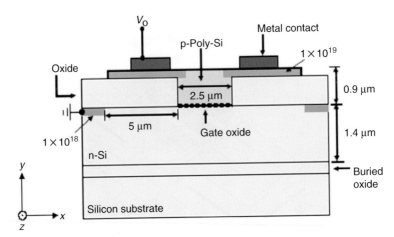

Fig. 4.23. The cross-sectional view of a MOS capacitor waveguide phase shifter in SOI technology, from [32]

an ~6.7 dB on-chip loss (active device length of 2.5 mm). Most of this loss is due to the doped polysilicon phase shifter and undoped polysilicon waveguide.

The authors suggest several methods that may improve the device performance even further. The first is replacing the p-type polysilicon with single crystal Si, where the latter is expected to reduce on-chip loss by ~5 dB. The n-type doped crystalline silicon layer thickness is 1.4 μm, while the p-type doped polysilicon thickness at the centre of the waveguide is 0.9 μm. The gate oxide thickness is 12 nm. Both the polysilicon rib and the gate oxide widths are 2.5 μm. The n-type silicon has an active doping concentration of 1.7×10^{16} cm^{-3} and the p-type polysilicon has an active doping concentration of 3×10^{16} cm^{-3}. The speed of the device depends on silicon and poly silicon resistances. More details on this device can be found in Paniccia's chapter in this book.

4.6 How to Fabricate and Characterise an Electro-Optical Modulator

In order to design, fabricate, and characterise a device, a modulator in our case, few rules must be kept in mind. A schematic of the different steps needed to fabricate a modulator is plotted in Fig. 4.24. The first thing is the definition of the operation principle. We have described all the modulation principles so far known for Si, and the free carrier absorption is nowadays the most efficient. The next step is the definition of the device layout. In particular, we decided to use a device very similar to the one proposed by Cutolo in 1997 [51], a BMFET. Typically, this device opportunely biased shows the modulation of the conductivity in the electrical channel due to a change in the

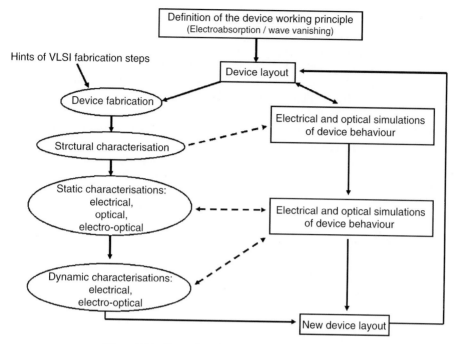

Fig. 4.24. Steps for the fabrication of a device

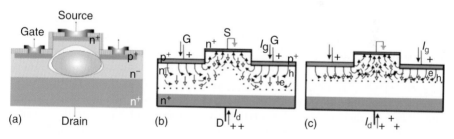

Fig. 4.25. **(a)** The cross-sectional view of the BMFET [58]; **(b)** distribution of the plasma of free carriers when the device is in the off state (light passes through the device); **(c)** plasma distribution when the device is in the on state (light is absorbed in the device length)

free carrier concentration. The device is integrated within a rib waveguide and opportunely designed in order to have the optical channel coincident with the BMFET electrical channel. A schematic of the modulator cross-sectional view is shown in Fig. 4.25. The device source (S) is placed on top of the rib, while the gates (G) are located on both sides of the rib, in a configuration similar to the symmetrical p–i–n shown in Fig. 4.16b. The drain region (D) is placed at the back of the device. Since a back contact is required, the waveguide lower cladding is realised using the variation in the refractive index of heavily doped Si (see (4.16)). This choice will cause an increase in the insertion loss.

In our configuration the device works as a pure-intensity modulator, hence no interferometric systems are needed, thus strongly reducing the overall device geometrical dimensions. This allows one to have a very small lateral dimension, below 50 μm. The working wavelength is ≥1.2 μm, corresponding to energies below the Si bandgap. The operation principle, as mentioned earlier is the free carriers dispersion/absorption effect achieved by the presence of a carriers plasma in the electrical/optical channel moved in order to have the light transmission (off state, Fig. 4.25b) or dispersion/absorption (on state, Fig. 4.25c). The holes injected from the device gate produce an excess in the n^- channel causing the free electrons generation (due to the action mass law) and the formation of quasi-neutral carrier plasma. By increasing the gate current, the free carrier concentration in the plasma is increased. The device is biased in common-source configuration with a constant gate current in order to have a constant carrier flux injected in the electro-optical device channel. An electric field, applied to the drain control terminal, allows one to move the plasma inside or outside the channel, thus switching the device between the two states.

Once the operation principle is defined, a set of electrical and optical simulations is needed (see Fig. 4.24) in order to define the best device structure. These studies have been carried out in literature [51, 57, 77] and the optimised device dimensions defined [51].

The following step (see Fig. 4.24) is the definition of the fabrication flowchart. The devices were fabricated using n-type epitaxial (epi) Si layers 10 μm thick, P doped with a resistivity of ~1 Ω cm (2×10^{15} P cm^{-3}) grown on n-type P doped Czochralski (CZ) Si wafers with a resistivity of ~0.003 Ω cm (2×10^{19} P cm^{-3}). The samples underwent standard clean room processing. This is a fundamental step if, in the end, the device will be passed in production. The rib waveguide is obtained by combining standard photolithography and reactive ion etching of Si. In this way 90° walls with a good quality are obtained [58]. The rib width is 10 μm, the Si core height (H) 10 μm, while the rib edge height is 0.75 μm. In the fabrication process the main issue to be considered is whether a discrete device is fabricated: the overall thermal budget experienced by the wafer must be as low as possible, with the highest temperature process at the beginning of the flow chart; the number of masks must be reduced, if possible, using self-aligned processes. For this reason the rib walls were passivated with a thin layer of thermally grown SiO_2, to ensure a good interface between Si and SiO_2, and the cladding was completed by a thick layer of deposited oxide on top of it.

The BMFET was fabricated in a limited region of the guide by implanting arsenic ions (at a dose of 5×10^{14} cm^{-2}) in the upper part of the rib to form the shallow n^+ source region. The p^+ gate region is obtained by implanting boron at a dose of 2×10^{14} cm^{-2} on the rib sides. The drain region (n^+) is the Si CZ substrate on the back of the device. The dopants are activated

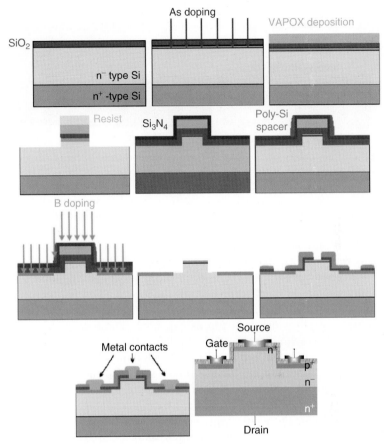

Fig. 4.26. The processing steps followed for the fabrication of the BMFET modulator

through thermal annealing at 940°C for 40 min (As activation) and 900°C for 15 min (B activation, also used for SiO_2 thermal growth). Finally, 3-µm thick Al contacts are deposited on top of the devices. A schematic of the processing steps is shown in Fig. 4.26. The modulator length is 100 µm, while the waveguide length is 1 cm.

At this point, the structural characterisation is required and a comparison with the simulations, previously performed, is needed (see Fig. 4.24). In particular, the structural characterisation was performed by combining different techniques, and also using "non-traditional" approaches. As an example, we combined scanning electron microscopy (SEM) analyses and electrochemical etching [78]. Thin sample sections underwent anodic oxidation in a $HF/HCl/H_2O$ solution (at a ratio of 1:3:2) applying a bias of 0.5 V for 30 s. Regions with doping concentration $\geq 10^{16}$ cm^{-3} are preferentially etched

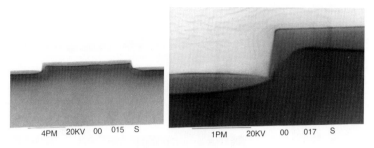

Fig. 4.27. SEM images of the BMFET rib after electrochemical etching [58]. **(a)** Full rib, **(b)** particulars of the rib edge

from the sample leaving a well-defined delineation of the bi-dimensional doping profile. The SEM image of the sample after the etch provides the doping profile. The results are shown in Fig. 4.27.

In this way it is possible to simultaneously study the structural quality of the rib waveguide and to determine both the vertical and the lateral distribution of dopants in different regions of the devices. A SEM photograph of the rib geometry is shown in Fig. 4.27a for a sample that underwent electrochemical etching. The rib width is $10\,\mu$m as expected. These data show the good quality of the dry etching process used to define the rib. Particulars of the rib edge are shown in Fig. 4.27b. The rib wall has an angle of 89° with respect to the surface and it does not exhibit roughness visible to the microscope (TEM analyses, not shown, confirm this result). A careful inspection of Fig. 4.27 evidences the dopant profile. In fact, in both figures a lighter grey layer close to the surface is visible. The source region extends for $\sim0.1\,\mu$m in depth on top of the rib and the gate regions on the sides, $0.3\,\mu$m deep, are clearly visible. This analysis shows a contact region between the n$^+$ and p$^+$ pockets at the edge of the rib. Nevertheless, the electrical characterisation of the gate–source diode (not shown) provides a breakdown voltage value of ~25 V, as expected for an avalanche diode. It demonstrates that the dopant concentration in the overlap region is very low. Finally, spreading resistance profiling measurements confirm the depth values of the gate and source pockets.

Once the structural characterisation is complete a feedback with simulations is required. Moreover, the electrical characterisation is mandatory (see Fig. 4.24). The electrical characterisation performed is quite standard for these devices, hence we do not discuss it here but refer to [58] for details. The only result we mention is the modulation of the electrical channel conductivity used to obtain the electro-optic modulation and typical of BMFET devices [77]. The electrical modulation was monitored measuring the channel resistance variation as a function of the gate bias (V_{G}) in common-source configuration. The resistance values derived from the linear region slope of the BMFET current–voltage characteristics are plotted in Fig. 4.28. The resistance ranges from 100 to $6\,\Omega$ for gate voltages from 0 to 1.4 V. A variation of a factor ~17

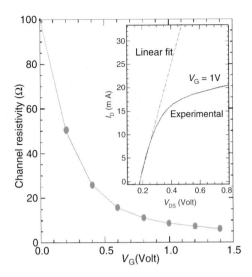

Fig. 4.28. Channel resistivity modulation as a function of the gate voltage. In the inset device electrical characteristic in common-source configuration for $V_G=0$ V. Linear fit of the characteristic (*dashed line*)

in the channel resistivity is observed by applying 1.4 V to the gate. This variation, even if it proves the possibility to use the charge injection to achieve the light modulation, is not the maximum modulation achievable with this structure [58].

The last routine characterisation is the static optical characterisation. The insertion loss of the passive device is evaluated using the cut back method on a structure much longer than the device (1 cm) but having the same structural and doping characteristics. Also in this case the measure is quite standard, and hence not reported. These waveguides exhibit losses of $\sim$5 dB cm^{-1}. The coupling losses between the fibre and the device have been measured and are $\sim$3 dB, corresponding to the Fresnel coupling loss ($\sim$1.6 dB at each Si/air interface). Recent experiments carried out on Si-based waveguides provided loss values below 2 dB cm^{-1} and we are confident similar values can also be achieved in such devices. Despite the fact that the loss is higher here than in other materials (e.g. LiNbO$_3$), it is still "reasonable" if the possibility to monolithically integrate all the optical components in Si is considered. The total insertion loss for the passive device is only 0.06 dB thanks to the device's small length, 100 µm. Therefore, the proposed device is a good solution when small dimensions are required and low optical power signals are used.

Finally, once the device is fully characterised a new set of simulations on the actual device structure is needed. The electrical simulations were performed using standard Silvaco programs ATHENA and ATLAS. Due to the perfect symmetry of the device structure in the simulations only half of the device has been taken into account. In this way it is possible to strongly reduce the simulation time. Since, also in this case, standard procedures were applied, a detailed description of the simulations [58] is recommended. Of course, the simulations provide some hints for the device optimisation.

To determine the device performance as a modulator, the electro-optical characterisation must be carried out (see Fig. 4.24). The device was bonded and electrically connected to a generator. The spectrum of the transmitted optical signal was monitored as a function of the drain polarisation conditions in the common-source configuration. In order to have a constant flux of carriers injected in the device channel, the gate current was set to a constant value. The optical absorption was monitored from negative (-25 V) to positive ($+25$ V) drain biases using gate current of either 5 or 10 mA. Once the voltage corresponding to the maximum transmitted power is identified, the on/off voltages can be defined and M calculated. (See [79] for more details). The experimental data demonstrate that good device performances are obtained even when applying low gate currents. This result is very important if low power consumption needs to be achieved.

Particular care must be taken when measuring the electro-optical performances of this device, due to the high current and voltages involved in the full device characterisation. As an example, when an electrical and thermal insulating package is used thermal dissipation is not allowed. We measured a significant device temperature increase during operation and obtained a working temperature of $140°$ C for $V_{DS} = +15$ V in static operation. Hence, an insulating package cannot be used. Such relevant temperature change determines an increase in the free carrier concentration, in both the modulator and waveguide regions. As a result, the absorption coefficient increases while an increase in the real part of the effective refractive index is achieved (see the thermo-optical effect theory). This last effect has an opposite trend with respect to the plasma dispersion effect, thus reducing its effectiveness. The increase in the waveguide absorption implies a significant reduction in the optical transmitted power also in the off state. As a result, a reduction in the M value is achieved when the modulation switching time is small with respect to the device thermal response time.

During the device testing, the output light intensity was monitored by an infrared camera. The camera faces the output cross-section of the tested device with a three-axis stage for a micrometric controlled positioning. It allows one to observe the intensity and the modal distribution of the transmitted optical signal. The optical characterisation indicates a static M of $90 \pm 5\%$ (at $I_G = 5$ mA, $V_{OFF} = +10$ V, $V_{ON} = +20$ V). Furthermore, the images collected by the infrared camera (not shown), confirm that for low positive drain bias (off state) the light intensity is higher and confined in a smaller area than in the unbiased configuration since the plasma under the gate terminals (see Fig. 4.25b) improves the waveguide lateral confinement. When increasing the drain polarisation (on state), the output light intensity is very low and it is confined in the lower region of the optical channel (see Fig. 4.25c). Since both the physics of plasma formation and its localisation have been theoretically studied as a function of the BMFET terminal biases [51] (trough experimental data are nowadays missing) the electro-optical measurements have also been used to experimentally verify the plasma behaviour under the electric field effect.

The dynamic electro-optical characterisation was carried out by applying different gate current values, in a wide range of drain voltage frequencies and amplitudes, in order to fully characterise the device. The measurements here reported were performed with $I_G = 5\,mA$ in common-source configuration and applying either a quasi-square or a sinusoidal signal as drain voltage (V_{DS}) with a duty cycle of 50%.

At low frequency, the optical transmitted signal follows the voltage applied to the drain control terminal in common-source configuration. The optical output power plotted in Fig. 4.29 is referred to as the left vertical axis (see the arrows) and is expressed in mW, while the reference voltage wave, with a frequency of 10 kHz and values between $\sim 10\,V$ and $\sim 22.5\,V$, is referred to as the vertical axis on the right and it is expressed in volts. Both are measured as a function of time. A modulation depth of $73 \pm 5\%$ was estimated at such a frequency [80]. Since the BMFET optical performances depend on the electrical ones, the device optical commutation speed is related to the electrical response time, hence the drain current I_D as a function of time with respect to the drain bias V_{DS} was measured and is shown in the inset of Fig. 4.30. I_D perfectly follows V_{DS} with a (very small) delay of about 10 ns. Such a value is not affected by the temperature in the monitored temperature range (up to 200°C). The data provide an electrical switching frequency above 100 MHz as foreseen by theoretical studies [51] and suggest an upper limit for the device optical performances below 10 ns commutation time.

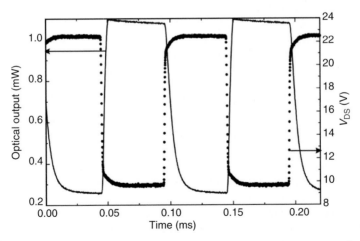

Fig. 4.29. Electro-optical dynamic performances in common-source configuration, $I_G = 5\,mA$ and a 10 kHZ square wave voltage. *Solid line*: optical power (left-hand scale); *dots*: V_{DS} (right-hand scale). The modulation depth is about 73% [80]

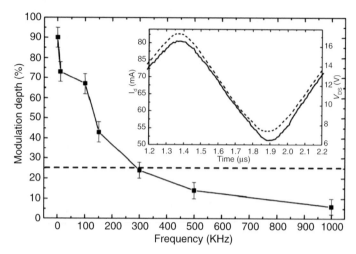

Fig. 4.30. Modulation depth vs. the driving frequency. In the inset, the drain current I_D follows the drain voltage V_{DS} with a delay of 10 ns [80]

The best M values, obtained for different voltages and/or wave shapes, are summarised in Fig. 4.30 where M, in percentage, is plotted as a function of the electrical modulation frequency. M strongly decreases when the V_{DS} frequency increases, approaching the 25% threshold at 300 kHz (as marked by the horizontal dashed line in the figure) and to about 8% at 1 MHz. We believe that the strong reduction observed is due to a change in the main factor determining the plasma position (1) free carrier generation/recombination and a small contribution of carriers drift at low frequency and (2) plasma drift at high frequency.

Optical simulations performed using commercial programs (Selene and Prometheus by Kymata BBV) allowed us to infer some information on the plasma concentration in our device. The results are shown in Fig. 4.31. When the free carrier concentration increases, the modulation depth, for a fixed length, increases. In our device static modulation depth of 90% has been achieved, suggesting a free carrier concentration of the order of 5×10^{19} cm^{-3}. Moreover, the simulations suggest that smaller devices could be fabricated, thus further reducing the device dimension, or in an optimised geometrical design the driving gate current is reduced. According to the schematic shown in Fig. 4.24 a new optimised device can now be designed.

Now, we focus our attention on the possibility, mentioned earlier, to use the electro-optical characterisation to better understand the physics involved in the device operation. Hence, the last part of this section is devoted to the description of a method to study the plasma behaviour as a function of the bias providing, for the first time, an experimental study of the plasma localisation in the device in both static and dynamic conditions [81]. We used standard emission microscopy (E.Mi.) analysis, and spectrally resolved photons counts,

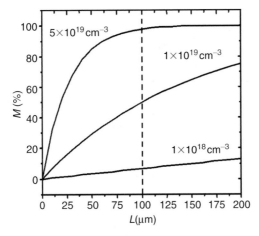

Fig. 4.31. Simulated modulation depth as a function of the device length assuming different free carrier concentrations uniformly filling the waveguide core

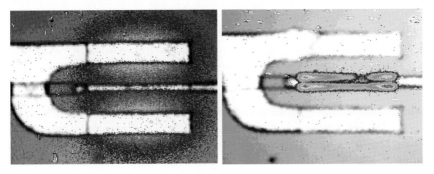

Fig. 4.32. E.Mi. images (1-s acquisition time) in (**a**) off state (light at the output); (**b**) on state (no light at the output)

to monitor the photons emitted by electron–hole recombination of the biased device, thus providing a bidimensional map of the plasma distribution in the BMFET.

E.Mi. measurements are widely used in failure or thermal analysis and for imaging of current conduction in dielectric films. E.Mi. images shown in Fig. 4.32a, b can be immediately compared with the theoretical analysis (shown in Fig. 4.25b, c), providing the experimental verification of the theory. The E.Mi. analysis consists of a time-integrated collection of the light emission from a polarized device from the visible to the infrared spectral region. The measurements provide a bidimensional map of the plasma distribution, since the electron–hole recombination within the plasma is monitored, through the photon emission at the Si band–edge, using spectrally resolved measurements.

When the drain terminal is grounded, in common-source configuration, the holes injected from the gate flow towards the drain and the source, as schematically described in Fig. 4.25b. Since the gate-facing area is larger for the drain than for the source, the carriers are preferentially localised in the area under the gate region and the n^+/n^- (source/epi-channel) transition region, as confirmed by the E.Mi. image of Fig. 4.32a. The emission intensity in figure is represented in false colours. Red indicates strong emission, blue, low emission, and white regions are metal contacts. They adsorb the light coming from the regions beneath them. In the off state the emission is due to two spatially separated regions, under and laterally to the gate terminals, in full agreement with the theoretical carrier distribution already described.

A positive current flows through the drain (I_D) when the device is in the on state. The holes injected from the gate reach the source due to this lower potential as schematically shown in Fig. 4.25c. The hole excess causes the plasma formation in a region whose extension depends on the drain voltage: by increasing the bias, the plasma localisation in the upper electro-optical channel region of the device increases as does its concentration. The E.Mi. picture in Fig. 4.32b, acquired in this configuration, clearly shows that the emission is localised in the region under the rib. The comparison between the E.Mi. images demonstrates that the plasma can be confined in a very small device region only, acting on the drain bias. Finally, in the open gate or zero gate current configurations no emission from the device is detected (not shown), even in presence of high drain currents. This result confirms the gate-injected holes the role in the plasma formation.

We used the measurements so far reported as starting point for the dynamical characterisation. In normal operation conditions, a modulated V_{DS} bias, at frequencies up to 5 MHz, is applied to switch the device between the on and off states. The plasma behaviour observation was again performed using the E.Mi. analysis. The measurement integration time (1 s) is much longer than the switching time; hence, the final image is the emission overlap of the two "final" states after switching. With "final" state we mean the plasma localisation at the end of the transition time (switch event) between the two operation states. These results must be compared with the images obtained in static conditions and are already presented.

The results in dynamic operation, when the device is switched between the two operation states at frequencies of 100 Hz, 500 kHz, and 5 MHz, are shown in Fig. 4.33a, b, and c, respectively. The plasma distribution at low operation frequencies (100 Hz, Fig. 4.33a) is the sum of the emission in the on and off, as clearly visible from the comparison of Fig. 4.33a with Fig. 4.32a, b. By increasing the frequency, the emission from the regions beneath the gate reduces while from the rib region increases (Fig. 4.33b). Such behaviour has been observed for frequencies up to ~500 kHz. At higher frequencies a relevant change in the emission shape is detected, i.e. at 5 MHz (Fig. 4.33c) it is localised in the rib region and it is very similar to the on static state of Fig. 4.32b.

The experimental data suggest that two different phenomena drive the plasma in the two extreme frequency conditions; otherwise exactly the same

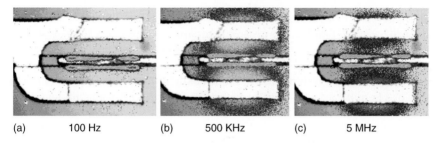

Fig. 4.33. E.Mi. images during dynamic device operation in common-source configuration at $I_G = 10\,\mathrm{mA}$: (**a**) 100-Hz, (**b**) 500-kHz, and (**c**) 5-MHz operation frequency [81]

image would have been detected by the E.Mi., regardless of the device operation frequency.

The estimated recombination lifetime is of $\sim 2\,\mu\mathrm{s}$ corresponding to a frequency of 100 kHz; hence the free carrier recombination plays an important role in the final plasma localisation only at frequencies below such value, as in a Fig. 4.33a. At low frequencies the drift effect, occurring in the first tenths of a microsecond, is overwhelmed by the generation/recombination phenomena and the final state is mainly due to this last one. On the other hand, at frequencies above the maximum for generation/recombination phenomena, the phenomenon driving the plasma is the drift due to the electric field. In fact, as shown in Fig. 4.33, during the measurement, only the emission under the rib is detected. Hence, the plasma moves under the source region due to the drift field effect. The behaviour at low and high frequencies is quite different and confirms the presence of a threshold for the carrier effective lifetime. A concomitant occurrence of both phenomena must be visible by E.Mi. at the threshold frequency. The experimental data provide a frequency threshold between the two effects at about 500 kHz, as shown in Fig. 4.33b. In fact, a combination of the two emissions already observed at (a) low and (c) high frequencies emerge as a result of the measurement. These results show that non-standard techniques, such as E.Mi., can be used to experimentally study the physics behind a device operation.

4.7 Conclusions

In this chapter we reviewed the physical mechanisms regulating light modulation in Si and the work reported so far in literature to design and fabricate Si-based optical modulators. The device performances reviewed clearly indicate that a winning approach to obtain fast modulation in Si (in a Si rib integrated device), above 1 GHz, is to reduce the channel waveguide dimensions below $1\,\mu\mathrm{m}$. Structures proposed more than 20 years ago (e.g. p–i–n devices) can operate at very high frequencies if their dimensions are reduced.

An improvement of the bandwidth above tens of GHz is predicted in the near future [62]. These bandwidths are low compared to those of "classical" electro-optical materials but are a great breakthrough considering that silicon does not exhibit a significant electric field-based modulation mechanism. Moreover, the possibility to integrate monolithically in electrical Si circuits optical functions is quite an attractive prospect for future generations of devices.

References

1. http://www.itrs.net/Common/2004Update/2004Update.html
2. J.C. Palais, *Fiber Optic Communications*, (5th edn), Prentice Hall, Upper Saddle River, NJ, 2004
3. R. Narayanaswamy, O.S. Wolfbeis (eds.), *Optical Sensors: Industrial, Environmental and Diagnostic Applications*, Springer series on Chemical Sensors and Biosensors 1, Springer, Berlin Heidelberg New York, 2004
4. S. Sudo, K. Okamoto (eds.), *New Photonics Technologies for the Information Age: The Dream of Ubiquitous Services*, Artech House Optoelectronic Library, Nippon Telegraph and Telephone Corporation, 2004
5. B.E.A. Saleh, M.C. Teich, *Fundamentals of Photonics*, Wiley, New York, 1991
6. J.K. Ebeling, *Integrated Optoelectronics*, Springer, Berlin Heidelberg New York, 1993
7. K. Noguchi, H. Miyazawa, O. Mitomi, Frequency-dependent propagation characteristics of coplanar waveguide electrode on 100 GHz Ti:LiNbO$_3$ optical modulator, Electron. Lett. **34** (7), 661–663, 1998
8. S.M. Sze, *Physics of Semiconductor devices*, Wiley, New York, 1987
9. A.E. Siegman, *Lasers*, In A. Kelly (ed), University Science Books, Sausalito, CA, 1986
10. J. Hecht, *The Laser Guidebook*, 2nd edn., McGraw Hill, New York, 1992
11. http://www.newfocus.com/Online_Catalog/Literature/apnote2.pdf
12. D. Dragoman, M. Dragoman, *Advanced Optoelectronic Devices*, Springer series in photonics, Springer, Berlin Heidelberg New York, 1999
13. K. Wakita, I. Kotaka, Multiple quantum well optical modulators and their monolithic integration with DFB lasers for optical communications, Microwave Opt. Technol. Lett. **7**, 120–128, 1994
14. http://www.micronic.se/site_eng/product/aom.shtml
15. Ch. Buchal, M. Siegert, *The Physics of Modulators*, In: O. Bisi, S.U. Campisano, L. Pavesi, F. Priolo, (eds.), Proceedings of the International School of Physics "E. Fermi", 369–395, Varenna, 1998
16. R.A. Soref, B.R. Bennet, Electro-optical effect in silicon, IEEE J. Quant. Electron. **23**, 123–129, 1987
17. T.S. Moss, G.J. Burrell, B. Ellis, *Semiconductor Opto-electronics*, Butterworths, London, 1973
18. P.H. Wendland, M. Chester, Electric field effects on indirect optical transitions in silicon, Phys. Rev. A **140** (4), 1384–1390, 1965
19. M. Bertolotti, V. Bogdanov, A. Ferrari, A. Jasow, Temperature dependence of refractive index in semiconductors, J. Opt. Soc. Am. B **7** (6), 918–922, 1990
20. M. Lipson, Overcoming the limitations of microelectronics using Si nanophotonics: solving the coupling, modulation and switching challenges, Nanotechnology **15**, S622–S627, 2004

21. G. Cocorullo, F.G. Della Corte, I. Rendina, Temperature dependence of the thermo-optic coefficient in crystalline silicon between room temperature and 550 K at the wavelength of 1523 nm, Appl. Phys. Lett. **74**, 3338–3340, 1999

22. G. Cocorullo, I. Rendina, Thermo-optical modulation at 1.5 μm in silicon etalon, Electron. Lett. **28**, 83–84, 1992

23. G.E. Jellison, H.H. Burke, The temperature dependence of the refractive index of silicon at elevated temperatures at several wavelengths, J. Appl. Phys. **60**, 841–843, 1986

24. R. Thalhammer, Internal laser probing techniques for power devices: analysis, modeling, and simulation, Ph.D. dissertation, University of Munich, 2000

25. P.E. Schmid, Optical absorption in heavily doped silicon, Phys. Rev. **B 23**, 5531–5536, 1981

26. A.A. Volfson, V.K. Subashiev, Fundamental absorption edge of silicon heavily doped with donor or acceptor impurities, Sov. Phys. Semicond. **1**(3), 327–332, 1967

27. N.W. Ashcroft, N.D. Mermin, *Solid State Physics*, Holt, Rinehart and Winston, Philadelphia, USA, 1976

28. H.Y. Fan, M. Becker, Infrared absorption of silicon, Phys. Rev. **78**, 178, 1950

29. W. Spitzer, H.Y. Fan, Infrared absorption in n-type silicon, Phys. Rev. **108**, 268–271, 1957

30. H.C. Huang, S. Yee, M. Soma, Quantum calculations of the change of refractive index due to free carriers in silicon with nonparabolic band structure, J. Appl. Phys. **67**, 2033–2039, 1990

31. J. Hecht, *Understanding Fiber Optics*, Prentice Hall, Upper Saddle River, NJ, 1999

32. A. Liu, R. Jones, L. Liao, D. Samara-Rubio, D. Rubin, O. Cohen, R. Nicolaescu, M. Paniccia, A high-speed silicon optical modulator based on a metal-oxide-semiconductor capacitor, Nature, **427**, 615–618, 2004

33. K. Yasumoto, Coupled-mode formulation of parallel dielectric waveguides, Opt. Lett. **18**, 503–506, 1993

34. H. Haus, W.P. Huang, Coupled-mode theory, Proceed. IEEE **79** (10), 1505–1518, 1991

35. A.K. Ghatak, K. Thyagarajan, *Integrated Optics. Optical Electronics*, Cambridge University Press, Cambridge, 421–460, 1989

36. K.C. Cadien, *Challenges for On-Chip Optical Interconnects*, invited talk at the Photonic West Conference – Optoelectronic Integration on Silicon III San Josě, California, January 2005

37. J.P. Lorenzo, R.A. Soref, 1.3 μm electro-optic silicon switch, Appl. Phys. Lett. **51**, 6–8, 1987

38. B.R. Hemenway, O. Solgaard, D.M. Bloom, All-silicon integrated optical modulator for 1.3 μm fibre-optic interconnects, Appl. Phys. Lett. **55**, 349–350, 1989

39. G.V. Treyz, P.G. May, J.-M. Halbout, Silicon optical modulators at 1.3 μm based on free-carrier absorption, IEEE Electron Device Lett. **12**, 276–278, 1991

40. G.V. Treyz, P.G. May, J.-M. Halbout, Silicon Mach-Zehnder waveguide interferometers based on the plasma dispersion effect, Appl. Phys. Lett. **59**, 771–773, 1991

41. X. Xiao, J.C. Sturm, K.K Goel, P.V. Schwartz, Fabry–Perot optical intensity modulator at 1.3 μm in silicon, IEEE Photon. Technol. Lett. **3**, 230–232, 1991

42. H.C. Huang, T.C. Lo, Simulation and analysis of silicon electro-optic modulators utilizing the carrier-dispersion effect and impact-ionization mechanism, J. Appl. Phys. **74** (3), 1521–1528, 1993

43. Y.L. Liu, E.K. Liu, S.L. Zhang, G.Z. Li, J.S. Luo, Silicon 1 × 2 digital optical switch using plasma dispersion, Electron. Lett. **30**, 130–131, 1994

44. C.C. Wang, M. Currie, S. Alexandrou, T.Y. Hsiang, Ultrafast, all-silicon light modulator, Opt. Lett. **19**, 1453–1455, 1994

45. Y. Liu, E. Liu, G. Li, S. Zhang, J. Luo, F. Zhou, M. Cheng, B. Li, H. Ge, Novel silicon waveguide switch based on total internal reflection, Appl. Phys. Lett. **64**, 2079–2080, 1994

46. C.Z. Zhao, G.Z. Li, E.K. Liu, Y. Gao, X.D. Liu, Silicon on insulator Mach-Zehnder waveguide interferometers operating at 1.3 μm, Appl. Phys. Lett. **67**, 2448–2449, 1995

47. M.Y. Liu, S. Chou, High-modulation-depth and short-cavity-length silicon Fabry–Perot modulator with two grating Bragg reflectors, Appl. Phys. Lett. **68**, 170–172, 1996

48. C.Z. Zhao, E.K. Liu, G.Z. Li, Y. Gao, C.S. Guo, Zero-gap directional coupler switch integrated into a silicon-on-insulator for 1.3 μm operation, Opt. Lett. **21**, 1664–1666, 1996

49. C. Marxer, M.A. Gretillat, V.P. Jaecklin, R. Baettig, O. Anthamatten, Megahertz opto-mechanical modulator, Sens. Actuators **A52**, 46–50, 1996

50. A. Cutolo, M. Iodice, A. Irace, P. Spirito, L. Zeni, An electrically controlled Bragg reflector integrated in a rib silicon on insulator waveguide, Appl. Phys. Lett. **71**, 199–201, 1997

51. A. Cutolo, M. Iodice, P. Spirito, L. Zeni, Silicon electro-optic modulator based on a three terminal device integrated in a low-loss single-mode SOI waveguide, J. Lightwave Technol. **15**, 505–518, 1997

52. C.Z. Zhao, A.H. Chen, E.K. Liu, G.Z. Li, Silicon-on-insulator asymmetric optical switch based on total internal reflection, IEEE Photon. Technol. Lett. **9**, 1113–1115, 1997

53. G. Breglio, A. Cutolo, A. Irace, P. Spirito, L. Zeni, M. Iodice, P.M. Sarro, Two silicon optical modulator realizable with a fully compatible bipolar process, IEEE J. Sel. Top. Quantum Electron. **4** (6), 1003–1010, 1998

54. P.D. Hewitt, G.T. Reed, Improved modulation performance of a silicon p–i–n device by trench isolation, IEEE J. Lightwave Technol. **19**, 387–390, 2001

55. A. Irace, G. Breglio, G. Coppola, A. Cutolo, Fast silicon-on-silicon optoelectronicrouter based on a BMFET device, IEEE J. Sel. Top. Quantum Electron. **6** (1), 14–18, 2000

56. G. Coppola, A. Irace, M. Iodice, A. Cutolo, Simulation and analysis of a high-efficiency silicon optoelectronic modulator based on a Bragg mirror, Opt. Eng. **40**, 1076–1081, 2001

57. G. Coppola, A. Irace, G. Breglio, A. Cutolo, All-silicon mode-mixing router based on the plasma-dispersion effect, J. Opt. A: Pure Appl. Opt. **3**, 346–354, 2001

58. A. Sciuto, S. Libertino, A. Alessandria, S. Coffa, G. Coppola, Design, fabrication and testing of an integrated Si-based light modulator, IEEE J. Lightwave Technol. **21** (1), 228–235, 2003

59. C.A. Barrios, V.R. Almeida, M. Lipson, Low-power-consumption short-length and high-modulation-depth silicon electro-optic modulator, IEEE J. Lightwave Technol. **21**, 1089–1098, 2003

60. C.A. Barrios, V.R. Almeida, R.R. Panepucci, M. Lipson, Electro-optic modulation of silicon-on-insulator submicron-size waveguide devices, IEEE J. Lightwave Technol. **21**, 2332–2339, 2003

61. G.T. Reed, C.E.J. Png, Silicon optical modulators, Mater. Today, 40–50, January 2005
62. http://www.intel.com/technology/silicon/sp/index.htm
63. K. Gustafsson, R. Hok, A silicon light modulator, J. Phys. E: Sci. Instrum. **21** 680–685, 1988
64. L. Que, G. Witjaksono, Y.B. Gianchandani, Modeling and Design of a Micromechanical Phase-Shifting Gate Optical Modulator, http://www.nsti.org/publ/MSM2000/W42.03.pdf
65. C.-L. Dai, H.-L. Chen, P.-Z. Chang, Fabrication of a micromachined optical modulator using the CMOS process, J. Micromech. Microeng. **11**, 612–615, 2001
66. G.V. Treyz, Silicon Mach-Zehnder waveguide interferometer operatine at 1.3 μm, Electron. Lett. **28**, 118–120, 1991
67. G. Cocorullo, F.G. Della Corte, I. Rendina, P.M. Sarro, Thermo-optic effect exploitation in silicon microstructures, Sens. Actuators **A 71**, 19–26, 1998
68. G. Cocorullo, M. Iodice, I. Rendina, All-silicon Fabry–Perot modulator based on the thermo-optic effect, Opt. Lett. **19** (6), 420–422, 1994
69. S.A. Clark, B. Culshaw, E.J.C. Dawney, I.E. Day, *Thermo-optic phase modulators in SIMOX material*, In: G.C. Righini, S. Honkanen (eds.), Integrated Optics Devices IV, Proc. SPIE **3936**, 16–24, 2000
70. R.A. Soref, B.R. Bennett, Kramers–Kronig analysis of E–O switching in silicon, Proc. SPIE Integrat. Opt. Circuit Eng. **704**, 32–37, 1986
71. M. Iodice, G. Coppola, R. C. Zaccuri, I. Rendina, Waveguide-vanishing-based optical modulator in embedded all-silicon structure, Proc. SPIE **5730** to be published
72. C.E. Png, S.P. Chan, S.T. Lim, G.T. Reed, Optical phase modulators for MHz and GHz modulation in silicon-on-insulator (SOI), J. Lightwave Technol. **22**, 1573–1582, 2004
73. G.T. Reed, A.P. Knights, Silicon Photonics: An Introduction, Wiley, New York, 2004
74. C.K. Tang, G.T. Reed, Highly efficient optical phase modulator in SOI waveguides, Electron. Lett. **31** (6), 451–452, 1995
75. T. Pirnat, L. Friedman, R.A. Soref, Electro-optic mode-displacement silicon light modulator, J. Appl. Phys. **70**, 3355–3359, 1991
76. A. Irace, G. Breglio, A. Cutolo, All-silicon optoelectronic modulator with 1 GHz switching capability, Electron. Lett. **39**, 232–233, 2003
77. P. Spirito, G.V. Persiano, A.G.M. Strollo, The bipolar mode FET: a new power device combinino FET with BJT operation, Microelectr. J. **24**, 61–74, 1993
78. G. D'Arrigo, C. Spinella, High resolution measurements of two-dimensional dopant diffusion in silicon, Microsc. Microanal. **6**, 237–245, 2000
79. A. Sciuto, S. Libertino, S. Coffa, G. Coppola, A miniaturizable Si-based electro-optical modulator working at 1.5 μm, Appl. Phys. Lett. 86, 201115, 2005
80. A. Sciuto, S. Libertino, Miniaturizable Si-Based Light Intensity Modulator for Integrated Sensing Applications, accepted for publication in J. of Lightwave Techn.
81. A. Sciuto, S. Libertino, S. Coffa, G. Coppola, M. Iodice, Experimental evidences of carrier distribution and behaviour in a frequency in a Bipolar Mode Field Effect Transistor light modulator, IEEE Trans. on Electron Dev., **52**, 2374–2378, 2005.

5

Silicon Photodetectors and Receivers

H. Zimmermann

Summary. The properties of photodiodes being exploitable in standard silicon bipolar, complementary metal oxide semiconductor (CMOS), and BiCMOS technologies are summarized. In addition, examples of advanced photodiodes are introduced in order to show how the properties of integrated photodiodes can be improved significantly by minor process modifications. Furthermore, examples of optoelectronic integrated circuits (OEICs) for such important applications like optical storage systems, optical fiber receivers, and optical interconnect receivers are described. New trends for the circuit topology of digital-video-disk (DVD) and digital-video-recording (DVR) read-OEICs are covered. Progress of OEIC receivers for optical data transmission and communication as well as low-cost optical interconnects is also summarized. BiCMOS optoelectronic application-specific integrated circuits (OPTO-ASICs) seem to be a good choice for low-cost, high-speed, high-sensitivity optical receivers especially for free-space, between-electronic-boards, and optical-backplane applications.

5.1 Photodetectors in Standard Silicon Technologies

Nowadays, there are three types of silicon process technologies for the fabrication of integrated circuits: bipolar, complementary metal oxide semiconductor (CMOS), and a combination of both in BiCMOS processes. These technologies are very mature and there is a trend towards integrated sensors. The integration of optical sensors especially is very interesting. Many semiconductor companies are investigating photodetectors in some of their processes. In the following, properties of photodiodes being available without process modifications are described.

5.1.1 Photodetectors in Bipolar Technology

Modern bipolar integrated circuits are manufactured in standard-buried-collector (SBC) technology [1]. Without any technological modifications, the buried N^+ collector can serve as the cathode (see Fig. 5.1), the N collector

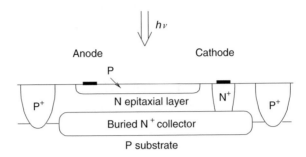

Fig. 5.1. Schematic cross-section of a base–collector photodiode

epitaxial layer can serve as the "intrinsic" layer of a PIN photodiode, and the base implant can serve as the anode in order to integrate PIN photodiodes with a thin "intrinsic" region in bipolar technologies [2, 3]. The process of [2] was described in [4]. The only difference in Fig. 5.1 compared to the cross-section of an NPN transistor is that the N^+ emitter (lying within the P-base) is omitted, which is a pure layout matter and does not need any technological modification.

The small thickness of the epitaxial layer of high-speed bipolar processes in the range of about $1\,\mu m$ causes a low quantum efficiency in the yellow to the infrared spectral region (580–1,100 nm). The rise and fall times of the photocurrent for light pulses are very short due to the small epitaxial layer thickness. In [2] a bit rate (BR) of $10\,Gb\,s^{-1}$ for the base–collector diode and a responsivity R of merely $48\,mA\,W^{-1}$ for 840 nm were reported. This low responsivity at wavelengths from 780 to 850 nm, which is widely used for optical data transmission on short fiber lengths of up to a few kilometers, is a major disadvantage of standard bipolar OEICs.

The base–collector diode with a sensitive area of $100\,\mu m^2$ fabricated in a $0.8\,\mu m$ silicon bipolar technology worked up to $3\,Gb\,s^{-1}$ for a wavelength of 850 nm [3]. A sensitivity of $0.045\,A\,W^{-1}$ was reported. A phototransistor with a very small sensitive area of $10\,\mu m^2$ reached a data rate of $5\,Gb\,s^{-1}$.

5.1.2 Photodetectors in CMOS Technology

The simplest way to build CMOS OEICs is to use the PN junctions available in CMOS processes: source/drain-substrate, source/drain-well, and well–substrate diodes. These PN photodiodes, however, possess regions which are free of electric fields. In these regions, the slow diffusion of photogenerated carriers determines the transient behavior of such PN photodiodes. Published PN CMOS OEICs are characterized by bandwidths of less than 15 MHz [5–7]. Another example is described in [8], where the OEIC was optimized for a dynamic range of six decades in illumination.

The source/drain-substrate and source/drain-well photodiodes are more appropriate for the detection of wavelengths shorter than about 600 nm,

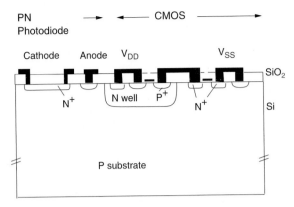

Fig. 5.2. Schematic cross-section of a PN photodiode integrated in a one-well CMOS chip

whereas the well–substrate photodiode is more appropriate for long wavelengths like 780 or 850 nm. Figure 5.2, for instance, shows an N^+/P-substrate photodiode.

In addition to carrier diffusion, the series resistance of the photodiodes due to lateral anode contacts at the silicon surface together with the relatively large junction capacitance of the photodiode may limit the dynamical PN photodiode behavior. In an N-well process, the anode of the N^+/P-substrate photodiode has to be at V_{SS} potential, which may be a restriction for circuit design.

A lateral N^+/P-substrate/P^+ photodiode was used as an optical detector into which light was coupled by an integrated waveguide [5]. This detector was realized in a 0.8 µm N-well CMOS process. A maximum bandwidth of 10 MHz was achieved with a transimpedance amplifier for a photocurrent of 1 µA with $\lambda = 675$ nm. The speed of the detector was limited by carrier diffusion due to photogeneration outside the diode [6].

The N-well/P-substrate diode in a 2 µm N-well CMOS process was used as a photodiode in [7]. A bandwidth of 1.6 MHz with a wavelength $\lambda = 780$ nm was reported for an unoptimized system. The leakage current density of the photodiode was 15 pA mm^{-2} at 5 V. The responsivity for $\lambda = 780$ nm was 0.5 A W^{-1} (quantum efficiency $\eta = 70\%$). The light was coupled into the photodiode via an integrated waveguide.

5.1.3 Photodetectors in BiCMOS Technology

In order to demonstrate the speed limitation caused by the rather high doping concentrations in modern standard technologies, we consider an N^+/P-substrate photodiode in a CMOS-based BiCMOS technology. The problem, however, is exactly the same in modern CMOS processes. Figure 5.3 shows the cross-section of this photodiode.

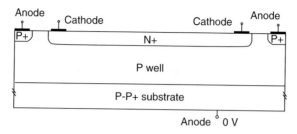

Fig. 5.3. Cross-section of a N$^+$/P-substrate photodiode in self-adjusting well BiC-MOS technology

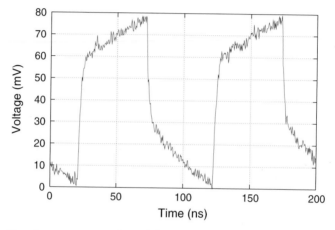

Fig. 5.4. Transient response of a N$^+$/P-substrate photodiode in a self-adjusting well BiCMOS technology

Modern CMOS and BiCMOS processes are optimized with respect to a minimum number of masks and, therefore, implement a self-aligned well processing scheme. Such a process scheme has the effect where only N wells can be defined in the layout (for a process using P-type substrate) and everywhere outside the N wells, P wells are present. Therefore, there is a P well with a doping concentration exceeding 10^{16} cm^{-3} incorporated in the N$^+$/P-substrate photodiode. This leads to a thin N$^+$/P-substrate space-charge region and the effect of slow carrier diffusion is very pronounced (see Fig. 5.4). Rise and fall times of 26 and 28 ns, respectively, for 638 nm light are determined due to the pronounced diffusion tail. The measured -3 dB bandwidth is 6.7 MHz [9].

A standard BiCMOS technology with a minimum effective channel length of 0.45 μm (corresponding to a drawn or nominal channel length of about 0.6 μm) without any modifications was used [10, 11] to exploit a fast photodiode. The buried N$^+$-collector in Fig. 5.5 was used for the cathode of the PIN photodiode. The P$^+$-source/drain island served for the anode. The intrinsic

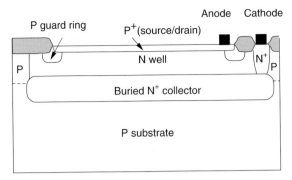

Fig. 5.5. PIN photodiode in a SBC-based BiCMOS technology [10]

zone of the PIN photodiode was formed by the N well and, therefore, had only a thickness of 0.7 μm. Consequently, the responsivity of the photodiode was only 0.07 A W^{-1} for a wavelength of 850 nm [10]. For a bias of 2.5 V across the 75×75 μm^2 PIN photodiode with a capacitance of 1.8 pF, a -3 dB bandwidth of 700 MHz was reported. The OEIC reached a BR of 531 Mb s^{-1} with a bit error rate of 10^{-9} and a sensitivity of -14.8 dBm. This BR was limited by the capacitance of the photodiode and the feedback resistor of 1.4 kΩ in the amplifier transimpedance input stage. The dark current of the photodiode was 10 nA for a reverse voltage of 2.5 V at room temperature.

In [11], a laser with a wavelength of 670 nm was used for the characterization of the same OEIC as in [10]. The data rate was increased to 622 Mb s^{-1} for this wavelength. No measured result for the responsivity was given in [11]. Instead, a three times larger responsivity value of approximately 0.2 A W^{-1} for $\lambda = 670$ nm was estimated due to the lower penetration depth than for $\lambda = 850$ nm. This estimation, however, may be doubtful, because possible destructive interference in the isolation and passivation stack is neglected. The main drawback for the photodiode used in [10,11] was its low responsivity due to the thin epitaxial layer in the SBC-bipolar based BiCMOS process.

5.1.4 Lateral PIN Photodiodes on SOI

Silicon-on-Insulator (SOI) was suggested to be the technology of future CMOS [12]. SOI uses a thin crystalline silicon film for the integration of devices.

A cross-section of the integration scheme used for SOI receivers [13] is shown in Fig. 5.6. The starting material was bonded and etched-back silicon-on-insulator (BESOI), and the active Si layer was N-type (100) with a resistivity of 50–80 Ω cm and a thickness of 3 μm. The buried oxide layer (BOX) was also 3 μm thick. The fabrication steps have been described in detail elsewhere [14]. All processing steps were fully compatible with standard CMOS processes. Indeed, this structure could easily be implemented in a standard process with no modifications other than the starting substrate.

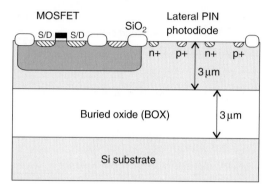

Fig. 5.6. Cross-section of lateral SOI PIN photodiode

The P- and N-fingers of the lateral, interdigitated PIN photodiode were formed during the P-well contact and source/drain implants, respectively. The finger width was $2\,\mu m$ and the spacing between the fingers was $5\,\mu m$. The total photodiode area was $51\,\mu m$ by $46\,\mu m$. The dark current of the SOI device was $2.1\,nA$ at $5\,V$ and $2.2\,nA$ at $20\,V$. The onset of breakdown occured at $40\,V$. The external quantum efficiency was measured as a function of wavelength using a white light source, spectrometer, chopper, and lock-in amplifier using a calibrated Si photodiode as a reference. At $850\,nm$, the efficiency of the SOI photodiode was $29\,\%$ corresponding to a responsivity of $0.20\,A\,W^{-1}$. The $-3\,dB$ bandwidth of the SOI photodiode with a bias $|V_{PD}| = 20\,V$ was $2.8\,GHz$ [13].

5.2 Advanced Photodetectors

5.2.1 Double Photodiode

In a standard CMOS or a CMOS-based triple-diffused (3D) BiCMOS technology [9], the so-called double photodiode (DPD) can be applied in order to eliminate slow carrier diffusion. In a CMOS-based (3D) BiCMOS process, an N well can be used as the collector of a bipolar NPN transistor. This N well can also be used in order to form the cathode of the DPD. Figure 5.7 shows the cross-section of such a DPD [15]. The two anodes are connected to ground. The cathode is connected to the amplifier in the OEIC. Two PN junctions are vertically arranged.

In addition to the two space-charge regions at the two vertically arranged PN junctions, an electric field is also present between the two space-charge regions due to the doping gradient of the N well. Therefore, there is no contribution of slow carrier diffusion from the region between the two space-charge regions. With an integrated polysilicon resistor of $1\,k\Omega$, rise and fall times of 3.2 and $2.8\,ns$, respectively, were measured for the DPD with $\lambda = 638\,nm$ and

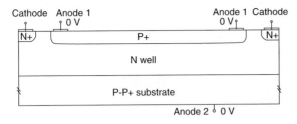

Fig. 5.7. Cross-section of a DPD in BiCMOS technology [15]

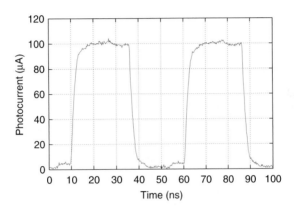

Fig. 5.8. Transient response of the DPD in BiCMOS technology shown in Fig. 5.7 [16]

$U_\mathrm{r} = 2.5\,\mathrm{V}$ [16]. With an integrated $500\,\Omega$ resistor, rise and fall times of 1.8 and 1.9 ns, respectively, were found [15]. There is no indication of a so-called diffusion tail in the photocurrent (Fig. 5.8).

5.2.2 Vertical PIN Photodiodes

Figure 5.9 shows the structure of an $\mathrm{N^-N^+}$ CMOS-OEIC in the twin-well CMOS approach [9]. Here the $\mathrm{N^+}$ substrate serves as the cathode and the $\mathrm{P^+}$ source/drain region forms the anode of the integrated PIN photodiode.

PIN-CMOS photodiodes with an area of $2{,}700\,\mu\mathrm{m}^2$, a standard doping concentration of $1 \times 10^{15}\,\mathrm{cm}^{-3}$ and reduced doping concentrations in the epitaxial layer down to $2 \times 10^{13}\,\mathrm{cm}^{-3}$, and an integrated polysilicon resistor of $500\,\Omega$, were fabricated in an industrial $1.0\,\mu\mathrm{m}$ CMOS process [17]. For the measurements, a laser with $\lambda = 638.3\,\mathrm{nm}$ was modulated with a commercial emitter-coupled logic (ECL) generator. The light pulses were coupled into the photodiodes on a wafer prober via a single-mode optical fiber. The rise (t_r) and fall (t_f) times of the photocurrent of the photodiodes were measured with a picoprobe (pp), which possesses a $-3\,\mathrm{dB}$ bandwidth of 3 GHz and an input capacitance of 0.1 pF, and with a 20 GHz digital sampling oscilloscope HP54750/51.

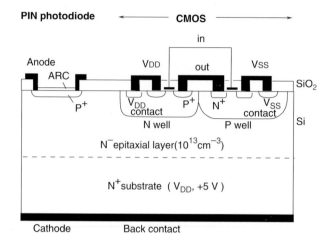

Fig. 5.9. Cross-section of a N^-N^+ PIN-CMOS-OEIC [9]

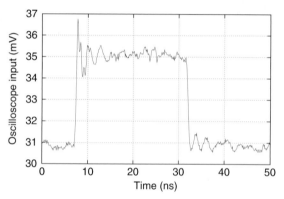

Fig. 5.10. Measured transient response of a CMOS-integrated PIN photodiode with an I-layer doping concentration of 2×10^{13} cm^{-3} for $\lambda = 638$ nm and $|V_{PD}| = 3.0$ V. The overshoot in the signal is due to the direct modulation of the laser [18]

For the PIN photodiode with a doping concentration in the epitaxial layer of 2×10^{13} cm^{-3}, the oscilloscope extracted $t_r^{osc,disp} = 0.37$ ns and $t_f^{osc,disp} = 0.57$ ns from the waveform shown in Fig. 5.10. The evaluation according $(t_r^{PIN})^2 = (t_r^{osc,disp})^2 - (t_r^{laser})^2 - (t_r^{pp})^2 - (t_r^{osc})^2$ with $t_r^{laser} = 0.30$ ns, $t_f^{laser} = 0.51$ ns, $t_{r/f}^{pp} = 0.1$ ns (pp stands for picoprobe) and $t_{r/f}^{osc} \approx 0.02$ ns results in $t_r^{PIN} = 0.19$ ns and $t_f^{PIN} = 0.24$ ns [19]. With $f_{3dB} = 2.4/(\pi(t_r + t_f))$, the -3 dB bandwidth can be estimated to be 1.7 GHz for $|V_{PD}| = 3.0$ V. With the conservative estimate BR $= 1/(1.5(t_r + t_f))$, a BR, of 1.5 Gb s^{-1} results for $|V_{PD}| = 3.0$ V. With an antireflection coating (ARC), the quantum efficiency η could be increased from 49 to 94%. To our knowledge this is the first time that such a high speed and such a high quantum efficiency have been achieved with an integrated silicon photodiode for a reverse voltage of only 3 V, whereby only a slightly modified standard CMOS process has been used. This slight modification consisted of one additional mask to block out the originally unmasked p-type threshold implantation. Later, however, it was shown [9] that this additional mask is not necessary for the structure shown in Fig. 5.9.

In contrast to transistors in bipolar OEICs, the electrical performance of the N- and P-channel MOSFETs is not degraded when the epitaxial layer is modified. This statement was verified by measurements (Table 5.1). The threshold voltages of the NMOS and PMOS transistors are practically independent of the doping concentration in the epitaxial layer, because these transistors are placed inside wells which possess a much higher doping level of several times 10^{16} cm^{-3} than the standard epitaxial layer with about 1×10^{15} cm^{-3} and because the threshold implants produce an even higher doping level ($\approx 10^{17}$ cm^{-3}) than the wells [20].

The reverse, i.e., the leakage current of the drain to well diodes is also listed in Table 5.1 for the NMOS and PMOS transistors [20]. The leakage current for the NMOS transistor in an epitaxial layer with reduced doping concentrations actually seems to be smaller than for the standard concentration.

These results confirm the superiority of the PIN-CMOS integration [21] over the PIN-bipolar integration [22] concerning process complexity. For the PIN-CMOS integration only epitaxial wafers with a low-doped top layer have to be used. This is not a process modification since these epi wafers can be bought from silicon wafer manufacturers. In contrast to the PIN-bipolar integration, the electrical transistor parameters of the standard twin-well CMOS process are completely unaffected for the PIN-CMOS integration and can be used for circuit simulations within the design of OEICs.

A photodiode reported in [22] is appropriate for the combination of high speed and high responsivity in bipolar and BiCMOS technologies. A thick intrinsic zone of this vertical pin photodiode allows the high responsivity. The other advantage of this photodiode was not reported or suggested in [22]: this photodiode allows to increase the reverse voltage far above the circuit supply voltage, which allows to enhance the speed of this photodiode by increasing the electric field strength and thereby the drift velocity of photogenerated

Table 5.1. Measured threshold voltages U_{Th} and drain leakage currents I_{D}^{r} for different doping levels in the epitaxial layer

I-Doping (cm^{-3})	$U_{\text{Th}}^{\text{NMOS}}$ (V)	$I_{\text{D}}^{\text{r,NMOS}}$ (pA)	$U_{\text{Th}}^{\text{PMOS}}$ (V)	$I_{\text{D}}^{\text{r,PMOS}}$ (pA)
Standard	0.79	2.19	−0.62	63.1
1×10^{14}	0.79	0.83	−0.60	66.1
5×10^{13}	0.79	0.81	−0.60	66.1
2×10^{13}	0.78	1.02	−0.60	67.6

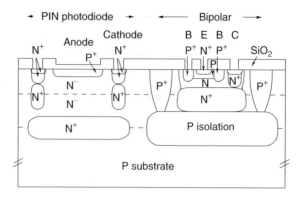

Fig. 5.11. Cross-section of a PIN-BiCMOS-OEIC [22]

electrons and holes leading to reduced rise and fall times of the photocurrent [23,24].

Figure 5.11 shows the cross-section of the OEIC proposed in [22]. The PIN cathode and a P isolation were implanted into the substrate. A three-step epitaxial growth (first $2\,\mu m$ as a buffer layer in order to avoid autodoping during the epitaxial process, then a $9\,\mu m$ I-layer, and finally a $4\,\mu m$ I-layer after the buried collector implant) was performed in order to supply the thick N-type "intrinsic" layer with a reduced doping level ($<3 \times 10^{13}\,cm^{-3}$) and a thin collector [22]. Assuming that the N$^+$-buried collector and the N$^+$-contact implant can be used for the N$^+$-cathode contact diffusions, that the P$^+$-isolation implant of the standard process can be used for the P isolation contact, and that the base contact implant is used for the PIN anode, the process complexity is increased by three additional lithography steps compared to the original bipolar process (without steps for ARC) (1) for the N$^+$-buried PIN cathode; (2) for the P isolation in order to isolate the buried collectors from the N$^-$ "intrinsic" region and to restrict the cathode potential to the photodiode area; and (3) for the N collector implant in order to provide the doping level of about $10^{16}\,cm^{-3}$. Here, the bias voltage of the photodiode is not limited by the circuit operating voltage U_{CC} or by the breakdown voltage BV$_{CE0}$ of the NPN transistors. Larger voltages, therefore, enable high-speed operation of the PIN photodiode. This possibility, however, was not exploited in [22]: A $-3\,dB$ bandwidth of $300\,MHz$ for $\lambda = 780\,nm$ was reported for the PIN photodiode with an area of $0.16\,mm^2$ at a bias of $3\,V$. The rise and fall times of the photocurrent were $1.6\,ns$ for these wavelength and bias values. Bandwidth values for higher bias values, were not reported in [22]. The responsivity of the PIN photodiode was $0.35\,A\,W^{-1}$ ($\eta = 57\%$) for $\lambda = 780\,nm$. This responsivity seems to be somewhat low, although an ARC is indicated in Fig. 5.11.

The speed enhancement possibility of this photodiode was verified in [23–25]. A photodiode with a similar cross-section was integrated in a $0.5\,\mu m$ industrial BiCMOS process [26,27]. By increasing the photodiode bias voltage from 5 to $11\,V$, the bandwidth of the photodiode was increased from 1.5 to

2.4 GHz [27]. To generate this high bias voltage from only one 5 V supply voltage for the photodiode a voltage-up-converter was integrated on the optical receiver chip.

5.3 Photo-Receiver Circuits

The design of microelectronic integrated circuits is a highly developed area. Circuits for advanced OEICs, however, are not yet investigated thoroughly. In the following, progress in this field will be summarized especially in the examples of OEICs for optical storage systems (OS-OEICs) like DVD and DVR as well as OEICs for receiver applications in optical interconnects, optical data transmission, and optical communication.

5.3.1 OS-OEICS for DVD Applications

PIN-CMOS OEICs for DVD applications were investigated [28], but were outperformed by BiCMOS OEICs [9]. The first BiCMOS approach presented here requires no process modification to implement a DPD as shown in Fig. 5.7. An OEIC with the DPD with a light-sensitive area of about $50 \times 50 \, \mu m^2$ was fabricated in a $0.8 \, \mu m$ BiCMOS process [29].

The circuit diagram of a fast channel (A–D) of the OEIC for optical storage systems (OS-OEIC) is shown in Fig. 5.12. An integrated DPD is connected to a transimpedance amplifier using an operational amplifier to obtain a low output offset voltage compared to a 2.5 V reference voltage as is required for applications in optical storage systems.

For a universal applicability, the gain is switchable by MOS elements among high (H, $R3$), medium (M, $R4 \| R3$), and low (L, $R5 \| R4 \| R3$) with a ratio of approximately 1/3 each (Fig. 5.12). Polysilicon–polysilicon capacitors can be produced in the $0.8 \, \mu m$ BiCMOS process and are used for frequency compensation with $C1$, $C2$, and $C3$. Only NPN transistors were applied in the signal path of operational amplifier to achieve high $-3 \, dB$ bandwidths.

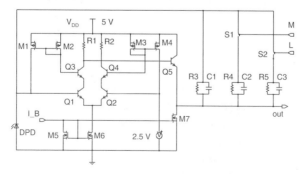

Fig. 5.12. Schematic of a fast channel amplifier (A–D) in a BiCMOS OS-OEIC [29]

The NPN transistors Q3 and Q4 are used to sense the base current of Q1 and Q2, respectively. These currents are mirrored by the current mirrors M1/M2 and M3/M4 back to the bases of the input transistors Q1 and Q2, respectively. This method is called bias-current cancelation. This biasing of the input transistors Q1 and Q2 reduces the systematic output offset of approximately 110 mV, which would result from the base current of Q1 across the largest resistor $R3 = 20\,k\Omega$, by more than one order of magnitude. The low-frequency open-loop gain of the operational amplifier was 27 dB and its transit frequency was 870 MHz with a $1\,k\Omega$ and 10 pF load. An OEIC was packaged, mounted on a printed circuit board together with these load elements, and a probe head with an input capacitance of 1.7 pF is used to measure the frequency response shown in Fig. 5.13.

The measured bandwidths exceeded 92 MHz, which was much larger than the bandwidth value 7.3 MHz of the circuit for $8\times$ speed CD-ROMs fabricated in 0.8 µm CMOS technology with off-chip photodiodes [30].

The four sensitive channels E–H have a ten times larger sensitivity for tracking control in the optical storage system. A DPD with approximately twice the size of the DPDs in the channels A–D was implemented here. A value of 90 mV µW^{-1} sensitivity in combination with a low offset voltage was realized. Each amplifier of channels A–H covered an active area of approximately 0.079 mm^2. The total OEIC die area was 3.25 mm^2. The power consumption of the OS-OEIC was less than 75 mW at 5.0 V. Table 5.2 summarizes the properties of the OS-OEIC channels A–H.

The next shown circuit topology implements transistor transimpedance amplifiers instead of operational amplifiers in transimpedance configuration. The innovative topology of this OEIC for optical storage systems is shown in Fig. 5.14.

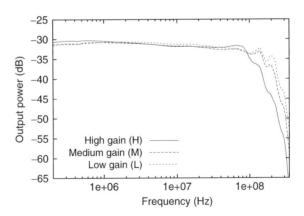

Fig. 5.13. Frequency response of a fast channel amplifier (A–D) in a BiCMOS OS-OEIC [29]

Table 5.2. Measured results of the high-bandwidth OS-BiCMOS-OEIC

	H	M	L
$f_{-3\,dB}$ (MHz) A–D	92.0	94.9	95.1
$f_{-3\,dB}$ (MHz) E–H	5.2	8.5	14.6
Sensitivity $(mV\,\mu W^{-1})$ A–D	8.8	2.9	0.9
Sensitivity $(mV\,\mu W^{-1})$ E–H	88.1	29.3	9.1
U_{Offset} (mV) A–D	<10.8	<9.5	<9.0
U_{Offset} (mV) E–H	<7.4	<6.4	<6.4
Noise (dBm) @10 MHz with 30 kHz RBW A–D	−81.5	−85.0	−85.2
Noise (dBm) E–H	−66.0	−67.5	−73.5

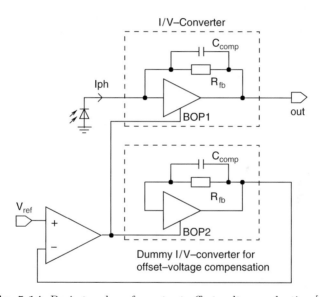

Fig. 5.14. Basic topology for output offset voltage reduction [31]

A current-to-voltage (I/V) converter is connected to the integrated photodiode. The BOP1 (bias operating point) input allows to correct the operating point of this transimpedance amplifier in order to cancel the output offset voltage.

For this purpose an identical dummy I/V converter with an open input and a low-offset operational amplifier are integrated together with the transimpedance amplifier connected to the photodiode. The operational amplifier measures the output offset voltage of the dummy I/V converter compared to $V_{ref} = 2.1\,V$ and changes its operation point via the BOP2 input in order to minimize the output offset voltage. The BOP1 input of the actual transimpedance amplifier is also connected to the output of the operational amplifier.

In such a way, the output offset voltage of the transimpedance amplifier can be reduced provided that a good matching between the two I/V converters and a low-offset voltage of the operational amplifier can be guaranteed. The first requirement can be dealt with a careful layout reducing the distance between corresponding devices in the two I/V converters. The second requirement can be fulfilled, because the operational amplifier can be designed to have a very slow response, i.e., transistors with a large emitter area or large gate lengths and widths for a good matching can be used.

An OEIC containing four fast channels (for data extraction) and four sensitive channels (for tracking control in the optical storage system) with $R_{fb} \approx 60\,k\Omega$ values was fabricated in a $0.8\,\mu m$ BiCMOS technology. The N-well/P-substrate photodiode was implemented with a light sensitive area of $2500\,\mu m^2$ in the fast channels and $9100\,\mu m^2$ in the sensitive channels.

The schematic diagram of the fast and sensitive channels of the BiCMOS OEIC is shown in Fig. 5.15. The transistor transimpedance amplifier is formed by Q1 in common-emitter configuration, the PMOS load element M1, the emitter follower Q3, and the feedback resistor R_{fb1}. Q3 provides a low output impedance of this input stage. The capacitor C_{fb1} is used for compensation. The emitter follower Q11 decouples the feedback path from the output and provides a low overall output impedance. The dummy I/V converter is formed by Q2, M2, Q4, R_{fb2}, C_{fb2}, and Q12. Q5–Q10, $R5$–$R10$, and M3/M4 are used

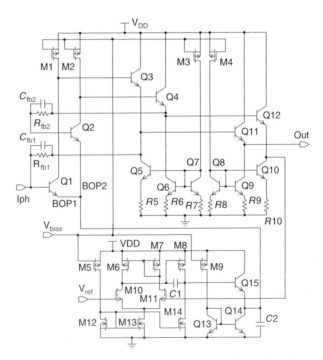

Fig. 5.15. Schematic diagram of one BiCMOS OS-OEIC channel

for biasing. The circuit diagram in Fig. 5.15 corresponds to the interlaced layout for good matching of the two I/V converters.

The BiCMOS operational amplifier is based on a CMOS Miller type topology with the NPN emitter follower Q15 at the output in order to supply a voltage source with a low output resistance for the emitter inputs BOP of the two I/V converters. The simulated open-loop gain of the operational amplifier was 77 dB and its transit frequency was 7.3 MHz.

A sensitivity for the fast channel of 25 and of $51\,\text{mV}\,\mu\text{W}^{-1}$ for the sensitive channels at 638 nm was found by determining the incident light power with a calibrated photodiode. A $-3\,\text{dB}$ frequency of 58.5 MHz was measured for a fast channel with a load of $10\,\text{k}\Omega$ and 10 pF (Fig. 5.16). The sensitive channels showed a $-3\,\text{dB}$ frequency of 26.7 MHz with the same load. All measured output offset voltages for photodiodes and OEICs in the dark were lower than 7.5 mV. This value is considerably lower than in the maximum simulated offset voltage of a reference amplifier without offset compensation of more than 72 mV for worst case transistor parameters at room temperature.

It should be mentioned that the BiCMOS operational amplifier and the emitter BOP inputs reduce the bandwidth of the I/V converters. The output resistance of the BiCMOS operational amplifier, however, is obviously rather low, since the bandwidth of 58 MHz is not considerably lower than that of a reference I/V converter of 82 MHz where the emitter of Q1 was connected to V_{ref} directly.

The bandwidth of the N-well/P-substrate photodiode may seem to be rather high compared to [7] where a bandwidth of 1.6 MHz was reported for a similar photodiode for $\lambda = 780\,\text{nm}$. The high bandwidth determined here, however, can be explained by the shorter wavelength and by an electric field, i.e., by a drift zone for photogenerated carriers, due to the doping gradient of the N-well similarly to that in the N-well of a DPD [32].

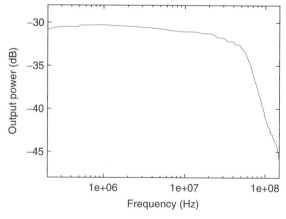

Fig. 5.16. Frequency response of fast channel of BiCMOS OS-OEIC [31]

112   H. Zimmermann

Noise measurements at 10 MHz with a resolution bandwidth of 30 kHz revealed values of −76 and −65 dBm for the fast and sensitive channels, respectively. The power consumption determined by simulation was 36 mW for a fast channel at a supply voltage of 5.0 V. The active die area of each channel was approximately $0.042\,\mu m^2$ and the total chip area was $3.1\,mm^2$.

When a high speed and a higher sensitivity are required in addition to a lower output offset voltage for the OS-OEICs, a two-stage optical receiver may be necessary. The circuit principle of such a two-stage amplifier [33, 34] is shown in Fig. 5.17.

The circuit consists of a transimpedance amplifier for the photocurrent and a reference I/V converter for offset compensation plus an operational amplifier in subtractor configuration. The subtractor can also be used as a voltage amplifier with the amplification factor RS2/RS1 enabling a high overall sensitivity of the OEIC. It has be mentioned, however, that offset voltages due to mismatch of the two transimpedance preamplifiers are also amplified. The circuit diagram of the complete circuit is shown in Fig. 5.18.

In order to achieve a high bandwidth, only NPN transistors are used in the signal paths of the preamplifiers. Q1 is used in common-emitter configuration, Q3 is used as emitter follower, and the feedback resistor R_{fb1} together with Q1 and Q3 represents a low input impedance for the photocurrent of the DPD.

Thereby, the effect of the DPD capacitance is minimized. The reference voltage V_{ref} is chosen as the emitter potential of Q1 in order to increase the reverse voltage of the DPD to $V_{BE,Q1} + V_{ref}$. The values for R_1, R_{fb1}, and C_{fb1} were $3\,k\Omega$, $27\,k\Omega$, and $25\,fF$, respectively. A second emitter follower (Q11) is implemented for level shifting and decoupling of output and feedback path. The second preamplifier consists of transistors Q2, Q4, and Q7 and the feedback resistor R_{fb2} plus the compensation capacitor C_{fb2}. At the outputs of the preamplifier, $C3$ and $C4$ are added as further compensation capacitors.

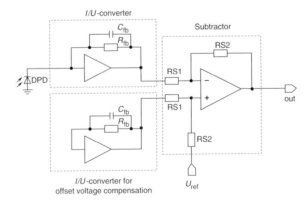

Fig. 5.17. Block diagram of a two-stage optical receiver for a fast channel of a high-speed OS-BiCMOS-OEIC

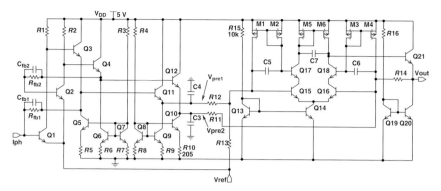

Fig. 5.18. Schematic of high-speed BiCMOS-OEIC for the fast channels A–D in an OS-BiCMOS-OEIC

Perfect matching of the two preamplifiers is necessary in order to obtain $V_{\mathrm{pre1}} = V_{\mathrm{pre2}}$ for a dark photodiode (Fig. 5.18). This perfect matching of the two preamplifiers requires a careful layout to achieve a low output offset voltage.

The preamplifiers are connected to the operational amplifier in subtractor configuration via R_{11} and R_{12}. The bias current cancelation introduced in Fig. 5.12 is applied to reduce the input currents of the operational amplifier enabling a low output offset voltage. A PMOS current mirror load with M5 and M6 is used here in order to obtain a higher open-loop gain of the operational amplifier. For $R_{11} = R_{12}$ and $R_{13} = R_{14}$, an analysis of the subtractor amplifier yields the transfer function

$$V_{\mathrm{out}} \approx V_{\mathrm{ref}} + \frac{R_{13}}{R_{12}} I_{\mathrm{ph}} R_{\mathrm{fb1}},$$

when we assume a large open-loop voltage gain of the operational amplifier. A $-3\,\mathrm{dB}$ frequency of $189\,\mathrm{MHz}$ was determined by numerical prelayout simulation for the complete amplifier with a load of $R_{\mathrm{L}} = 1\,\mathrm{k}\Omega$ and $C_{\mathrm{L}} = 10\,\mathrm{pF}$. The complete two-stage amplifier was designed for a sensitivity of $10\,\mathrm{mV}\,\mu\mathrm{W}^{-1}$ and an offset voltage less than $10\,\mathrm{mV}$ for $R_{11} = R_{12} = R_{13} = R_{14}$.

The OEIC with the DPD and the two-stage amplifier was fabricated in a $0.8\,\mu\mathrm{m}$ BiCMOS technology. The measured frequency response of this two-stage optical receiver is shown in Fig. 5.19. A $-3\,\mathrm{dB}$ frequency of $147.7\,\mathrm{MHz}$ is determined from this frequency response. The power consumption of the two-stage optical receiver was $35\,\mathrm{mW}$ at a supply voltage of $5\,\mathrm{V}$. The active die area of the two-stage optical receiver was $340\,\mu\mathrm{m} \times 140\,\mu\mathrm{m}$.

Current amplifiers instead of transimpedance amplifiers were suggested in [35, 36]. A bandwidth of $250\,\mathrm{MHz}$ was achieved with this new approach. Meanwhile, however, the performance of two-stage amplifiers like in [34] were improved considerably by supplying both preamplifiers with one common

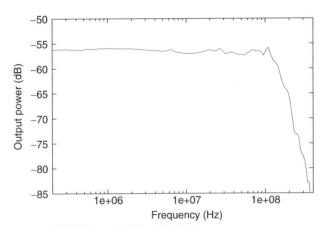

Fig. 5.19. Frequency response of a two-stage optical BiCMOS receiver for the fast channels A–D of a high-speed BiCMOS-OEIC for optical storage systems [34]

current source, whereby this measure allows to double the sensitivity. A bandwidth of 380 MHz at a sensitivity of $100\,\mathrm{mV\,\mu W^{-1}}$ at 660 nm was achieved [37, 38].

5.3.2 Fiber and Interconnect Receivers

First, we demonstrate an innovative monolithic integration of vertical PIN photodiodes in a twin-well CMOS process (Fig. 5.9), which uses epitaxial wafers, in order to combine the high speed and the large quantum efficiency of the photodiode [19]. In contrast to the OEICs of [39, 40], here, only a single power supply of 2.5, 3.3, of 5 V is needed. The integration of PIN photodiodes in a twin-well CMOS technology requires much less additional process complexity than the published approaches to SBC-based bipolar OEICs [22, 41]. Three additional masks were necessary for the PIN-bipolar integration and for the avoidance of the Kirk effect. Only one photodiode protection mask is added for the PIN-CMOS integration in order to block out an originally unmasked threshold implantation from the photodiode area [19, 42]. This additional mask seemed to be justified due to the expected high volumes of OEICs and OPTO-ASICs. Finally it was shown that this additional mask is not necessary [9].

The PIN-CMOS photodiodes with an area of $2{,}700\,\mu\mathrm{m}^2$, a doping concentration C_e in the epitaxial layer of $2\times10^{13}\,\mathrm{cm}^{-3}$, and an integration polysilicon resistor of $500\,\Omega$, as well as complete OEICs, were fabricated in an industrial $1.0\,\mu\mathrm{m}$ CMOS process. Reference samples both for the photodiodes and OEICs with the standard doping concentration $C_e = 10^{15}\,\mathrm{cm}^{-3}$ were also fabricated. The measurements on the PIN photodiodes and their results were described in Chap. 2.

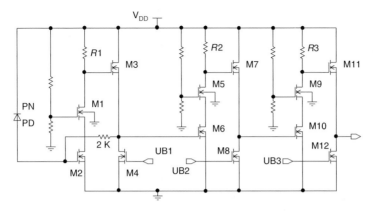

Fig. 5.20. Circuit of a fiber receiver OEIC [19]

The circuit of a high-speed preamplifier with an integrated PIN photodiode is shown in Fig. 5.20. The cascode transistors M1, M5, and M9 reduce the Miller effect and increase the bandwidth correspondingly. The input stage with transistors M1–M4 is a transimpedance configuration, which converts the photocurrent change in the integrated PIN photodiode to a voltage change. The source followers M3, M7, and M11 as well as the current sources M4, M8, and M12 are used for level shifting. The threshold voltage of transistors M3, M7, and M11 was reduced intentionally to about 0.4 V by the photodiode protection mask in order to obtain lower V_{GS} values. Due to the feedback across the $2 \, k\Omega$ resistor and to the identical dimensions of the transistors in the different stages, a good independence from process deviations within the relatively large specified process tolerances of the used digital CMOS process was obtained. Three identical biasing circuits (UB1=UB2=UB3) were used instead of one in order to avoid parasitic coupling between the stages. The sensitivity of the PIN amplifier OEIC was $4.7 \, \text{mV} \, \mu\text{W}^{-1}$ increasing to $9.0 \, \text{mV} \, \mu\text{W}^{-1}$ with ARC, which corresponds to an overall transimpedance of $18.4 \, k\Omega$. Its power consumption was 44 mW at 5.0 V reducing to 17 mW at 3.3 V and to 9 mW at 2.5 V. The photodiode together with its metal shield around covered an area of approximately $150 \, \mu\text{m} \times 150 \, \mu\text{m}$. The preamplifier occupied an active area of less than $190 \, \mu\text{m} \times 200 \, \mu\text{m}$.

The rise and fall times for the OEIC with the epitaxial doping concentration $C_e = 10^{15} \, \text{cm}^{-3}$ were $t_r = 15.5 \, \text{ns}$ and $t_f = 17.6 \, \text{ns}$. These large values for t_r and t_f were due to the slow carrier diffusion in the standard epitaxial layer of the photodiode. The corresponding values for the doping concentration $2 \times 10^{13} \, \text{cm}^{-3}$ in the epitaxial layer, where the depletion region spread through the whole epitaxial layer and carrier diffusion in the photodiode was eliminated, were 0.53 and 0.69 ns, respectively. These values indicate that CMOS OEICs with a reduced doping concentration in the epitaxial layer having an appropriate output buffer can be used as receivers for optical data transmission via fibers or for optical interconnects on a board level up to a BR of

$622\,\mathrm{Mb\,s^{-1}}$ in the nonreturn-to-zero (NRZ) mode, verified by a measured eye diagram with pseudo random bit sequences (PRBS) of $2^{23} - 1$ (Fig. 5.21).

The next fiber receiver described used PIN photodiodes in the same $1.0\,\mu\mathrm{m}$ CMOS process also with a reduced doping concentration in the epitaxial layer. The high-speed preamplifier circuit is shown in Fig. 5.22.

Only N-channel MOSFETs with a minimum channel lengths were used in order to achieve a high bandwidth. The transistors M1–M5 operate in common-source configuration. The input stage with M1 in transimpedance configuration converts changes in the photocurrent of the photodiode to voltage changes. The feedback resistor in the input stage is formed by gate polysilicon. Two further stages with two transistors each are used as voltage amplifiers. The polysilicon feedback resistor in each of the two stages (Rg1, Rg2) limits the gain and therefore boosts the bandwidth. Due to the feedback resistors, a good independence from process deviations within the relatively large

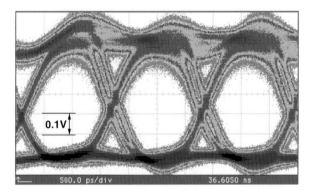

Fig. 5.21. Eye diagram of the CMOS preamplifier OEIC for a data rate of $622\,\mathrm{Mb\,s^{-1}}$ ($500\,\mathrm{ps\,div^{-1}}$; $0.1\,\mathrm{V\,div^{-1}}$)

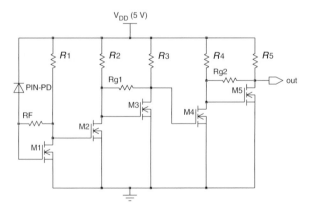

Fig. 5.22. Circuit of a fiber receiver OEIC [42]

specified process tolerances of the used digital CMOS process is obtained. The sensitivity of the PIN preamplifier OEIC was $13.8\,\mathrm{mV\,\mu W^{-1}}$ without ARC, which corresponds to an overall transimpedance of $45.9\,\mathrm{k\Omega}$. Compared to the above-mentioned receiver, the sensitivity of the new OEIC has been increased by a factor of more than 2.5. Its power consumption is $2.5\,\mathrm{mW}$ at $5.0\,\mathrm{V}$. The photodiode has an area of $2{,}700\,\mathrm{\mu m^2}$. The preamplifier occupies an active area of $245\,\mathrm{\mu m} \times 140\,\mathrm{\mu m}$ due to the large area necessary for the resistors formed by low-resistivity (gate) polysilicon.

The rise and fall times measured at the OEIC output with the picoprobe were 0.72 and $0.57\,\mathrm{ns}$, respectively, with a supply voltage of $5.0\,\mathrm{V}$. With these values a bandwidth of $f_{-3\mathrm{dB}} = 2.2/(\pi(t_\mathrm{r} + t_\mathrm{f})) = 550\,\mathrm{MHz}$ can be estimated for the OEIC. The higher data rate of this OEIC is due to less transistors in the preamplifier than in the typical high-frequency circuit with cascode transistors and source followers described before. In the above-mentioned receiver, 12 NMOS transistors were used in the preamplifier, which therefore contained more parasitic capacitances. Here, the number of parasitic capacitances is considerably lower. In addition, the gain per amplifying transistor is lower here, allowing for a higher data rate of the new preamplifier.

The so-called Q-factor is defined as $(B_1 - B_0)/(\sigma_1 + \sigma_0)$ [43], where B_1 and B_0 are the mean values and σ_1 and σ_0 are the standard deviations of the output signal for a "1" and "0." B_1, B_0, σ_1, and σ_0 can be determined from eye diagram measurements.

The bit error rate (BER) then is available from BER $\approx \frac{e^{-Q^2/2}}{Q\sqrt{2\pi}}$ [43]. The BER of the OEIC has been determined with an HP54570/51 digital sampling oscilloscope for the NRZ data rates of $622\,\mathrm{Mb\,s^{-1}}$ and $1\,\mathrm{Gb\,s^{-1}}$ in dependence on the optical input power. The results for the BER are shown in Fig. 5.23. From this diagram for $\lambda = 638\,\mathrm{nm}$ being especially interesting for data transmission via plastic optical fibers and for optical interconnects, it can be seen that an optical input power of $-15.4\,\mathrm{dBm}$ is necessary for a BER of 10^{-9} at a data rate of $1\,\mathrm{Gb\,s^{-1}}$. At $622\,\mathrm{Mb\,s^{-1}}$ the sensitivity of the OEIC for the same BER is $-17.2\,\mathrm{dBm}$. The sensitivity for a BER of 10^{-10} at a data rate of $1\,\mathrm{Gb\,s^{-1}}$ is $-14.8\,\mathrm{dBm}$. The sensitivity for a BER of 10^{-11} at a data rate of $622\,\mathrm{Mb\,s^{-1}}$ is $-16.5\,\mathrm{dBm}$. For $\lambda = 850\,\mathrm{nm}$, a sensitivity of $-15.3\,\mathrm{dBm}$ for a BER of 10^{-9} at a data rate of $622\,\mathrm{Mb\,s^{-1}}$ has been verified.

These values are very good results for an OEIC in a $1.0\,\mathrm{\mu m}$ CMOS technology. The overall transimpedance of $46\,\mathrm{k\Omega}$ and the data rate of $1\,\mathrm{Gb\,s^{-1}}$ result in a $46\,\mathrm{Tbit\Omega/s}$ transimpedance data rate product. With the bandwidth of $550\,\mathrm{MHz}$, a $25\,\mathrm{THz\,\Omega}$ effective transimpedance product results, which exceeds the value of $18\,\mathrm{THz\,\Omega}$ reported in [44]. The sensitivity of the PIN CMOS OEIC for a data rate of $1\,\mathrm{Gb\,s^{-1}}$ and a BER of 10^{-9} with a value of $-15.4\,\mathrm{dBm}$ for $638\,\mathrm{nm}$ exceeds the sensitivity of $-6.3\,\mathrm{dBm}$ of a $0.35\,\mathrm{\mu m}$ OEIC for $850\,\mathrm{nm}$ [40] by a factor of 8. It shall be mentioned that the implementation of an ARC will improve the sensitivity of the OEICs by a value of 2–3 dB. The minimum sensitivity of $-17\,\mathrm{dB m}$ specified in the Gigabit Ethernet networking standard,

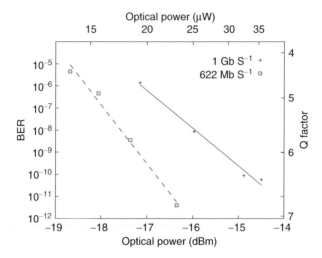

Fig. 5.23. BER of PIN OEIC without ARC in dependence on the optical input power at 638 nm

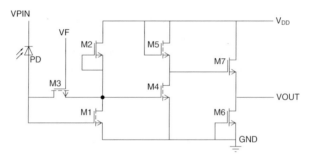

Fig. 5.24. Circuit diagram of the SOI receiver [13]

therefore, may be achievable with PIN-CMOS-OEICs. The data rate of the OEIC is limited by the amplifier in a 1.0 μm technology. With sub-micrometer PIN-CMOS-OEICs data rates in excess of $1\,\mathrm{Gb\,s^{-1}}$ are possible [36].

The circuit diagram for a receiver with a SOI PIN photodiode (Fig. 5.6) is shown in Fig. 5.24. Optimum results were obtained for input transistors (M1–M2) with 50 μm widths for the SOI receiver. The functionality of the three-stage amplifier has been described in [14]. The small-signal transfer function of this design is discussed in [13].

At $V_{\mathrm{DD}} = 5\,\mathrm{V}$ and $V_{\mathrm{PD}} = 20\,\mathrm{V}$, for 850 nm and a BER $= 10^{-9}$, the sensitivity of the SOI receiver was $-26.1\,\mathrm{dBm}$ at $622\,\mathrm{Mb\,s^{-1}}$, $-20.2\,\mathrm{dBm}$ at $1.0\,\mathrm{Gb\,s^{-1}}$ as well as $-12.2\,\mathrm{dBm}$ at $2.0\,\mathrm{Gb\,s^{-1}}$. The dependence of the sensitivity on the photodiode voltage was also investigated. The SOI receiver demonstrated a penalty that was typically less than 1 dB as the photodiode supply was reduced to 10 V. For single-supply 5 V operation, the sensitivity degraded by 2.8–3.9 dB and the maximum data rate was 1.5 Gb s^{-1} with a sensitivity of −12 dBm.

The dynamic range of the receivers was measured with both a constant voltage supply and with a variable feedback voltage (V_F) as a simple form of automatic gain control (AGC). The photodiode bias has a large impact on the dynamic range, although the feedback voltage can be used very effectively to compensate for this degradation. At the optimum bias point and a constant V_F, the SOI receiver exhibited dynamic ranges of >23.5 dBm, >17.7 dBm, and 12.0 dBm at 622 Mb s^{-1}, 1.0 Gb s^{-1} and 2.0 Gb s^{-1}.

This receiver was not optimized for the best sensitivity at the higher data rates. The dominant noise term was thermal noise of the feedback resistor (M_F) and not the channel noise of the input transistor. Better performance is possible with an optimized circuit design in a more advanced technology.

An optical receiver with the Yamamoto type photodiode (see Fig. 5.11) in a 0.5 μm BiCMOS technology achieved a data rate of 3 Gb s^{-1} and to our knowledge the best sensitivity of −24.3 dBm for 660 nm ever reported for an silicon OEIC at this wavelength. The circuit diagram of this optical receiver is shown in Fig. 5.25.

The receiver circuit contains only bipolar transistors with a maximum transit frequency of about 25 GHz to achieve a high data rate. At Gb s^{-1} data rates, bipolar amplifiers cause less noise than MOS amplifiers. A pseudo-differential amplifier topology allows implementation of optical receivers in analog–digital circuits. The frequency response of the OEIC is shown in

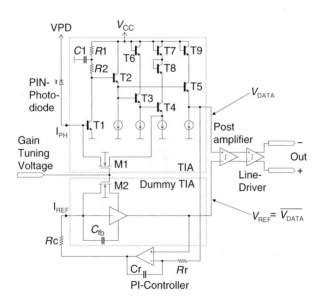

Fig. 5.25. Circuit diagram of the PIN-BiCMOS receiver [27]

Fig. 5.26. The eye diagram for a data rate of $3\,\mathrm{Gb\,s^{-1}}$ with the average optical input power of $-24.3\,\mathrm{dBm}$ can be seen in Fig. 5.27.

Table 5.3 summarizes more results of the 2.4 GHz BiCMOS-OEIC. New measurements at $5\,\mathrm{Gb\,s^{-1}}$ show a sensitivity of $-20.5\,\mathrm{dBm}$ for 660 nm light [45]. Besides the very good sensitivity very large values for the dynamical range of the average optical input power were obtained. This very good performance is due to the thick intrinsic zone of the integrated pin photodiode and the drift-field-enhancement with the high reverse voltage of the photodiode. A considerable advantage of the presented OEIC is that the user does not have to

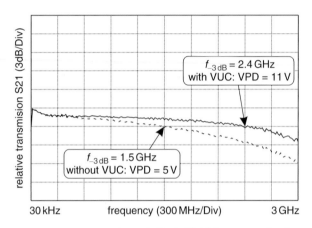

Fig. 5.26. Frequency response of the PIN-BiCMOS receiver for $\lambda = 660\,\mathrm{nm}$ [27]

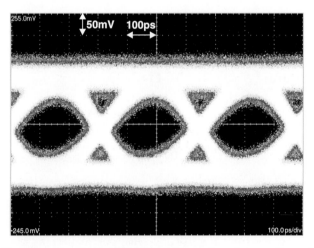

Fig. 5.27. Eye diagram of the PIN-BiCMOS receiver for av. P_{opt} $-24.3\,\mathrm{dBm}$ at $\lambda = 660\,\mathrm{nm}$, BER $= 10^{-9}$ and PRBS $= 2^{31} - 1$ [27]

Table 5.3. Measured results of the 2.4 GHz bandwidth BiCMOS OEIC for $\lambda = 660$ nm, BER $= 10^{-9}$ and PRBS $= 2^{31} - 1$ [27]

Data rate (Gb s^{-1})	Sensitivity (dBm)	Transimpedance (kΩ)	VT (V)	av. $P_{\mathrm{opt}}^{\max}$ (dBm)	Opt. dyn. range (dB)
1.0	−29.3	248	2.1	−7	22.3
1.5	−27.8	–	2.25	−7.2	20.6
2.5	−24.9	126	2.5	−7.8	17.1
3.0	−24.3	88	3.0	−7.9	16.4

provide this high photodiode bias voltage, because it is generated on the OEIC chip. It was also shown that the generation of this photodiode bias on chip does not degrade the sensitivity of the OEIC [45]. This PIN-BiCMOS technology allows fabrication of low-cost optical receivers in OPTO-ASIC technology for optical interconnect applications e.g., between electronic boards and for plastic optical fiber (POF) receivers.

5.4 Conclusion

Double photodiode, vertical PIN photodiode, and lateral SOI photodiode avoid slow carrier diffusion effects. The performance of integrated photodiodes can be improved significantly when minor process changes are made to obtain these advanced integrated photodiodes. Advanced OEICs for optical storage systems have been realized. These low-offset solutions have bandwidths ranging from 58 to 380 MHz. Furthermore, high-speed fiber receiver OEICs have been presented. PIN-CMOS-OEICs in 1.0 µm technology reach sensitivities of −17.2 dBm at 622 Mb s^{-1} and −15.4 dBm at 1 Gb s^{-1} (BER $= 10^{-9}$) with a single 5 V supply voltage. Using higher photodiode bias voltages, the SOI receiver reaches sensitivities of −20.2 dBm at 1 Gb s^{-1} and −12.2 dBm at 2 Gb s^{-1}. The speed of integrated photodiodes and of amplifiers can be increased with sub-µm technologies. A 0.6 µm BiCMOS-OEIC, which generated a high bias voltage for the pin photodiode from its single-supply voltage of 5 V, achieved a sensitivity of −24.3 dBm with 660 nm light at 3 Gb s^{-1} and −20.5 dBm at 5 Gb s^{-1}. Si PIN-BiCMOS low-cost highly reliable receivers for Gigabit Ethernet, the Fiber Channel, as well as optical interconnects working with 850 nm light, for instance, can be realized in OPTO-ASIC technology. In contrast, standard deep-sub-µm CMOS technologies suffer from the photodiode properties and, furthermore, are not interesting for OPTO-ASIC production due to high mask costs. It, however, seems feasible to achieve almost as fast CMOS optical receivers in deep-sub-µm CMOS technologies, which achieve a much worse sensitivity compared to the described PIN-BiCMOS OEIC.

References

1. Landolt-Börnstein: *Numerical Data and Functional Relationships in Science and Technology*, vol. 17c (Springer, Berlin Heidelberg New York, 1984), p. 474

2. J. Popp, H. v. Philipsborn: 10 Gbit/s on-chip photodetection with self-aligned silicon bipolar transistors, Proceedings of ESSCIRC, 1990, pp. 571–574

3. J. Wieland, H. Duran, A. Felder: Two-channel 5 Gbit/s silicon bipolar monolithic receiver for parallel optical interconnects, Electronics Letters **30**, 358 (1994)

4. H. Kabza, K. Ehinger, T.F. Meister, H.-W. Meul, P. Weger, I. Kerner, M. Miura-Mattausch, R. Schreiter, D. Hartwig, M. Reisch, M. Ohnemus, R. Köpl, J. Weng, H. Klose, H. Schaber, L. Treitinger: A 1-µm polysilicon self-aligned bipolar process for low-power high-speed integrated circuits, IEEE Electronics Device Letters **10**, 344–346 (1989)

5. E. Braß, U. Hilleringmann, K. Schumacher: System integration of optical devices and analog CMOS amplifiers, IEEE Journal of Solid-State Circuits **29**, 1006–1010 (1994)

6. U. Hilleringmann, K. Goser: Optoelectronic system integration on silicon: waveguides, photodetectors, and VLSI CMOS circuits on one chip, IEEE Transactions on Electronic Devices **42**, 841–846 (1995)

7. E. Fullin, G. Voirin, M. Chevroulet, A. Lagos, J.-M. Moret: CMOS-based technology for integrated optoelectronics: a modular approach, IEEE International Electronic Device Meeting, 1994, pp. 527–530

8. R. Kauert, W. Budde, A. Kalz: A monolithic field segment photo sensor system, IEEE Journal of Solid-State Circuits **30**, 807–811 (1995)

9. H. Zimmermann: *Integrated Silicon Optoelectronics* (Springer, Berlin Heidelberg New York, 2000)

10. P.J.-W. Lim, A.Y.C. Tzeng, H. L. Chuang, S.A.S. Onge: A 3.3 V monolithic photodetector/CMOS preamplifier for 531 Mb s^{-1} optical data link applications, Proceedings ISSCC, 1993, pp. 96–97

11. D.M. Kuchta, H.A. Ainspan, F.J. Canora, R.P. Schneider: Performance of fiber-optic data links using 670 nm CW VCSELs and a monolithic Si photodetector and CMOS preamplifier, IBM Journal of Research and Development **39**, 63–72 (1995)

12. J.-P. Colinge: *Silicon-on-Insulator Technology: Materials to VLSI*, (Kluwer, Boston, 1991)

13. J. Schaub, R. Li, J. Csutak, J. Campbell: High-speed monolithic silicon photo-receivers on high resistivity and SOI substrates, Journal of Lightwave Technology **19**, 272–278 (2001)

14. C. Schow, J. Schaub, R. Li, J. Qi, J. Campbell: A 1-Gb/s monolithically integrated silicon NMOS optical receiver, IEEE Journal of Selected Topics in Quantum Electronics **4**, 1035–1039 (1998)

15. H. Zimmermann, K. Kieschnick, M. Heise, H. Pless: BiCMOS OEIC for optical storage systems, Electronics Letters **34**, 1875–1876 (1998)

16. H. Zimmermann: Full custom CMOS and BiCMOS OPTO-ASICs, Proceedings of the 5th International Conference on Solid-State and Integrated-Circuit Technology, 1998, pp. 344–347

17. H. Zimmermann, U. Müller, R. Buchner, P. Seegebrecht: Optoelectronic receiver circuits in CMOS-technology, in: *Mikroelektronik'97, GMM-Fachbericht 17* (VDE, Berlin, Offenbach, 1997), pp. 195–202

18. H. Zimmermann, T. Heide, A. Ghazi, K. Kieschnick: PIN-CMOS-receivers for optical interconnects, Extended Abstracts of 2nd IEEE Workshop on Signal Propagation on Interconnects, 1998, pp. 88–89
19. H. Zimmermann, T. Heide, A. Ghazi: Monolithic high-speed CMOS photo-receiver, IEEE Photonics Technology Letters **9**, 254–256 (1999)
20. H. Zimmermann, A. Ghazi, T. Heide, R. Popp, R. Buchner: Advanced photo integrated circuits in CMOS technology, Proceedings of 49th Electronic Components and Technology Conference (ECTC), 1999, pp. 1030–1035
21. H. Zimmermann: Monolithic Bipolar-, CMOS-, and BiCMOS-receiver OEICs, Proceedings of the International Semiconductors Conference (CAS'96), 1996, pp. 31–40
22. M. Yamamoto, M. Kubo, K. Nakao: Si-OEIC with a built-in pin-photodiode, IEEE Transactions on Electronic Devices **42**, 58–63 (1995)
23. R. Swoboda, H. Zimmermann: A low-noise monolithically integrated 1.5 Gbps optical receiver in 0.6 μm BiCMOS technology, IEEE Journal of Selected Topics in Quantum Electronics **9**, No. 2, pp. 419–424 (2003)
24. C. Seidl, J. Knorr, H. Zimmermann: Single-stage 378MHz–178kΩ-transimpedance amplifier with capacitive coupled voltage dividers, IEEE International Solid-State Circuits Conference (ISSCC), Digest of Technical Papers, vol. 47, 2004, pp. 470–471, 540.
25. R. Swoboda and H. Zimmermann: A 2.5-Gpbs receiver OEIC in 0.6-μm BiCMOS technology, IEEE Photonics Technology Letters, **16**(7), pp. 1730–1732 (2004)
26. J. Sturm, M. Leifhelm, H. Schatzmayr, S. Groiss, H. Zimmermann: Optical receiver IC for CD/DVD/blue-laser application, Proceedings of 30th European Solid-State Circuits Conference ESSCIRC, 2004, pp. 267–270
27. R. Swoboda, H. Zimmermann: A 2.4GHz OEIC with voltage-up-converter, Proceedings of 30th European Solid-State Circuits Conference (ESSCIRC), 2004, pp. 223–226
28. A. Ghazi, T. Heide, H. Zimmermann, P. Seegebrecht: DVD OEIC and 1 GBit/s Fiber Receiver in CMOS Technology, Proceedings of 7th IEEE International Symposium on Electronic Devices for Microwave and Optoelectronic Applications (EDMO), 2000, pp. 224–229
29. H. Zimmermann, Kieschnick, M. Heise, H. Pless: High-bandwidth BiCMOS OEIC for optical storage systems, Proceedings of IEEE International Solid-State Circuits Conference, 1999, pp. 384–385
30. B. Kim, M. Jeong, D. Cho, J. Kim, J. Lee, S. Kim: 0.8 μm CMOS Analog Front-End Processor for 8× Speed CD-ROM, IEEE Transactions on Consumer Electronics **42**, 826–831 (1996)
31. H. Zimmermann, K. Kieschnick: Low-offset BiCMOS OEIC for optical storage systems, Electronics Letters **36**, 1223–1224 (2000)
32. H. Zimmermann, K. Kieschnick, T. Heide, A. Ghazi: Integrated high-speed, high-responsivity photodiodes in CMOS and BiCMOS technology, Proceedings of 29th European Solid-Stage Device Research Conference, 1999, pp. 332–335
33. T. Takimoto, N. Fukunaga, M. Kubo, N. Okabayashi: High speed Si-OEIC (OPIC) for optical pickup, IEEE Transactions on Consumer Electronics **44** 137–142 (1998)
34. K. Kieschnick, T. Heide, A. Ghazi, H. Zimmermann, P. Seegebrecht: High speed photonic CMOS and BiCMOS receiver ICs, Proceedings of 25th European Solid-State Circuits Conference, 1999, pp. 398–401

35. G. de Jong, J. Bergervoet, J. Brekelmans, J. van Mil: A DC-to-250MHz current pre-amplifier with integrated photodiodes in standard CBiMOS, for optical-storage system, Proceedings of IEEE International Solid-State Circuits Conference, 2002, pp. 362–363, 474

36. H. Zimmermann: *Silicon Optoelectronic Integrated Circuits* (Springer, Berlin Heidelberg New York, 2003)

37. J. Leeb, K. Schneider, H. Zimmermann: A 380 MHz two-stage OEIC for the use in DVD pick-up units, Proceedings of IEEE International Symposium on Consumer Electronics, 2004, pp. 381–384

38. J. Leeb: Ph.D. thesis, Vienna University of Technology, 2005

39. C. Schow, J. Qi, L. Garrett, J. Campbell: A silicon NMOS monolithically integrated optical receiver, IEEE Photonics Technology Letters **9**, 663–665 (1997)

40. T. Woodward, A. Krishnamoorthy: 1 Gbit/s CMOS photoreceiver with integrated detector operating at 850 nm, Electronics Letters **34**, 1252–1253 (1998)

41. M. Kyomasu: Development of an integrated high speed silicon PIN photodiode sensor, IEEE Transactions on Electronic Devices **42**, 1093–1099 (1995)

42. H. Zimmermann, T. Heide: A monolithically integrated 1-Gb/s optical receiver in 1-μm CMOS technology, IEEE Photonics Technology Letters **13**, 711–713 (2001)

43. G. Agrawal: *Fiber-Optic Communication Systems*, Wiley, New York, 1997

44. M. Ingels, G. Plas, J. Crols, M. Steyaert: A CMOS 18 THz 240 Nb/s transimpedance amplifier and 155 Mb/s LED-driver for low-cost optical fiber links, IEEE Journal of Solid-State Circuits **29**, 1552–1559 (1994)

45. R. Swoboda, H. Zimmermann: A 5 Gbps OEIC with voltage-up-converter, IEEE Journal of Solid-State Circuits, **40**, 7, 1521–1526 (2005)

6

Active SiGe Devices for Optical Interconnects

E. Cassan, S. Laval, D. Marris, M. Rouvière, L. Vivien, M. Halbwax, A. Lupu, and D. Pascal

Summary. The realization of integrated active devices is a key point for on-chip optical interconnects at 1.3–1.5 µm wavelengths. The required devices are high-speed, low-noise, and highly sensitive detectors, as well as efficient switching devices and optical modulators. In this context, as material compatibility must be insured, the group IV silicon–germanium technology is important due to the energy band engineering that it allows, with the integration of Si/SiGe heterostructures into the existing silicon technology. The aim of this chapter is to give a review of the recent progress made in the field of active SiGe devices for optical interconnects. Attention is focused on strained-layer superlattice SiGe/Si and pure Ge photodetectors, as well on SiGe-based switching devices and optical modulators.

6.1 Introduction

In 1965, Gordon Moore predicted that the number of transistors in an integrated circuit would double every two years. This tendency has appeared as a remarkable prediction since then. The dramatic increase in the number of transistors per chip and the shrinkage of device dimensions have allowed a spectacular increase of operating frequencies as well as circuit functionalities. Silicon CMOS technology is dominating the electronic market. Transistor gate length enters the sub-50 nm range, and the transistor count per chip is nearly exceeding the billion. Nevertheless, with the shrinkage of device dimensions down to the nanometer scale and the huge growth of metallic interconnect lengths to several tens of kilometers on a chip, it is expected that silicon circuits will encounter an interconnection bottleneck in the near future [1]. The main problems that arise are linked to increasing delays of global on-chip interconnects, signal distortion, and timing uncertainty (skew and jitter), as well as power consumption.

In a parallel fashion, lightwave communications have received considerable interest for 30 years with the fabrication of ultra-low-loss single-mode optical fibers, IR laser diodes, and optical amplifiers. Optical fiber communications have now been deployed all over the word, enabling modern high-speed communications. Due to the increasing demand for bandwidth in all applications, the general tendency has been the reduction of distances over which optical communications are employed. With the progressive drift from the long-haul to inter-chip optical links, and even to on-chip optical interconnects, as recently proposed [1], optical communications could meet microelectronics in a near future.

On-chip optical interconnects require the development of silicon-based materials and devices for the generation, guidance, control, and detection of light. Monolithic silicon microphotonics at the wavelength of 1.3–1.55 μm has received a growing interest for several years [2]. Silicon-on-insulator (SOI) integrated optical components have been demonstrated for light distribution at the chip scale [3], including submicron SOI waveguides, 90°-turns, beam splitters, and H-tree distributions. The development of high-speed and low-capacitance CMOS-compatible modulators and detectors is now among the major challenges for realizing on-chip interconnects. Toward efficient integrated active devices for optical interconnects, the group IV silicon–germanium technology is important, due to the possibilities of energy bandgap engineering that it allows. Material compatibility has allowed Si/SiGe heterostructures to be integrated into the existing silicon technology – both in bipolar and MOS applications. The SiGe HBT in BiCMOS process has been commercialized for high-frequency and low-noise applications. Si/SiGe modulation field effect transistors (MODFETs) have been reported with oscillation frequencies up to 158 GHz [4]. Other works suggest that SiGe technology is to be used for advanced CMOS applications [5].

The energy bandgap engineering brought by the SiGe technology is a key element for the optoelectronic integration of active devices dedicated to optical interconnects. The aim of this chapter is to give a review of the recent progress made in this field. Attention will be focused on SiGe and pure Ge photodetectors, as well on SiGe-based optical switches and modulators.

6.2 Silicon–Germanium Technology for Optical Interconnects

The Si and Ge band structures have been described in detail elsewhere [6–9]. Si and Ge are miscible over the complete range of compositions. One of the main features for optoelectronic applications is the indirect bandgap of both materials. Indirect bandgap energies of Si and Ge are 1.12 and 0.66 eV at 300 K, respectively. SiGe alloys maintain this indirect bandgap structure whatever be the alloy composition.

Following the Vegard's law, the evolution of the $Si_{1-x}Ge_x$ crystal lattice parameter is nearly linear from the Si parameter ($x = 0$) to the Ge one ($x = 1$):

$$a = (5.431 + 0.20x + 0.027x^2)\,\text{Å} \quad \text{at } T = 300\,\text{K}.$$

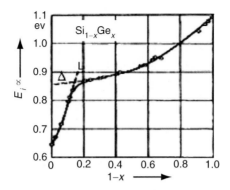

Fig. 6.1. Si_xGe_{1-x} indirect energy gap vs. composition at 296 K. At about $x = 0.15$ a crossover of the Ge-like [111] conduction band minima and the Si-like [100] conduction band occurs (after [6])

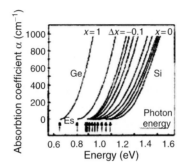

Fig. 6.2. $Si_{1-x}Ge_x$ (bulk alloys) absorption coefficient α vs. energies between 0.5 and 1.4 eV (with x varying by increments of 0.1 between Ge ($x = 1$) and Si ($x = 0$)). Vertical *arrows* show the respective indirect bandgaps (after [7])

The Ge/Si lattice parameter misfit at 300 K is thus about 4.2%. Figure 6.1 shows the evolution of the indirect bandgap E_G of bulk SiGe alloys. It is noticeable that two regions can be distinguished depending on the Ge mole fraction. $Si_{1-x}Ge_x$ presents a Si-like behavior for $x < 85\%$ and a Ge-like for $x > 85\%$. The sharp variation of E_G as a function of x suggests that optical absorption of bulk SiGe alloys at photon energies smaller than 1 eV (i.e., wavelengths larger than 1.24 µm) is low as soon as the Si percentage is more than a few percents. Figure 6.2, which is a plot of the SiGe optical absorption coefficient as a function of photon energy, shows that optical absorption at 1.3 µm (0.95 eV) is as low as a few cm^{-1} for Ge mole fractions up to 90–95%.

At the same time, the variation of the optical refractive index of bulk SiGe alloys is smoother (Fig. 6.3). The index contrast between SiGe and Si opens the possibility of light guidance using the SiGe/Si system as the core/cladding materials of optical waveguides.

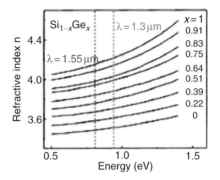

Fig. 6.3. $Si_{1-x}Ge_x$ (bulk) refractive index n vs. energy between 0.5 and 1.4 eV (after [7])

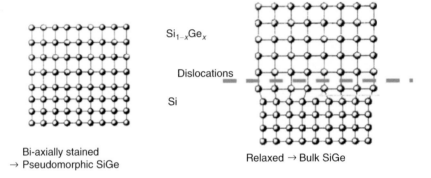

Fig. 6.4. Schematic description of the partition between pseudomorphic and relaxed SiGe layers

With the technical development of growth techniques, such as molecular beam epitaxy (MBE) or chemical vapor deposition (UHV-CVD, RT-CVD, RP-CVD), it has been soon recognized in the 1985s that it could be possible to take benefit of the Ge/Si lattice misfit by growing thin strained SiGe films. This option opened the way of energy bandgap engineering in silicon technology, which has first emerged for III–V materials in the 1970s. These thin SiGe films are usually deposited on an Si substrate. Because of the lattice mismatch of around 4.2% between pure Si and pure Ge, such films are tetragonally distorted, when grown to a thickness below the critical value for the onset of misfit dislocations. As illustrated in Fig. 6.4, these films begin to relax to their intrinsic cubic lattice constant, once the critical thickness is exceeded.

Hence, depending on the thickness of the $Si_{1-x}Ge_x$ film at a given composition, such films can be either biaxially strained, or strain-relaxed. The strain-relaxed films can be considered as a sort of virtual SiGe substrate. With the art of epitaxial growth rapidly advancing, it has been soon recognized that strain was an as important material parameter as composition in the $Si_{1-x}Ge_x$ heterostructure system. Many parameters, such as bandgaps, band offsets, effective masses, and so on, are indeed strongly strain-dependent, making strain control a vital necessity for any kind of energy bandgap engineering conceivable in these materials. The strain in the pseudomorphic active layer (either tensile or compressive), includes a hydrostatic component which shifts the average band energy level and a uniaxial component which splits the degenerate bands [8]. In case of $Si_{1-x}Ge_x/Si$ grown alloys on unstrained Si, which are of particular interest for the realization of active SiGe devices for optical interconnects, the heterostructure is of type I, with a very small conduction band offset ($\Delta E_C < 10$–$20\,meV$), and a valence band offset ΔE_V mainly due to the splitting between the light and heavy hole levels and is given by [8]

$$\Delta E_V = 0.74.x.$$

Figure 6.5 shows the evolution of the indirect bandgap energy of pseudomorphic SiGe alloys grown on $\langle 100 \rangle$ Si. It is noticeable that the bandgap energy of strained SiGe layers with 65% of Ge is lower than the energy bandgap of pure Ge. This result opens very interesting opportunities for the realization of integrated SiGe detectors on Si substrates.

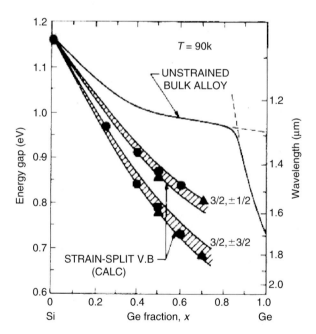

Fig. 6.5. Reduction of the indirect bandgap energy of pseudomorphic SiGe on $\langle 100 \rangle$ Si as a function of the Ge mole fraction at $T = 90\,K$

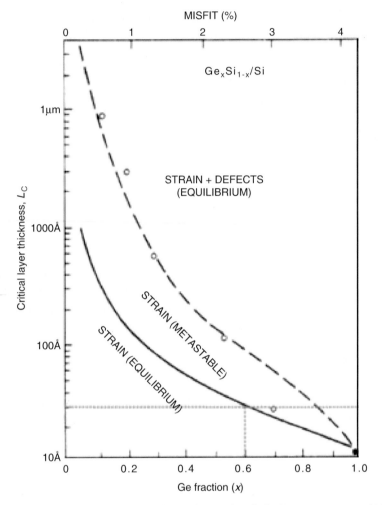

Fig. 6.6. Critical layer thickness of pseudomorphic SiGe layers grown on ⟨100⟩ Si as a function of the Ge mole fraction

Unfortunately, the interest of this bandgap energy decrease must be significantly moderated due to the order of magnitude of the SiGe/Si critical thickness, which is limited to less than a few tens of nanometers for Ge mole fractions above 50% (Fig. 6.6).

6.3 SiGe and Ge Photodetectors

As the final blocks dedicated to the optical/electrical conversion of signals, photodetectors are key elements for on-chip optical interconnects. In addition to the material compatibility with the mainstream silicon technology, the desired properties of detectors are a high responsivity ($R_\lambda \gg 0.1\,\mathrm{A\,W^{-1}}$), a high bandwidth ($f_C > 10\,\mathrm{GHz}$), and a low dark current to enhance the signal over noise ratio. Efficient optical coupling of light into the absorbing region is important to guaranty a high external yield. Surface- or waveguide illuminations have been proposed. For integration with optical micro-waveguides in a silicon-compatible planar technology, the second option is preferred.

6.3.1 SiGe Photodetectors

Due to the reduction of the energy bandgap if compared with silicon, strained SiGe materials are naturally suited to the realization of detectors in the 1.3–1.5 μm wavelength window. Nevertheless, the design of SiGe detectors is the result of a series of compromises.

Optimization of the SiGe Layer Stack

The design of SiGe detectors is first based on an optimization of the Ge mole fraction and the acceptable thickness of SiGe layers. The higher the Ge mole fraction, the higher the absorption coefficient α, but the lower the layer critical thickness h_C : α is less than $100\,\mathrm{cm^{-1}}$ for $x = 60\%$ while h_C is then about 5 nm [10]. In order to circumvent this severe layer thickness limitation, the use of multiple quantum well structures has been proposed, leading to strained-layer superlattices (SLS) composed of a series of SiGe/Si bi-layers. It was shown that the growth criterion is that the SLS thickness is lower than the critical thickness of the SiGe pseudomorphic layer with a Ge content equal to the SLS average Ge one. With the notations of Fig. 6.7, this condition can be written as

$$h_{\mathrm{SLS}} < h_C(x_{\mathrm{av}})$$

with

$$x_{\mathrm{av}} = \frac{h_{\mathrm{SiGe}}}{(h_{\mathrm{SiGe}} + h_{\mathrm{Si}})} \times x.$$

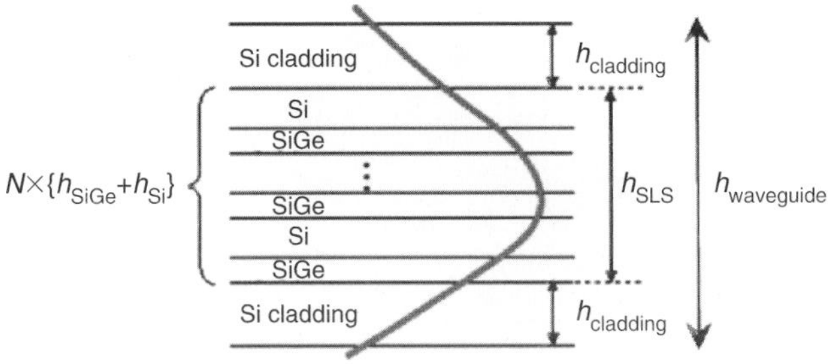

Fig. 6.7. Schematic view of an SiGe SLS grown on Si (slab waveguide optical field is depicted in *continuous line*)

Effective absorption is then estimated by taking into account the overlap with the optical field in optical waveguides.

Optimization is made by finding the best compromise between ultra-thin $Si_{1-x}Ge_x$ layers with high x values and thicker $Si_{1-x}Ge_x$ layers with lower x values. This optimization is also complicated by the increase of the effective energy bandgap due to wave-mechanical confinement effects [11]. Results showed that whatever the number of quantum wells and the associated possible Ge mole fractions, the overall limited content of SiGe limits the maximum effective absorption coefficient to less than $100\,cm^{-1}$ for Ge-rich SiGe layers (%Ge up to 60%) and optical waveguide heights up to $1\,\mu m$ [12].

Figure 6.8 is the schematic view of the 2 mm long SLS SiGe waveguide detector reported in [13]. Light guidance is ensured by a $2.5\,\mu m$ thick $Si_{0.98}Ge_{0.02}$ rib waveguide with $2.6\,dB\,cm^{-1}$ propagation loss. The SLS is composed of 20 $\{30\,nm\ Si + 5\,nm\ Si_{0.55}Ge_{0.45}\}$ periods, which corresponds to an average Ge mole fraction of about 6.43%. With such a low Ge mole fraction, the associated critical thickness is about $3\,\mu m$, which is higher than the layer stack thickness ($\sim 700\,nm$).

Optical loss due to free carrier absorption was estimated to be lower than $1\,dB\,cm^{-1}$. Quantum efficiency is about 12%, which corresponds to a $0.125\,A\,W^{-1}$ responsivity at $\lambda = 1.3\,\mu m$. The reported dark current and device capacitance are 200 nA and 1.3 pF, respectively, leading to a maximum operation frequency of 2.44 GHz for a $50\,\Omega$ resistive load. Carrier transit time experiments were also carried out. Time response under a reverse bias of 5V showed an impulse width at half maximum of about 400 ps, but also a slow component in the time response ($\sim 50\,ns$) probably due to trapping of holes in the valence band quantum wells.

Other works have reported surface-illuminated SiGe SLS devices with higher Ge average mole fractions (up to 10%), which exhibited low dark currents ($< 60\,mA\,cm^{-2}$), and quantum efficiencies of a few percent near

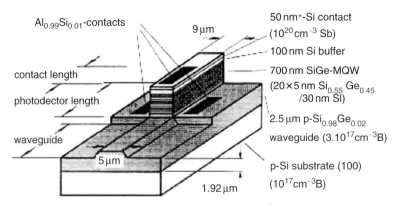

Fig. 6.8. Perspective view of the SiGe SLS detector reported in [13]

$\lambda = 1.3\,\mu m$. Two strategies have been proposed to compensate for the low absorption coefficient of SiGe SLSs, and thus to improve quantum efficiencies and lower the device lengths: avalanche process and resonant detectors.

Avalanche SiGe Photodetectors

The first option to enhance the SiGe detector responsivity is to use the photocarrier multiplication mechanism proper to avalanche photodiodes (APDs) [14]. In these devices, the reverse bias applied to the semiconductor region is high enough so that accelerated carriers initially created by interband optical absorption can be responsible for impact ionization. The device internal gain G depends on the electron (α) and hole (β) impact ionization coefficients, which are temperature and field dependent. The higher $k = \alpha/\beta$, the higher the bandwidth and the lower the device excess noise. This condition is well filled in Si and SiGe alloys. Advanced structures are required to increase the APD absorption (thick absorbing layer) while lowering the excess noise: structures with separated absorption and carrier multiplication region (SAM) have been proposed for this purpose both in III/V semiconductors and Si technologies [15, 16].

Several works have reported a significant enhancement of SiGe detectors responsivities using an APD approach. External quantum efficiencies up to 400% have been reported ($R_\lambda \approx 4\,\mathrm{A\,W^{-1}}$ at $\lambda = 1.3\,\mu m$) for millimeter long waveguide SiGe detectors [17]. High efficiency is then made possible by the long absorption length in waveguide configuration while optical absorption remains about $20\,\mathrm{cm^{-1}}$ at $\lambda = 1.3\,\mu m$. In the pioneering works mentioned here [17], quite high voltages were required to reach the avalanche regime (~ 20–$30\,\mathrm{V}$).

An advanced SiGe-based waveguide SAM APD structure has been recently proposed for $1.3\,\mu m$ operation [18]. Thanks to an accurate optimization of the transport and carrier multiplications layer thicknesses, and using travelling-wave electrodes to match the optical and radio-frequency signals all along the device length (150–300 μm), a $-3\,\mathrm{dB}$ bandwidth of 30 GHz and a gain of 10 were theoretically obtained for a 12 V reverse bias. Further improvements could lead to high-speed and highly sensitive SiGe APDs with lower bias voltages in the future.

Resonant Cavity Enhanced SiGe Photodetectors

The other way which has been proposed to enhance the external efficiency of SiGe detectors is to use of an optical Fabry–Perót cavity, so that light suffers backward and forward reflections inside the absorbing material. Distributed Bragg reflectors (DBR) can typically be obtained using $\lambda/4$ layers-based stacks. Resonant cavity enhanced (RCE) photonic devices have been extensively studied [19]. In a detector design, it appears that the device external quantum efficiency η_{EXT} presents a periodic evolution as a function of wavelength due to Fabry–Perót resonances.

For well-chosen mirror reflection coefficients ($R_2 = R_1 \exp(-\alpha d)$), where R_1, R_2, α, and d are the two mirror reflection coefficients, the material absorption coefficient, and the cavity length, respectively), η_{EXT} is multiplied by a cavity-enhancement factor, so that the maximum quantum efficiency can approach 100% even with poorly absorbing layers.

Figure 6.9 is a plot of the external quantum efficiency that can be reached with an RCE approach as a function of the normalized absorption coefficient $\alpha \times d$ for different values of R_2. It is noticeable that with a material with an absorbing coefficient as low as $10\,\mathrm{cm}^{-1}$, η_{EXT} reaches about 68% for $R_2 = 99\%$ ($R_\lambda \approx 0.71\,\mathrm{A\,W}^{-1}$ at $\lambda = 1.3\,\mu\mathrm{m}$). It thus appears that short devices (low capacitance) devices are feasible even with poorly absorbing SiGe materials.

Figure 6.10 shows the schematic view of the back-side-illuminated RCE SiGe/Si MQW photodiode reported in [20]. The Fabry–Perót cavity is composed of the interface between Si and the buried SiO_2 layer as bottom mirror and the SiO_2–Si DBR as top mirror. The obtained responsivity of the device was $31\,\mathrm{mA\,W}^{-1}$ at $1.305\,\mu\mathrm{m}$. Comparable responsivities have been reported in other works, such as in [21]. An SiGe/Si RCE SLS detector was fabricated using wafer bonding techniques for the realization of Bragg mirrors. It should be mentioned that fabrication of DBR is somewhat difficult in the $Si_{1-x}Ge_x$/Si system due to the lattice parameter misfit between Ge and Si (4.2%) which limits the layer stack thickness as well as due to the low index contrast between Si and SiGe alloys. Practical reflection coefficients are often about 50% when mirrors are only fabricated with SiGe/Si multi-layer stacks [19].

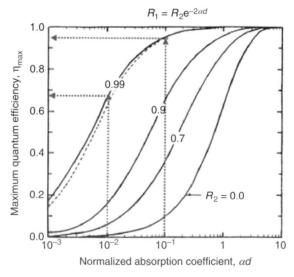

Fig. 6.9. Improvement of detector efficiency using an RCE approach (see [19])

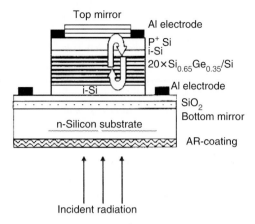

Fig. 6.10. Schematic of the $Si_{0.65}Ge_{0.35}/Si$ back-side-illuminated RCE detector (see [20])

Conclusion on SiGe Detectors

The main challenge in obtaining a high efficiency 1.3 μm SiGe detector is the lattice strain. SiGe detectors have layers with moderate absorption coefficients ($<100\,cm^{-1}$). Solutions such as RCE and APD are thus required, yet with the cost of additional complexity. Rather large voltages ($>10\,V$) are needed for the moment to obtain internal gain through the avalanche mechanism, although progress could be made on this point in a near future. RCE devices could also lead to high-sensitivity SiGe detectors. SiGe devices have long interaction lengths in most cases (>100–$200\,\mu m$). Travelling-wave electrodes are thus required (RF electrode engineering needed).

6.3.2 Ge Photodetectors

Pure germanium photodetectors have been proposed to circumvent some of the limitations of their SiGe counterparts. With the shift from strained SiGe materials to pure Ge, a strong increase of the optical absorption coefficient α is obtained due to the low direct bandgap energy of Ge ($0.8\,eV \leftrightarrow 1.55\,\mu m$): α is about $9,000\,cm^{-1}$ at $\lambda = 1.3\,\mu m$, which corresponds to a 95% absorption length of about $3.3\,\mu m$. On the other hand, the main challenge is the 4.2% lattice misfit between Ge and Si, which imposes specific strategies for the growth of thick Ge layers on Si substrates with heights comparable with optical waveguide thickness. Germanium layers also suffer from other drawbacks. Due to the low indirect bandgap energy of Ge ($\sim 0.66\,eV$), the realization of good Schottky contacts on Ge is difficult, which yields fairly large dark currents in the case of metal–semiconductor–metal (MSM) detector configurations [22]. Instability of the Ge surface also requires the use of advanced passivation techniques [23].

Germanium Growth on Si

Three main strategies have been proposed for the growth of high-quality Ge layers on Si substrates. The first one follows a very intuitive approach. The guiding idea is to take it the lattice parameter misfit by growing an SiGe graded layer on an Si substrate with an increasing Ge mole fraction up to 100%. Growth of a very high-quality active Ge layer is then possible. It has been yet a general practice that for each 10% Ge mole fraction, about 1 μm of linearly graded buffer is required to grow a high-quality SiGe alloy with a low dislocation density [24]. As a result, the Ge growth on an Si substrate requires a graded SiGe layer with a thickness of about 10 μm.

Figure 6.11 shows a TEM view of a thick Ge layer grown on a ⟨100⟩ Si substrate by this approach [24]. A surface-illuminated Ge PIN photodiode with interdigitated electrodes was reported with a 3 dB bandwidth of 3.8 GHz, a dark current of 5 μA, and a responsivity of 0.52 A W^{-1} at the wavelength of 1.3 μm. The main drawback of this approach relies in the rather large thickness of Ge layers, which limits the possibility of device integration in optical micro-waveguides.

In the effort towards thinner active layers (<1 μm) another approach based on the use of thin SiGe buffers has been recently reported [25]. This growth strategy is based on the optimization of a couple of two thin SiGe buffer layers with different Ge mole fractions (total SiGe height of 1 μm), that allows trapping most of threading dislocations at the heterojunction interface. It follows a significant reduction of the dislocation density in the top Ge absorbing layer. Responsivities up to 0.5 A W^{-1} at 1.3 μm have been recently reported using this approach with 3 dB bandwidth of 8 GHz under a reverse bias of 10 V [25]. This growth technique represents a significant improvement if compared with the growth of Ge on thick SiGe buffers although the required thickness to trap the dislocations could be still relatively large (∼1 μm) in the perspective view of the detector integration with submicron optical waveguides.

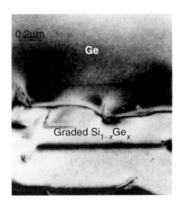

Fig. 6.11. Cross-sectional TEM image of a 1 μm-thick Ge layer grown on a 10 μm-thick SiGe buffer layer on a ⟨100⟩ Si substrate (after [24])

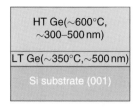

Fig. 6.12. Schematic view of the layer stack for the direct growth of Ge layers using the two-step CVD process described in [10]: the thickness of the Ge layer is not limited to 300–500 nm but such values are well suited to the integration of Ge detectors with SOI submicron optical waveguides

A third method was proposed for the direct heteroepitaxial growth of Ge on Si despite the challenging lattice parameter misfit between these two materials [10]. This method involves the growth of a thin (a few 10 nm) low-temperature Ge buffer layer between the underlying Si substrate and the Ge film grown at a higher temperature (see Fig. 6.12).

The growth of the thin Ge buffer allows avoiding 3D growth, and permits concentrating most of misfit dislocations in the Ge interfacial layer. This particular method is usually combined with a cyclic annealing process (~780°/900°) to reduce the threading dislocation density within the Ge active film. Due to high degree of the latitude in the choice of the active Ge layer (>200 nm), this growth method is well suited to the realization of both surface-illuminated or waveguide-illuminated devices.

Recent Advances on Ge Photodetectors

Due to the key role of detectors in all-optical communications and to the possibility of direct growth of Ge on Si, enabling a monolithic approach of light detection within the Si technology, germanium detectors have received a considerable and growing interest for a few years [10, 26–31]. Several configurations have been proposed, including MSM devices and PIN photodiodes. Results have been obtained at wavelengths up to 1,550 nm. Several groups have indeed pointed out measured absorption coefficients of Ge layers on Si at higher wavelengths than those expected for bulk Ge [26–28]. The absorption increase was attributed to bandgap narrowing induced by tensile strain within the Ge layer, resulting from the difference in the thermal expansion coefficients of Ge and Si. Although the Ge layer is relaxed during growth, strain is introduced as the structure is cooled down. For the moment, attention has essentially been paid on surface-illuminated devices.

The schematic cross-section of the RCE Ge detector reported in [29] is shown in Fig. 6.13. This lateral PIN photodiode with interdigitated Ti/Al electrodes was fabricated using the two-step temperature growth process described in the previous paragraph (low temperature Ge layer: 350°C, 50 nm and high-temperature Ge layer: 600°C, 350 nm), with a thermal cycling

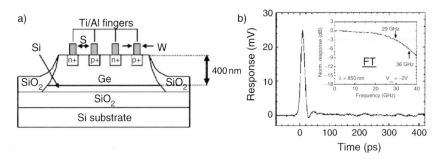

Fig. 6.13. (a) Schematic cross-section of the lateral PIN photodiode reported in [29]. (b) Impulse response of a $10\,\mu m \times 10\,\mu m$ device with finger spacing of $0.4\,\mu m$ at a bias of $-2\,V$, with the Fourier transform of the impulse response in the *inset*

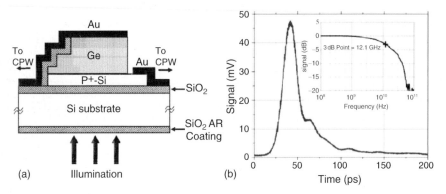

Fig. 6.14. (a) Cross-sectional view of the back-side-illuminated RCE Ge–SOI Schottky photodiode, (b) Temporal response of a $10\,\mu m$ diameter detector at $-3\,V$ and $\lambda = 1,550\,nm$. *Inset*: Fourier transform of the temporal response (after [30])

annealing (TCA) (780–900°) to lower the dislocation density. With a special design to minimize the device capacitance and a finger space ranging from 0.4 to 0.6 μm, the device exhibits a 3 dB bandwidth of 29 GHz under a reverse bias of 2 V, a dark current of 10 mA cm^{-2} and a quantum efficiency of 34% at $\lambda = 0.85\,\mu m$ (responsivity $= 0.233\,A\,W^{-1}$).

This work showed that the direct heteroepitaxial growth of thin Ge layers ($<500\,nm$) is possible with devices exhibiting low dark current and high frequency operation. These are important points for the future realization of waveguide detectors. The lack of response near 1.3 μm was due to Si diffusion (~5–10%) into the Ge layer induced by TCA and, which causes a rapid decrease of optical absorption with increasing wavelengths, but progress in this field are foreseeable in a near future.

Other works have related an NIR response near 1.5 μm, although with thicker Ge layers. Figure 6.14 shows the cross-sectional view of the back-side-illuminated RCE Ge–SOI Schottky photodiode reported in [30]. This RCE

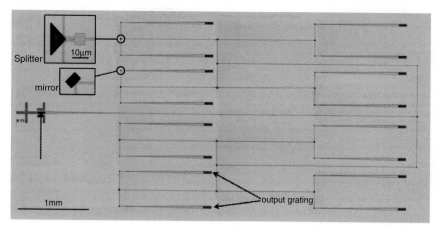

Fig. 6.15. Photograph of the 1 cm-long 1–16 light distribution based on SOI rib waveguides, corner mirrors, and beam splitters reported in [3]

device was realized after the growth of 1.45 μm thick Ge layers, and exhibited a 380 nA dark current at -5 V, a responsivity of 0.73 A W^{-1} at $\lambda = 1.55$ μm, and a high -3 dB bandwidth of 12.8 GHz at -4 V.

Waveguide Ge Photodetectors for On-Chip Optical Interconnects

Works reported in [29, 30] can be considered as important milestones towards the integration of efficient Ge detectors for optical interconnects: devices with high responsivities and high bandwidths have been demonstrated using direct growth of Ge on Si substrates. The integration of Ge detectors with optical micro-waveguides and the demonstration of high sensitivity devices with thin (<500 nm) Ge layer configurations and 1.3 μm operation are now among the most important points.

The possible distribution of light at the scale of a silicon chip was demonstrated using low-loss (0.4 dB cm^{-1}) submicron slightly etched rib SOI micro-waveguides, compact beam splitters and corner mirrors (Fig. 6.15).

Attention is focused here on the integration of Ge detectors in such optical waveguides. This geometry, which is shown in Fig. 6.16, was proved to be favorable to the placement of the electrodes required for biasing the active device.

The first challenge for the integration of Ge waveguide detectors is the possibility of selective growth of thin (<500 nm) Ge layers at the end of SOI waveguides. Growth of Ge layers was realized following a two-step temperature process in a UHV-CVD growth system [28, 31]. Figure 6.17 shows the cross-sectional view of Ge on Si selective epitaxy realized on masked substrates [31]. It is noticeable that Ge growth in Si recesses is possible without any lattice discontinuity, which is an important point for the detector integration.

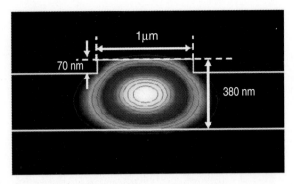

Fig. 6.16. Cross-sectional view of the SOI rib waveguide geometry reported in [3], with the single-mode optical field profile calculated at $\lambda = 1.3\,\mu m$

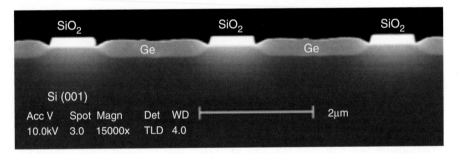

Fig. 6.17. SEM view of a Ge selective growth on (001) Si substrates (after [31])

Several TCA were tested to highlight the role of TCA on thin Ge layers (<500 nm) [28]. It was proved that annealing at a constant temperature (720°C) is enough to reduce the threading dislocations in Ge layers of 500 nm thickness. Such thermal annealing does not change the initial crystalline order and optical properties of Ge layers, contrary to TCA which tend to induce a noticeable diffusion of Si within the Ge layer, thus strongly reducing the absorption of active layers.

Two configurations of detector integration were investigated to couple the guided light from a rib SOI waveguide to the integrated Ge photodetector: either butt coupling where the Ge layer is grown in a recess etched in the waveguide, or vertical coupling with a Ge layer grown on top of the waveguide [32]. Three-dimensional finite-difference time-domain (3D-FDTD) and field mode matching simulations were performed to evaluate the efficiency of both photodetector integration configurations. Examples of the obtained field maps at $\lambda = 1.3\,\mu m$ are plotted in Fig. 6.18.

The main conclusions are given hereafter. About 100% of the incident light is transmitted from the Si waveguide to the Ge layer in case of butt coupling, i.e., losses by reflection and diffraction at the waveguide discontinuity are negligible. The main advantage of this option is the very short absorption

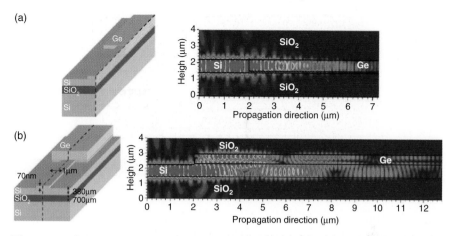

Fig. 6.18. Schematic views and fields obtained using 3D-FDTD simulations in a longitudinal cross-section of an integrated Ge photodetector: (**a**) butt coupling; (**b**) vertical coupling

length that is obtained: more than 95% of the light intensity is absorbed after less than a 4 μm propagation length. However, this configuration requires an accurate control of the etched depth of the silicon film in order to keep a thin Si layer (around 30 nm) to allow the Ge growth above the buried silicon oxide. In case of vertical coupling, the coupling between the Ge and Si waveguides induces a periodic transfer of the light between both materials, whose strength depends on the Ge layer thickness. The general trend is that light absorption increases with the Ge thickness, but this evolution is not monotonous due to a Fabry–Perót effect in the silicon–germanium bi-layers. For a well-chosen Ge layer thickness around 310 nm, more than 95% of light absorption then occurs within a 7 μm device length.

Such ultra-short devices are greatly beneficial to high-bandwidth operation. Several electrical configurations have been studied including vertical and lateral PIN Ge photodiodes and MSM detectors with both butt and vertical coupling of light into the Ge active layers. Small-signal and transient simulations were performed with a physical device simulator [33] to extract the device capacitance, dark current, and impulse response of a large set of waveguide Ge detectors. Figure 6.19, which shows the carrier collection process in an MSM Ge detector with a 1 μm spacing between two top-surface electrodes, demonstrates the important role of the displacement current in the temporal response of MSM Ge detectors. Depending on the device configuration, capacitances ranging from 0.2 to 5 fF were obtained, with FWHM temporal responses ranging from 2.6 to 15 ps.

These results show the high potentiality of ultra-short Ge waveguide detectors for high-bandwidth operation in the regime of tens of GHz at 1.3–1.5 μm.

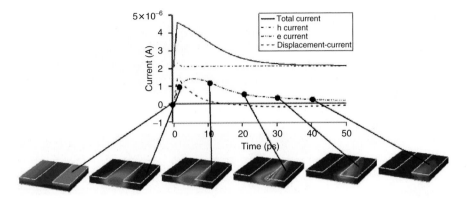

Fig. 6.19. Photo-current impulse response under IV on the positive contact of a Ge MSM detector with 1 μm spacing between the metallic electrodes

6.4 SiGe Switching Devices and Optical Modulators

Significant efforts have been reported for a few years towards silicon-based light emitters. Due to the indirect bandgap structure of silicon and to the high-speed operation needed for optical interconnects, silicon-compatible sources are likely to be used with a continuous emission. Such waves carry no data or information. An optical modulator is needed to encode data onto this continuous light wave. Considering the integration of optics within silicon chips, one of the targeted applications is the generation of global signals for on-chip optical interconnects, including the global clock. Optical modulators with speeds higher than 1 GHz are fabricated from either LiNbO$_3$ electro-optic crystal or III/V semiconductors including multiple quantum wells such as GaAlAs/GaAs and InGaAsP/InP using the quantum confined stark effect (QCSE). These devices have demonstrated modulation frequencies beyond 40 GHz but are not compatible with the Si technology.

With the rapid introduction of optics within shorter and shorter distances, silicon-based optical interconnects could also rapidly include technologies first introduced for long distance lightwave communications. Among them, photonic components for switching and routing, including wavelength-selective components to exploit the wavelength division multiplexing (WDM) technique, could be key elements. Such components are also traditionally fabricated using III/V compounds, which are not compatible with the mature silicon technology.

The realization of switching devices and modulators in silicon technology is a challenging task if compared with III/V semiconductor technology or LiNbO$_3$ due to the limited physical effects available to obtain refractive index and/or optical absorption variations. In this context, the introduction of silicon–germanium alloys is welcome due to the additional degree of freedom it allows.

6.4.1 Electro-Optical Effects in SiGe Materials

Although the refractive index and the optical absorption coefficient are linked one to each other by the Kramers–Krönig relationships, one of these two effects is generally favored. Optical modulators can be thus classified in two main categories: either electro-absorption or electro-refractive modulators. Photonic switches exploit the same physical effects.

Electro-optical effects in Si have been described elsewhere [34]. Here, we will focus on SiGe compounds. SiGe materials do not differ significantly from Si. Thermo-optical index variations ($\Delta n \sim 2 \times 10^{-4}\,\mathrm{K}^{-1}$ at $T = 300\,\mathrm{K}$) can be used yet with very slow time responses ($\sim 1\,\mathrm{ms}$). Electro-optical effects have a moderate amplitude: due to the crystal symmetry, there is no linear-field (Pockels) effect in Si, while the quadratic (Kerr) effect is very low ($\Delta n \sim 8 \times 10^{-7}\,\mathrm{K}^{-1}$ for a $10^{5}\,\mathrm{V\,cm}^{-1}$ electric field). As in Si, the more efficient effect to obtain a rapid refractive index variation is obtained by the variations of free carrier concentrations as described by Soref et al. [35].

Another approach is to exploit the electro-absorption properties of SiGe materials and heterostructures. Electro-absorption in bulk SiGe (Franz–Keldysh effect) is of the same order of magnitude as in Si: the obtained refractive index variation at $1.3\,\mathrm{\mu m}$ is about 10^{-6} for an applied field of about $10^{5}\,\mathrm{V\,cm}^{-1}$. A more interesting option is to use the possibility of energy band engineering offered by the silicon–germanium technology to exploit the quantum-confined Stark effect (QCSE) as used in III/V-based modulators.

6.4.2 SiGe Switching Devices

SiGe switching devices proposed up to now are based on free-carrier-induced refractive index variations using injection structures. SiGe alloys are mainly used for light guiding, which is allowed by the refractive index increase with the Ge mole fraction x. The energy bandgap of pseudomorphic SiGe alloys grown on Si is approximately $1.12\,\mathrm{eV}$–$0.74x$ [8], so that $x < 23\%$ ensures that there is no absorption at $\lambda = 1.3\,\mathrm{\mu m}$. In the limit of low Ge mole fractions ($x < 20\%$), the refractive index contrast Δn of SiGe layers with respect to Si is about $0.3x$ [36], so that $\Delta n \approx 0.012$ for $x = 4\%$. Photonic switching devices are composed of a solid-state structure for carrier injection (typically a forward-biased PN junction) and an optical integrated structure with input and output waveguides. The desirable properties of a photonic switch are low insertion loss (IL), low crosstalk, low switching voltage, high switching speed, low power consumption, high integration, and a good reliability.

Several similar structures have been proposed either based on a Y-branch, a two-mode interference directional coupler, a Mach–Zehnder interferometer, or waveguide optical cross-connects. As a typical example, an integrated 3×2 photonic switch in SiGe alloy for near $1.5\,\mathrm{\mu m}$ operation is reported in [37]. This 3 mm-long structure, which is based on a multi-wavelength cross-connect structure, was fabricated using UHV-CVD process to realize $\mathrm{Si}_{0.96}\mathrm{Ge}_{0.04}/\mathrm{Si}$

few-micrometer-scale single-mode rib waveguides at $\lambda = 1.55\,\mu m$. Depending on the chosen input wavelength (1,540, 1,550, or 1,560 nm) and on the ON/OFF bias states of two PN junctions, this structure can act as an optical power splitter, a one-wavelength switch, an optical multiplexer, an optical combiner, and optical add-drop multiplexer, and a three-wavelength switch. The device insertion loss was 2 dB, with a crosstalk of $-22\,dB$, a bias current of 120 mA for each of the two PN junctions, and a switching speed of 200 ns.

Other SiGe-based photonic switches recently published are characterized by comparable properties [38, 39]. These devices exploit rather large single-mode SiGe/Si optical waveguides ($\gg 1\,\mu m$). For the moment, the proposed components have DC currents around 100–200 mA, moderate switching speeds ($\sim 100\,ns$), and device lengths of a few millimeters, although further improvements could appear in the future. Such structures, allowing the control of light within the silicon technology, are important devices for optical interconnects.

6.4.3 SiGe Optical Modulators

Two types of SiGe optical modulators are described in this section. Electro-absorption modulators are described first. Electro-refractive SiGe modulators are then described, with an emphasis on depletion-based structures coupled with an integrated optical interferometer.

SiGe Electro-Absorption Modulators

The basic principle of electro-absorption modulators is to obtain a variation of light intensity at the device output through the variation of the optical absorption coefficient of the modulator material. ON/OFF states correspond to the situation of low absorption coefficient (α_{ON}) and high absorption (α_{OFF}), respectively. A schematic view of a waveguide multiple quantum well SiGe/Si electro-absorption modulator is given in Fig. 6.20. Due to the waveguide geometry, a large interaction length with the absorbing layers can

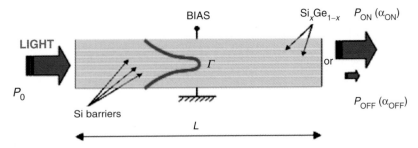

Fig. 6.20. Schematic view of a waveguide multiple quantum well SiGe/Si electro-absorption modulator (the optical field of the slab waveguide is plotted in red: the overlap factor with SiGe layers is Γ)

be achieved. The interaction with the active material is optimized through the optical factor Γ $(0 < \Gamma < 1)$ between the optical mode and the absorbing layers.

The desirable properties of an integrated modulator are a high contrast ratio (CR), low IL, high-speed operation, low voltage and low power consumption, and a high potential for monolithic integration. With the notations of Fig. 6.20, CR and IL are given by the following relationships:

$$\mathrm{CR} = P_{\mathrm{OFF}}/P_{\mathrm{ON}} = \exp[-\Gamma \times (\alpha_{\mathrm{OFF}} - \alpha_{\mathrm{ON}}) \times L],$$

$$\mathrm{IL}_{\mathrm{dB}} = 10 \times \log(P_0/P_{\mathrm{ON}}) = 4.34\Gamma \times \alpha_{\mathrm{ON}} \times L.$$

Large values for CR are obtained with large differences of optical absorption between the ON and OFF states.

QSCE is an electro-absorption effect based on the wave-mechanical confinement of electrons and holes in quantum well (QW) structures [14]. In absence of applied electric field, the exciton oscillator strength is reinforced due to the electron and hole confinement, so that an exciton peak absorption is present. When a transverse field is applied, the confined conduction and valence band energy levels decrease so that the photon energy undergoes a red-shift and the exciton peak is lowered due to electron and hole separation [14]. Contrary to what holds for III/V semiconductors, it was proved that QSCE is very weak in case of SiGe QW structures due to the indirect bandgap structure of SiGe materials and to the very low confinement of electrons in conduction band QWs [40]. A modified approach was proposed in [41]. Considering an SiGe QW with a well chosen Ge mole fraction, photons with energy equal to the electron/heavy hole transition are absorbed in absence of field due to the strong overlap between the electron and hole wavefunctions. A small transverse field then reduces the oscillator strength corresponding to this transition due to the delocalization of the carrier wavefunctions, thus reducing the optical absorption of layers. It is noticeable that, unlike in the case of QSCE electro-absorption modulators, the photon energy in this scheme should be higher than the bandgap of the SiGe QW ($x \gg 23\%$ for $1.3\,\mu\mathrm{m}$ operation) for optimal operation.

The strength of this electro-absorption effect was theoretically estimated in [41]. The interband absorption was calculated in the unique SiGe/Si QW depicted in Fig. 6.21 as a function of the QW width and the applied transverse electric field. In each case, all energy levels and carrier wavefunctions were calculated, and the interband absorption coefficient was then evaluated using the Fermi Golden Rule and taking into account all transitions between the conduction and valence levels.

Figure 6.22 shows the result obtained at $\lambda = 1.2\,\mu\mathrm{m}$ under flat-band conditions as a function of a square $\mathrm{Si}_{0.6}\mathrm{Ge}_{0.4}/\mathrm{Si}$ QW width. It shows then an optimum is found for a QW width of about $3.5\,\mathrm{nm}$, with an optical absorption coefficient of about $445\,\mathrm{cm}^{-1}$.

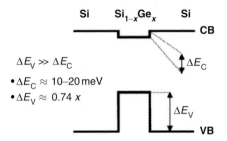

Fig. 6.21. Conduction and valence band energy profiles under flat-band conditions of an $Si_{0.6}Ge_{0.4}$ square quantum well – the thermal energy ($\sim$25 meV at $T = 300$ K) is higher than the conduction band quantum well offset

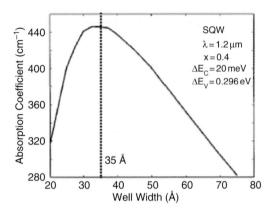

Fig. 6.22. Calculated optical absorption coefficient at λ=1.2 μm as a function of the width of the square quantum well (SQW) depicted in Fig. 6.20 (see [41])

The calculated effect of the transverse field is shown in Fig. 6.23 for this optimized QW. Figure 6.23 shows that with moderate fields involved around $2\,kV\,cm^{-1}$, the reachable optical absorption ON/OFF contrast is about $80\,cm^{-1}$.

An experimental demonstration of an $Si_{0.6}Ge_{0.4}$–Si electro-absorption modulator was reported in [42]. The device length was $100\,\mu m$, with IL $= 28\,dB$ and CR $= 2.9$ at $\lambda = 1.15\,\mu m$. This wavelength was chosen due to the high Ge mole fractions needed for $1.3\,\mu m$ operation. The requirement of an accurate control of the doping profiles to control the CB and VB shapes was mentioned in this work. These VB and CB profiles have indeed a direct impact on the carrier wavefunctions, and thus on the device operation.

SiGe Electro-refractive Modulator

Due to the relative weakness of electro-absorption effects in SiGe heterostructures, the use of free carrier concentration variations was considered for the realization of fast modulators. Contrary to the electro-absorption-based mod-

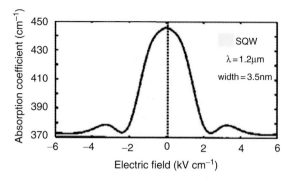

Fig. 6.23. Absorption coefficient as a function of the transverse field in case of a single square $Si_{0.6}Ge_{0.4}/Si$ quantum well of 3.5 nm thickness (see [41])

ulators, an integrated interferometer is then needed to convert the refractive-index induced phase modulation into an intensity one. Injection-based devices have generally rather large DC currents ($\gg$10 mA) and moderate response times ($\gg$1 ns) due to electron/hole recombination processes. For this reason, a modulation-doped (MD) multi-layer SiGe/Si device integrated in a reversed-biased PIN diode was proposed [43,44]. In this case, only one kind of carrier is involved, so that no recombination process takes place and a high-frequency operation can be expected ($>$1 GHz). Moreover, the current density is reduced in the active region, if compared with carrier injection in forward-biased PIN diode, and this limits power dissipation.

Figure 6.24 is a schematic view of the MD SiGe/Si multiple well structure from [44]. The active region of the SiGe/Si modulator is constituted of a stack of $Si_{1-x}Ge_x$ thin layers surrounded by Si non-intentionally doped (nid) layers ($\sim 10^{16}$ cm^{-3}). As shown in Fig. 6.24, silicon highly doped P$^+$ δ-layers are introduced in the core of the nid Si layers as a source of free holes.

To avoid optical absorption of the electro-magnetic field propagating through the device, the material bandgap must be kept above the incident photon energy. $Si_{1-x}Ge_x$ bandgap can then be estimated from the relation: $E_{SiGe} = 1.12–0.74x$. Optical absorption at $\lambda = 1.3\,\mu m$ of the $Si_{1-x}Ge_x$ alloy is negligible for $x < 0.23$. The Ge content was then chosen equal to 0.2. SiGe/Si layers were epitaxially grown on an SOI substrate. Due to the lattice parameter difference, $Si_{1-x}Ge_x$ is compressively strained on Si(001) thus leading to the energy band profile depicted in Fig. 6.21. SiGe layers thus acts as wells for holes. Due to the rather low Ge content (20%), the growth of such multi-layer SiGe/Si stacks easily satisfies the critical layer thickness condition introduced in "Optimization of the SiGe Layer Stack" for the realization of SLS SiGe photodetectors.

When a reverse bias voltage is applied to the diode, holes are swept out from the active region thanks to the induced bending of the valence band. Calculation of hole distributions were performed using a physical device simulator

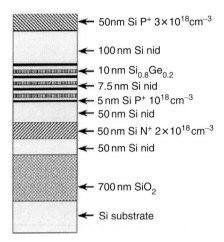

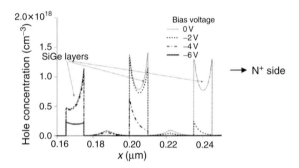

Fig. 6.24. Schematic view of the active region of the MD SiGe/Si multi-layer structure grown on an SOI substrate

Fig. 6.25. Evolution of the hole density distribution in a three MD $Si_{0.8}Ge_{0.2}$/Si structure as a function of the applied reverse bias

that performs the numerical resolution of Poisson, carrier density continuity, and drift-diffusion equations [45]. The hole emptying process is illustrated in Fig. 6.25, which shows the evolution of the hole density distribution in a structure comprising three $Si_{0.8}Ge_{0.2}$ (10 nm)/Si (25 nm) VB wells for various values of the applied reverse bias. At zero bias, holes are confined in the three SiGe wells, whereas for a -2 V reverse bias the first well is depleted. At -4 V, the second well is partially depleted, whereas at -6 V only a few holes remain in the last well. The variation of the free carrier density induces a refractive index change inside the active region, and therefore a phase shift of the optical guided mode. The refractive index variation and the absorption losses at $\lambda = 1.3\,\mu m$ due to free carriers were calculated from hole density distribution using the works of Soref [35].

Experiments were performed around 1.55 μm on SiGe/Si multi-layer stack to determine the index variation. They consisted of measuring the shift of Fabry–Perót resonance fringes as a function of the applied voltage in a 2.3 mm long SiGe modulator waveguide cavity [46]. Special care was taken to clearly identify the contribution of the electro-refractive refractive index variations from the thermo-optic induced variations. This was performed by recording both the electrical dissipated power and the optical absorption of the structure as a function of the applied bias. Figure 6.26 is a plot of the obtained result for the structure depicted in Fig. 6.24.

The main conclusion was that a noticeable free-carrier induced index variation can be obtained in such SiGe/Si reverse-biased structures. The effective index variation of the optical mode of the slab is larger than 10^{-4} for reverse bias higher than 3 V, in close agreement with the performed simulations [44, 45, 47]. The factor of merit of the device, defined as the product of the length to ensure a π phase shift by the bias voltage is about 2.3 V cm. This is about three times larger than the value of the MOS capacitor silicon-based optical modulator proposed in [48].

The integration of such a modulation multi-layer structure in a single-mode SOI waveguide, which is a key point for on-chip optical interconnects, was considered in [47].

As shown in Fig. 6.27, the device dimensions were chosen in accordance with 380 nm high and 1 μm wide slightly etched rib waveguides that were used for the demonstration of a 1–16 H-tree 1 cm-long light distribution at 1.3 μm [3]. The main objective for the structure optimization was to maximize the effective index variation of the guided mode while maintaining optical propagation losses at an acceptable level. The optimization consisted in finding the number and the location of SiGe wells needed to maximize the hole density variation and their influence on the guided mode propagation. For example, increasing the total charge quantity inside the structure improved the effective index variation. However, the voltage needed to deplete

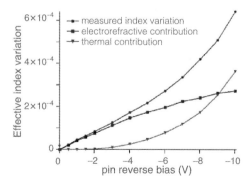

Fig. 6.26. Measured effective index variation vs. voltage and determination of the electro-refractive and thermo-optical index variations

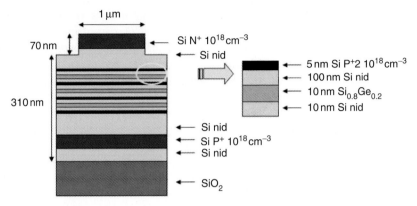

Fig. 6.27. Schematic view of the SiGe/Si MD structure integrated in an SOI slightly etched rib waveguide for 1.3 μm operation

all the hole population then also increased. A multi-factor simulation analysis was realized to understand the effects and interactions of the different structure parameters for the integration of the multi-layer stack into a slab optical waveguide. The local refractive index variations determined from calculated hole density profiles using a drift-diffusion simulator [33] were introduced in a full-vectorial optical mode solver [49] in order to evaluate the effective index variation of the optical mode propagating through the multi-layer stack. The conclusion was that the optimum SiGe multi-layer stack comprises three SiGe 10–15 nm VB wells, with 5 nm thick Si P^+ (2×10^{18} cm^{-3}) δ-doped layers. The predicted electro-refractive effect then results in an effective index variation $\Delta n = 1.7 \times 10^{-4}$ under a -6 V bias voltage. The associated absorption variation is 3 dB cm^{-1}. Special care was taken to minimize optical losses induced both by free carrier absorption in the P^+ and N^+ regions of the PIN diode and by the metallic electrodes needed for biasing the device. An overall optical loss level of 20 dB cm^{-1} was obtained.

The intrinsic frequency response of the MD SiGe/Si structure was estimated in [45] using transient simulations to quantitatively evaluate the global response time of the entire device. It mainly depends on the time needed for carriers to escape from and to be captured into the QWs. The delays needed for holes to be collected by electrical contacts when the bias voltage is switched on, and to come back in the SiGe wells of the active region when the bias voltage is switched off, were estimated using transient drift-diffusion simulation including a description of thermionic emission and tunnelling of carrier captures/escapes into/from VB wells.

As a reverse bias step from 0 to -6 V (rise time <1 fs) was applied to the PIN diode, simulation was performed until a new equilibrium was obtained. The hole density distributions obtained at several times ranging from 0 to 50 ps are reported in Fig. 6.28. Holes are initially confined into the three SiGe QWs (as in Fig. 6.25 for a 0 bias). When time increases, holes begin to drift,

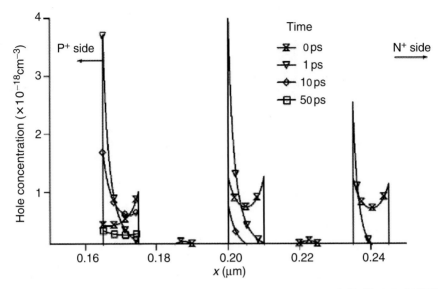

Fig. 6.28. Time evolution of the hole depletion process in the SiGe/Si MD-MQW structure following a reverse-bias step from 0 to $-6\,\mathrm{V}$ (rise time $<1\,\mathrm{fs}$) applied to the PIN diode

due to the electric field, and then tend to escape from the first quantum well to be captured by the following ones. After $1\,\mathrm{ps}$, holes are confined near the left side of each SiGe well. It can be seen that the time needed for holes to escape from the well nearest to the $\mathrm{N^+}$ region is only a few ps, according to the strong electrical field in this space charge region. After $10\,\mathrm{ps}$, the first well is emptied out but the hole concentration is still rather large into the two other QWs, from which they will also have to escape. As the electrical field in the third QW (far from the $\mathrm{N^+}$ region) is lower than in the previous ones, the escape time is longer than for the first two wells. At $t = 50\,\mathrm{ps}$, holes are only present in the third QW, and the hole distribution is very close to the equilibrium one under $V = -6\,\mathrm{V}$.

The reverse transition process was also studied, when a bias step from $-6\,\mathrm{V}$ to 0 is applied to the PIN diode (rise time $<1\,\mathrm{fs}$). The inhomogeneous space charge effect, which is related to the electrical potential via Poisson equation, is now to attract holes back into the QWs. Results are given in Fig. 6.29. Holes have first to fill the QW nearest to the $\mathrm{P^+}$ region. They get over the first Si barrier by thermionic emission and tunnelling. This does not need a long time, because of a high value of the attractive electrical field. As shown in Fig. 6.29, after $2\,\mathrm{ps}$ a large number of holes are confined into the first two wells, and after $10\,\mathrm{ps}$ holes are present in the three QWs. The equilibrium state is reached for $t = 50\,\mathrm{ps}$.

The time evolution of the optical effective index n_{eff} was calculated from the time-varying hole concentration inducing local variations of the layer

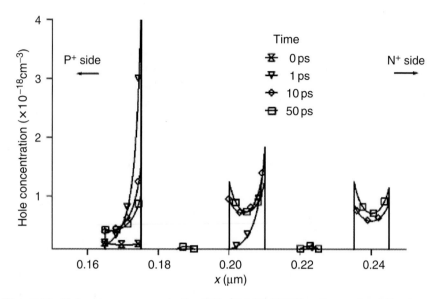

Fig. 6.29. Hole return process in the SiGe/Si MD-MQW active region following a reverse-bias step from $-6\,$V to 0 applied to the PIN diode (rise time $<1\,$fs)

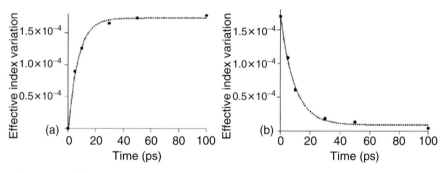

Fig. 6.30. Effective index evolution of the optical mode propagating in the modulator in response to: (**a**) a $0 \rightarrow -6\,$V bias voltage input, (**b**) a $-6\,$V$\rightarrow 0$ bias voltage input

refractive indices. Figure 6.30 shows the evolution of n_{eff} for a 0 to $-6\,$V and from a $-6\,$V to 0 voltage transitions, respectively.

The rise time and fall time that were extracted are 7.2 and 9.5 ps, respectively, leading to about a 16 GHz intrinsic bandwidth. An estimate of RC time constant of the device was also performed. It was found that the device RC is roughly independent of its length L. This result is due to the linear dependence of C as a function of L, while R scales as $1/L$. The modulator capacitance per unit length was estimated using an electrical device simulator in AC mode [33]. The conclusion was that RC ranges from 20 to 100 ps depending on the electrode sheet resistances and on the modulator geometry.

These results showed the potentiality of SiGe depletion-based optical modulators for high-speed operation ($>10\,$GHz).

To get intensity modulation from phase modulation, the use of an integrated interferometer device is necessary. Two types of interferometers can be considered: either a Fabry–Perót resonant cavity or a Mach-Zehnder interferometer [47]. These two interferometer structures were compared in terms of two parameters: the modulation depth D and the insertion loss IL defined as $1 - T_{OFF}/T_{ON}$, and $10 \times \log(1/T_{ON})$, respectively, T_{ON} and T_{OFF} being the optical power device transmittance in the "ON" and "OFF" states. Phase modulation properties described above were used ($\Delta n = 1.7 \times 10^{-4}$, total optical loss $= 20\,$dB cm^{-1}).

An efficient way to realize an integrated resonant Fabry–Perót cavity is to insert the active region between two Braggg mirrors, as shown in Fig. 6.31. Periodic microstructures deeply etched into an SOI waveguide offer high refractive-index contrasts leading to fairly high reflection coefficients with short structures.

Transmission vs. wavelength of a Fabry–Perót resonant cavity exhibits maxima when the optical length of the cavity matches the resonance condition (Fig. 6.32). In order to increase the modulation depth, the index and

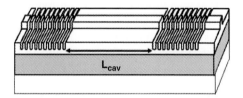

Fig. 6.31. Schematic view of a waveguide Fabry–Perót cavity

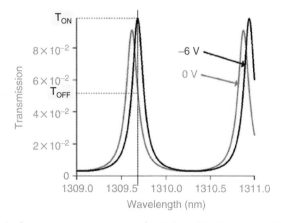

Fig. 6.32. Optical power transmittance of a Fabry–Perót cavity with the definition of the modulator "ON" and "OFF" states. Parameters used: $\Delta n = 1.7 \times 10^{-4}$, total intra-cavity optical loss of $20\,$dB cm^{-1}, cavity length of $200\,\mu$m, Bragg mirror reflection and transmission: $R = 0.62$ and $T = 0.1$

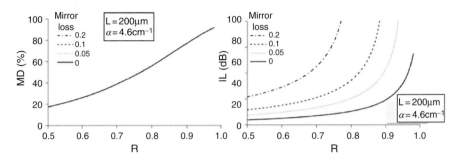

Fig. 6.33. Modulation depth and insertion loss vs. mirror reflection coefficient. Parameters used: $\Delta n = 1.7 \times 10^{-4}$, total intra-cavity optical loss of $20\,\mathrm{dB\,cm^{-1}}$, cavity length of $200\,\mu\mathrm{m}$, Bragg mirror reflection and transmission: $R = 0.62$ and $T = 0.1$)

absorption variations are combined: the ON state is defined when a reverse bias is applied, i.e., when holes are depleted which corresponds to a minimum of light absorption, and the OFF state corresponds to the zero bias condition.

Insertion loss depends on the two sources of optical loss: diffraction in the Bragg mirror and absorption in the cavity. The optimization of the Fabry–Perót cavity showed that a compromise must be found between low IL and high MD due to the cavity losses ($20\,\mathrm{dB\,cm^{-1}} \leftrightarrow 4.6\,\mathrm{cm^{-1}}$). Short devices are indeed desirable to lower the IL but are detrimental to the MD. The overall efficiency of the integrated interferometer was found to be closely related to the Bragg mirror properties. The calculated MD and IL for a $200\,\mu\mathrm{m}$ long cavity are plotted in Fig. 6.33 as a function of the mirror reflection coefficient R for different values of mirror losses.

Low IL ($<5\,\mathrm{dB}$) and high CR ($>50\%$) values can be simultaneously obtained provided that mirror reflection (R) and loss (Loss) are optimized. As a typical example, $\mathrm{IL} = 5\,\mathrm{dB}$ and $\mathrm{CR} = 60\%$ with $R = 0.8$ and $\mathrm{Loss} = 0.05$. A careful optical tapering is needed at the cavity input and output to minimize the out-of-plane scattering loss traditionally encountered in waveguide Bragg mirrors. Previous works reported some ways for reducing the loss of integrated Bragg mirrors in SOI submicron waveguides for the realization of efficient Fabry–Perót cavities on SOI substrates [50].

Despite the interesting properties of Fabry–Perót cavities for the realization of short modulators ($<300\,\mu\mathrm{m}$), it is worth noting that such interferometers tend to be sensitive to temperature variations and to process fluctuations. For this reason, the use of Mach-Zehnder interferometers (MZI) was proposed as an alternative solution [47]. A schematic view is shown in Fig. 6.34.

The operating principle is to divide the input electro-magnetic field and then to recombine the two guided waves. When a reverse bias is applied on the modulation doped SiGe/Si heterostructure in one arm of the MZI, the induced index variation is responsible for a phase shift of the electro-magnetic wave propagating in this arm. At the output of the MZI, if the phase difference

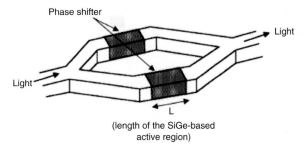

Fig. 6.34. Schematic view of the integration of the MD multiple SiGe/Si heterostructure in the two arms of a Mach-Zehnder interferometer

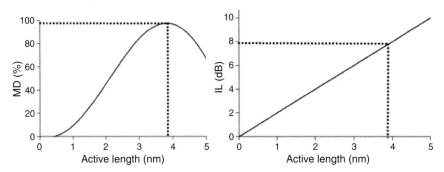

Fig. 6.35. Modulation depth and insertion loss of an SiGe/Si multiple well-structure integrated into a reverse-biased PIN device and placed in one arm of an MZI as a function the structure active length (after [47])

between the two electro-magnetic waves is a λ multiple, the MZI transmission is maximal (ON state). If the phase difference between the two electromagnetic waves equals $(2p+1)\lambda/2$, the MZI transmission is minimal (OFF state). For the minimal transmission to be nearest to zero, the two wave amplitudes should be as close as possible. The SiGe/Si MD-MQW active region must be thus included in each arm to balance the optical loss. The length g required for a π phase shift of the optical mode between the two arms, which allows the maximum contrast between the ON and OFF states is given by $L_\pi = \lambda/(2\Delta n)$. Due to the fairly low index variation ($1.7.10^{-4}$), L_π is, however, about 3.8 mm. It should yet emphasized that, as the modulator RC constant does not depend on its length, device lengths in the range of several mm are not directly detrimental to high-speed operation. A main advantage of this structure relies in its intrinsic geometrical symmetry, which nearly cancels the influence of temperature variations.

Figure 6.35 shows the evolutions of the MD and the IL as a function of the MZI length. It shows that MZI can perform very high modulation depth (MD > 98%) without excessive losses (IL < 8 dB). This option may yield high-performance integrated optical Si-based modulators. Properties of photonics crystals could also be beneficial to design shorter MZI by using slow-wave optical propagation based on low group velocity in periodically corrugated structures [51].

6.5 Conclusion

The realization of integrated active devices is a key point for optical interconnects at 1.3–1.5 μm wavelengths. The required devices are high-speed, low noise, and highly sensitive detectors as well as efficient switching devices and optical modulators. In this context, with a material compatibility allowing the Si/SiGe heterostructures to be integrated into the existing silicon technology, the group IV silicon–germanium technology is important due to the energy bandgap engineering that it allows.

The two main options to realize detectors are either based on optimized SiGe/Si SLS or on pure germanium layers. In both cases, due to the 4.2% of lattice mismatch between Si and Ge, accurate epitaxial growth methods are needed. SiGe/Si SLS detectors have moderate absorption coefficient ($<100\,\mathrm{cm}^{-1}$ at 1.3 μm) due to the reachable Ge mole fraction and SiGe layer thickness before the appearance of misfit dislocations. They thus require the use of enhancement mechanisms. Resonant cavity enhancement of the detector sensitivities is possible. SiGe avalanche detectors can yield very high sensitivities (up to $4\,\mathrm{A\,W}^{-1}$). New progress in the lowering of the reverse bias needed to exploit the avalanche mechanism in SiGe APDs could appear in a near future. An interesting approach is also based on the integration of pure germanium detectors. Previous works have proved high responsivities ($>0.7\,\mathrm{A\,W}^{-1}$) and high bandwidth ($>13\,\mathrm{GHz}$). Further works are currently under progress towards the integration of compact Ge detectors in SOI micro-waveguides for on-chip optical interconnects.

The realization of switching devices and optical modulators in the silicon technology is a challenging task if compared with III/V semiconductor technology or LiNbO$_3$ due to the weakness of linear electro-optical effects. Electro-absorption have a rather low efficiency due to the indirect bandgap of SiGe materials, and the most efficient effect is to obtain a refractive index variation by inducing variations of free carrier concentrations. SiGe technology is then used either as a platform for the realization of few-micrometers waveguides or for the valence band engineering that it allows.

SiGe switching devices based on carrier-injection in forward-biased PN junctions have been proposed and realized. These interesting components can act as optical power splitters, multi-wavelength optical switches, and add-drop multiplexers. For the moment, DC power consumption is in the range of 100–200 mW, and response times are around 100–200 ns.

Electro-refractive SiGe/Si modulators integrated in a reverse-bias PIN junction have been proposed to get free some of the limitations of forward-biased optical modulators. These devices can be integrated in SOI submicron waveguides for on-chip optical interconnects and have short intrinsic response times ($<10\,$ps). Optimized structures exhibit fairly low RC response times (20–$100\,$ps) and can be placed either in waveguide Fabry–Perót cavities or Mach-Zehnder interferometers.

References

1. ITRS (2003), International Technology Roadmap for semiconductors: 2003, http://public.itrs.net/
2. L.C. Kimerling, L. Dal Negro, S. Sanai, Y. Yi, D. Ahn, S. Akiyama, D. Cannon, J. Liu, J.G. Sandlan, D. Sparacin, J. Michel, K. Wada, and M.R. Watts: Monolithic Silicon Microphotonics, L. Pavesi, D.J. Lockwood: Silicon Photonics, Top. Appl. Phys. **94**, 89, Springer, Berlin Heildelberg New York (2004)
3. L. Vivien, S. Lardenois, D. Pascal, S. Laval, E. Cassan, J.L. Cercus, A. Koster, J.M. Fédéli, M. Heitzmann: Appl. Phys. Lett. **85**, 703 (2004)
4. F. Aniel, M. Enciso-Aguillar, P. Crozat, R. Adde, T. Hackbarth, U. Seiler, H.J. Herzog, U. König, and H.V. Känel: Proceedings of ESSDERC, Editor G. Bacarani, 167 (2002)
5. J.L. Hoyt, H.M. Nayfeh, S. Eguchi, I. Aberg, G. Xia, T. Drake, E.A. Fitzerald, A.D. Antoniadis, Proceedings of Technical Digest International Electron Devices Meeting, **23** (2002)
6. R. Braunstein, A.R. Moore, F. Herman: Phys. Rev. **109**, 695 (1958)
7. J. Humlicek: Properties of Strained and Relaxed Silicon Germanium Editor K. Kasper, EMIS Datareviews Series, No. 12, INSPEC, London, **4.6**, 116–131 (1995)
8. L. Yang, J.R. Watling, R.C.W. Wilkins, M. Boriçi, J.R. Barker, A. Asenov, and S. Roy: Semiconductor Science and Technology **19**, 1174 (2004)
9. R. People: IEEE J. Quant. Electron. **22(9)**, 1696 (1986)
10. L. Colace, G. Masini, and G. Assanto: IEEE J. Quant. Electron. **35(12)**, 1843 (1999)
11. L. Masarotto, J.M. Hartmann, G. Bremond, G. Rolland, M. Papon, and M.N. Séméria : J. Cryst. Growth **255**, 8 (2003)
12. L. Naval, B. Jalali, L. Gomelsky, and J.M. Liu: J. Lightwave Technol. **14(5)**, 788 (1996)
13. A. Splett, T. Zinke, K. Petermann, H. Kibbel, H.J. Herzog, and H. Presting: IEEE Photon. Technol. Lett. **6(1)**, 59 (1994)
14. K.J. Ebeling: Integrated Opto-electronics **chapter 11**, Springer, Berlin Heidelberg New York (1992)
15. K. Kato: IEEE Trans. Microwave Theory Technol. **47**, 1265 (1999)
16. J. Wei, F. Xia, and S.R. Forest: IEEE Photon. Technol. Lett. **14**, 1590 (2002)
17. H Temkin, A. Antreasyan, N.A. Olsson, T.P. Pearsall, and J.C. Bean: Appl. Phys. Lett. **49(13)**, 809 (1986)
18. J.W. Shi, Y.H. Liu, and C.W. Liu: J. Lightwave Technol. **22(6)**, 1583 (2004)
19. M.S. Ünlu and S. Strite: J. Appl. Phys. **78(2)**, 607 (1995)
20. C. Li, Q. Yang, H. Wang, J. Yu, Q. Wang, Y. Li, J. Zhou, H. Huang, and X. Ren: IEEE Photon. Technol. Lett. **12(10)**, 1373 (2000)
21. C. Li, C.J. Huang, B. Cheng, Y. Zuo, L. Luo, J. Yu, and Q. Wang: J. Appl. Phys. **92(3)**, 1718 (2002)

22. C.O. Chui, A.K. Okyay, and K. Saraswat: IEEE Photon. Technol. Lett. **15(11)**, 1585 (2003)
23. D. Bodlaki, H. Yamamoto, D.H. Waldeck, and E. Borguet: Surf. Sci. **543**, 63 (2003)
24. J. Oh, J.C. Campbell, S.G. Thomas, S. Bharatan, R. Thoma, C. Jasper, R.E. Jones, and T.E. Zirkle: IEEE J. Quant. Electron. **38(9)**, 1238 (2002)
25. Z. Huang, J. Oh, and J. Campbell: Appl. Phys. Lett. **85(15)**, 3286 (2004)
26. D.D. Cannon, J. Liu, Y. Ishikawa, K. Wada, D.T. Danielson, S. Jongthamma-nurak, J. Michel, and L.C. Kimerling: Appl. Phys. Lett. **84(6)**, 906 (2004)
27. J.M. Hartmann, A. Abbadie, A.M. Papon, P. Holliger, G. Rolland, T. Billon, J.M. Fédéli, M. Rouvière, L. Vivien, and S. Laval: J. Appl. Phys. **95(10)**, 5905 (2004)
28. M. Halbwax, M. Rouvière, Y. Zheng, D. Debarre, L.H. Nguyen, J.L. Cercus, C. clerc, V. Yam, S. Laval, E. Cassan, and D. Bouchier: Opt. Mater. **27**, 822 (2005)
29. G. Dehlinger, S.J. Koestler, J.D. Schaub, J.O. Chu, Q.C. Ouyang, and A. Grill: IEEE Photon. Technol. Lett. **16(1)**, 1041 (2004)
30. O.I. Dosunmu, D.D. Cannon, M.K. Emsley, L.C. Kimerling, and M.S. Ünlu: IEEE Photon. Technol. Lett. **17(1)**, 175 (2005)
31. M. Halbwax, M. Rouvière, Y. Zheng, D. Debarre, L.H. Nguyen, J.L. Cercus, C. Clerc, V. Yam, S. Laval, E. Cassan, and D. Bouchier: Proceedings of E-MRS, Symposium A1, Strasbourg (2004)
32. M. Rouvière, M. Halbwax, J.L. Cercus, E. Cassan, L. Vivien, D. Pascal, M. Heitzmann, J.M. Hartmann, and S. Laval, Integration of Germanium Waveguide Photodetectors for Intra-chip Optical Interconnects, to be published in Opt. Eng.
33. DESSIS simulator: http://www.ise.ch
34. A. Cutolo, M. Iodice, P. Spirito, and L. Zeni: J. Lightwave Technol. **15(3)**, 505 (1997)
35. R. Soref: Proc. SPIE **704**, 32 (1986)
36. S. Janz, J.M. Baribeau, A. Delâge, H. Lafontaine, S. Mailhot, R.L. Wiiliams, D.X. Xu, D.M. Bruce, P.E. Jessop, and M. Robillard: IEEE J. Quant. Electron. **4(6)**, 990 (1998)
37. B. Li, J. Li, Y. Zhao, and X. Lin, S.J. Chua, L. Miao, E.A. Fitzgerald, M.L. Lee, and B.S. Chaudharai: Appl. Phys. Lett. **84(13)**, 2241 (2004)
38. B. Li, S.J. Chua: Appl. Phys. Lett. **80(2)**, 180 (2002)
39. B. Li and S.J. Chua: J. Lightwave Technology **21(7)**, 1685 (2003)
40. Y. Miyake, J.Y. Kim, Y. Shiraki, and S. Fukatsu: Appl. Phys. Lett. **68**, 2097 (1996)
41. O. Qasaimeh, J. Singh, and P. Bhattacharya: IEEE J. Quant. Electron. **35(9)**, 1532 (1997)
42. O. Qasaimeh, J. Singh, and E.T. Croke: IEEE Photon. Technol. Lett. **10(6)**, 807 (1998)
43. A. Vonsovici and L. Vescan: IEEE J. Quant. Electron. **4(6)**, 1011 (1998)
44. D. Marris, A. Cordat, D. Pascal, A. Koster, E. Cassan, L. Vivien, and S. Laval: IEEE J. Select. Top. Quant. Electron. **9(3)**, 747 (2003)
45. D. Marris, E. Cassan, L. Vivien: Journal of Applied Physics 96, 6109 (2004)
46. A. Lupu, D. Marris, D. Pascal, J.-L. Cercus, A. Cordat, V. Le Thanh, and S. Laval: Appl. Phys. Lett. **85(6)**, 887 (2004)

47. D. Marris., E. Cassan, L. Vivien, D. Pascal, A. Koster, S. Laval: Design of a Modulation-Doped SiGe/Si Optical Modulator Integrated in a Sub-micrometer Silicon-On-Insulator Waveguide, to be published in Opt. Eng.
48. A Liu, R. Jones, L. Lia, D. Samara-Rubio, D. Rubin, O. Cohen, R. Nicolaescu, and M. Paniccia: Nature **427**, 615 (2004)
49. Fimmwave simulator: http://photondesign.com
50. D. Peyrade, E. Silberstein, Ph. Lalanne, A. Talneau, Y. Chen : Appl. Phys. Lett. **81(5)**, 829 (2002)
51. A. Martinez, A. Griol, P. Sanchis, J. Marti: Opt. Lett., **28**(6), 405 (2003)

7

An Introduction to Silicon Photonics

G.T. Reed

Summary. This chapter introduces the reader to the technology of silicon photonics as well as provide some insight to waveguides and devices in general. The chapter leads the reader through the fundamentals of modes of an optical waveguide, propagation constants, reflection coefficients and electric field profiles. Both planar and rib waveguides are discussed in the context of modal properties, and the discussion of propagation constants leads on to the concept of effective index. The contributions of loss to optical waveguides are discussed together with ways to measure optical loss. Finally a series of fundamental devices that make up many optical circuits are described together with their operational characteristics. These notes have been adapted from [G.T. Reed, A.P. Knights, Silicon Photonics: An Introduction, Wiley, UK, January 2004, ISBN 0-470-87034-6] with the permission of John Wiley and Sons, UK.

7.1 The Ray Optics Approach to Describing Planar Waveguides

The study of light is a study of electromagnetic waves. Consequently, the photonics engineer will inevitably encounter electromagnetic theory during his or her career. However, the rigours of Maxwell's equations are not always required for all applications, and one can make significant progress in understanding the basics of guided wave propagation with more simple models. Therefore, this section of this chapter introduces the well-known ray optical model, and uses it to investigate a number of important phenomena in simple optical waveguides. Later in this chapter, we will build on these concepts to describe more complex structures.

First consider a light ray (E_i) propagating in a medium with refractive index n_1, and impinging on the interface between two media, at an angle θ_1, as shown in Fig. 7.1.

At the interface the light is partially transmitted (E_t) and partially reflected (E_r). The relationship between the refractive indices n_1 and n_2, and the angles of incidence (θ_1) and refraction (θ_2), is given by Snell's law

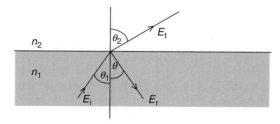

Fig. 7.1. Light rays refracted and reflected at the interface of two media

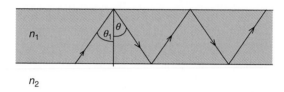

Fig. 7.2. Total internal reflection at two interfaces demonstrating the concept of a waveguide

$$n_1 \sin \theta_1 = n_2 \sin \theta_2, \tag{7.1}$$

In Fig. 7.1, the refractive index of media 1 (n_1) is clearly higher than the refractive index of media 2 (n_2), as the angle θ_2 is greater than θ_1. Consequently as θ_1 is increased, θ_2 will approach 90°. For some angle θ_1, the corresponding angle θ_2 will reach 90°, and hence Snell's law simplifies to

$$n_1 \sin \theta_1 = n_2. \tag{7.2}$$

Hence we can define the critical angle θ_c as

$$\sin \theta_c = \frac{n_2}{n_1}. \tag{7.3}$$

For angles of incidence greater than this critical angle, no light is transmitted and total internal reflection (TIR) occurs. If we now consider a second interface below the first, at which the wave also experiences TIR, we can understand the concept of a waveguide, as the light is confined to the region of refractive index n_1, and propagates to the right (Fig. 7.2).

This simplistic approach suggests, however, that the waveguide will support propagation at any angle greater than the critical angle. In Sect. 7.2, we shall see that this is not the case.

7.2 Reflection Coefficients

To fully understand the behaviour of the optical waveguide it is necessary to use electromagnetic theory. However, by enhancing our ray model a little we can improve our understanding significantly. Thus far we have implicitly considered reflection/transmission at an interface to be "partial" reflection and "partial" transmission, without defining the term "partial". With reference to Fig. 7.1, consider the reflection and transmission of the wave at a single interface. It is well known that the reflected and transmitted waves can be described by the Fresnel formulae. For example, the reflected wave will have complex amplitude, E_r, at the interface, related to the complex amplitude, E_i, of the incident wave by

$$E_r = r \cdot E_i, \tag{7.4}$$

where r is a complex reflection coefficient. The reflection coefficient is a function of both the angle of incidence and the polarization of light. Hence before proceeding further we must define the polarization with respect to the interface. The electric and magnetic fields of an electromagnetic wave are always orthogonal to one another, and both are orthogonal to the direction of propagation. Hence propagating electromagnetic waves are also referred to as transverse electromagnetic waves, or TEM waves. The polarization of a wave is deemed to be the direction of the electric field associated with the wave. For the purposes of this discussion we wish to consider cases where either the electric field, or the magnetic field, is perpendicular to the plane of incidence (i.e. the plane containing the wave normal and the normal to the interface). The transverse electric (TE) condition is defined as the condition when the electric fields of the waves are perpendicular to the plane of incidence. This is depicted in Fig. 7.3. Correspondingly, the transverse magnetic (TM) condition occurs when the magnetic fields are perpendicular to the plane of incidence. As previously mentioned, the Fresnel formulae describe the reflection coefficients r_{TE} and r_{TM}, usually written as follows:

For TE polarization

$$r_{TE} = \frac{n_1 \cos \theta_1 - n_2 \cos \theta_2}{n_1 \cos \theta_1 + n_2 \cos \theta_2}. \tag{7.5a}$$

Similarly for TM polarization, the reflection coefficient is

$$r_{TM} = \frac{n_2 \cos \theta_1 - n_1 \cos \theta_2}{n_2 \cos \theta_1 + n_1 \cos \theta_2}. \tag{7.6a}$$

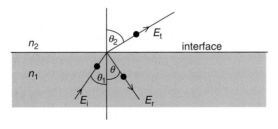

Fig. 7.3. Orientation of electric fields for TE incidence at the interface between two media. Circles indicate that the electric fields are vertical (i.e. coming out of the plane of the paper)

Using Snell's law (7.1), (7.5a) and (7.6a) can be rewritten as

$$r_{\mathrm{TE}} = \frac{n_1 \cos\theta_1 - \sqrt{n_2^2 - n_1^2 \sin^2\theta_1}}{n_1 \cos\theta_1 + \sqrt{n_2^2 - n_1^2 \sin^2\theta_1}}, \qquad (7.5\mathrm{b})$$

$$r_{\mathrm{TM}} = \frac{n_2^2 \cos\theta_1 - n_1\sqrt{n_2^2 - n_1^2 \sin^2\theta_1}}{n_2^2 \cos\theta_1 + n_1\sqrt{n_2^2 - n_1^2 \sin^2\theta_1}}. \qquad (7.6\mathrm{b})$$

When the angle of incidence is less than the critical angle, only partial reflection occurs and the reflection coefficient is real. However, when the critical angle is exceeded we have TIR. We can see from (7.5b) and (7.6b) that the term inside the square root becomes negative. Clearly this implies that

$$|r| = 1 \qquad (7.7)$$

and that r is also complex and hence a phase shift is imposed on the reflected wave. Hence we may denote this by

$$r = \exp\left(\mathrm{j}\,\phi\right), \qquad (7.8)$$

where ϕ_{TE} and ϕ_{TM} are given by

$$\phi_{\mathrm{TE}} = 2\tan^{-1}\frac{\sqrt{\sin^2\theta_1 - \left(\frac{n_2}{n_1}\right)^2}}{\cos\theta_1} \qquad (7.9)$$

and

$$\phi_{\mathrm{TM}} = 2\tan^{-1}\frac{\sqrt{\frac{n_1^2}{n_2^2}\sin^2\theta_1 - 1}}{\frac{n_2}{n_1}\cos\theta_1}. \qquad (7.10)$$

Note, however, that we have now defined reflection coefficients without too much discussion of the incident and reflecting waves. In fact it is important to note that the reflection coefficients as defined, relate the relative amounts of the fields, not the powers, which are reflected. In practical terms the field is difficult to measure directly, so we are more interested in the power (or intensity) that is reflected or transmitted. In an electromagnetic wave, the propagation of power is described by the Poynting vector, usually denoted as S ($\mathrm{W\,m^{-2}}$). The Poynting vector is defined as the vector product of the electric and magnetic vectors, and hence indicates both the magnitude and direction of power flow. One of the most common ways of expressing the Poynting vector is

$$S = \frac{1}{Z}E^2 = \sqrt{\frac{\varepsilon_{\mathrm{m}}}{\mu_{\mathrm{m}}}}E^2 \qquad (7.11)$$

where E is electric field, ε_m is the permittivity of the medium, μ_m is the permeability of the medium, and Z is the impedance of the medium. We can then define a reflectance, R, which relates the incident and reflected powers associated with the waves as

$$R = \frac{S_r}{S_i} = \frac{E_r^2}{E_i^2} = r^2. \tag{7.12}$$

7.3 Phase of a Propagating Wave and its Wavevector

If we enhance our simple ray model a little further we can understand more about the propagation of light in simple waveguides. Thus far the ray has simply represented the direction of propagating light. To do this we need briefly consider a propagating electromagnetic wave. In common with electrical circuit theory, it is convenient to use the exponential form of the propagating wave, to make a sufficiently general point. Let the electric and magnetic fields associated with a propagating wave be described, respectively, as

$$E = E_0 \exp[j(kz \pm \omega t)] \tag{7.13}$$

and

$$H = H_0 \exp[j(kz \pm \omega t)], \tag{7.14}$$

where z is merely the direction of propagation (for the sake of argument). This means that the phase of the wave is

$$\phi = kz \pm \omega t. \tag{7.15}$$

It is clear that the phase varies with time (t), and with distance (z). These variations are quantified by taking the time derivative and the spatial derivative, i.e.

$$\left| \frac{\partial \phi}{\partial t} \right| = \omega = 2\pi f, \tag{7.16}$$

where ω is angular frequency (rad s^{-1}), and f is frequency (Hz). Both angular frequency and frequency in hertz are familiar to engineers, and describe how the phase of the wave varies with time. A similar expression is found to relate phase to propagation distance by taking the spatial derivative of (7.15), i.e.

$$\frac{\partial \phi}{\partial z} = k, \tag{7.17}$$

where k is the wavevector or the propagation constant in the direction of the wavefront. It is related to wavelength, λ, by the relation

$$k = \frac{2\pi}{\lambda}. \tag{7.18}$$

In free space k is usually designated k_0. Hence k and k_0 are related by the refractive index of the medium, n, i.e.

$$k = nk_0. \tag{7.19}$$

Hence in free space

$$k_0 = \frac{2\pi}{\lambda_0}. \tag{7.20}$$

7.4 Modes of a Planar Waveguide

Having defined the wavevector we can now further consider propagation in a waveguide. The simplest optical waveguide is the planar waveguide, depicted in Fig. 7.2. This figure is reproduced in Fig. 7.4a, with the addition of axes x, y and z.

Let the waveguide height be h, and propagation be in the z direction, with the light being confined in the y direction by TIR. The zig-zag path indicated in the figure now means more than it did previously. It represents the direction of the wavenormals as the waves propagate through the waveguide, with wavevector k ($=k_0 n_1$). We can use an associated figure, Fig. 7.4b to explain this further. If we decompose the wavevector k, into two components, in the y and z directions. By simple trigonometry, referring to Fig. 7.4b

$$k_z = n_1 k_0 \sin \theta_1 \tag{7.21}$$

and

$$k_y = n_1 k_0 \cos \theta_1 \tag{7.22}$$

Having determined the propagation constant in the y direction, we can imagine a wave propagating in the y direction. Since this theoretical wave will be reflected at each interface, there will potentially be a standing wave across the waveguide in the y direction. Hence we can sum all the phase shifts introduced in making one complete "round trip" across the waveguide and back again. For a waveguide thickness h (and hence a traversed distance of $2h$), using (7.17), a phase shift is introduced of

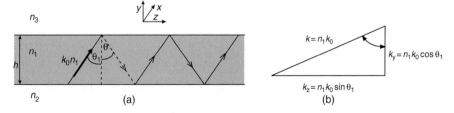

Fig. 7.4. (a) Propagation in a planar waveguide. (b) The relationship between propagation constants in the y, z and wavenormal directions

$$\phi_h = 2k_y h = 2k_0 n_1 h \cos\theta_1. \tag{7.23}$$

From (7.8) to (7.9), it is clear that phase changes are introduced upon reflection at upper and lower waveguide boundaries. Let us refer to these phase shifts as ϕ_u, and ϕ_ℓ, respectively. Hence the total phase shift is

$$\phi_t = 2k_0 n_1 h \cos\theta_1 - \phi_u - \phi_\ell. \tag{7.24}$$

For self-consistency (i.e. the preservation of a wave across the waveguide), this total phase shift must be a multiple of 2π, hence

$$2k_0 n_1 h \cos\theta_1 - \phi_u - \phi_\ell = 2m\pi, \tag{7.25}$$

where m is an integer. This equation can be reduced to an equation in θ, by substituting for ϕ_u and ϕ_ℓ from (7.9) or (7.10). Before we do that, however, note that because m is an integer, there will be a series of discrete angles θ, for which (7.25) can be solved, corresponding to integral values of m. For each solution there will therefore be a corresponding propagation constant in the y and z directions (for each polarization). This shows us that light cannot propagate at any angle θ, but only at one of the allowed discrete angles. Each allowed solution is referred to as a mode of propagation, and the mode number is given by the value of integer m. The modes are identified using a notation utilising the polarization and the mode number. For example, the first TE mode, or fundamental mode will be described as TE_0. Higher order modes are correspondingly described using the appropriate value of m. There is also a further limit on m, indicating that there is a limit to the number of modes that can propagate in a given waveguide structure. The limiting conditions correspond to the propagation angle, θ_1 becoming less than the critical angle at either the upper or lower waveguide interface. We will consider this further in the following few sections of this chapter, as we discuss the modes in the waveguides more comprehensively.

7.4.1 The Symmetrical Planar Waveguide

Having established (7.25), which describes the discrete nature of waveguide modes, let us now discuss solving the equation. The planar waveguide we have discussed thus far was described by Fig. 7.4, having an upper cladding of refractive index n_3, and a lower cladding refractive index of n_2. In the case where $n_2 = n_3$, the waveguide is said to be symmetrical since the same boundary conditions will apply at the upper and lower interfaces. In terms of solving (7.25), this means that $\phi_u = \phi_\ell$. Therefore, using (7.9) (assuming TE polarization), (7.25) becomes

$$2k_0 n_1 h \cos\theta_1 - 4\tan^{-1}\left[\frac{\sqrt{\sin^2\theta_1 - \left(\frac{n_2}{n_1}\right)^2}}{\cos\theta_1}\right] = 2m\pi. \tag{7.25}$$

This can be rearranged as

$$\tan\left[\frac{k_0 n_1 h \cos\theta_1 - m\pi}{2}\right] = \left[\frac{\sqrt{\sin^2\theta_1 - \left(\frac{n_2}{n_1}\right)^2}}{\cos\theta_1}\right]. \tag{7.26}$$

The only variable in (7.26) is θ_1, and therefore solving the equation yields the propagation angle, but also it is a straightforward calculation to find the propagation constants associated with the mode in question.

Using (7.10) for the phase shift, the corresponding TM equation is

$$\tan\left[\frac{k_0 n_1 h \cos\theta_1 - m\pi}{2}\right] = \left[\frac{\sqrt{\left(\frac{n_1}{n_2}\right)^2 \sin^2\theta_1 - 1}}{\left(\frac{n_2}{n_1}\right)\cos\theta_1}\right]. \tag{7.27}$$

It is often convenient to have an approximate idea of the number of modes supported by a waveguide. Consider (7.26), the eigenvalue equation for TE modes. We know that the minimum value that θ_1 can take corresponds to the critical angle θ_c. From (7.3),

$$\sin\theta_c = \frac{n_2}{n_1}. \tag{7.28}$$

Hence the right-hand side of (7.26) reduces to zero. Since θ_1, decreases with mode number, the minimum value of θ_1 ($=\theta_c$), corresponds to the highest possible order mode. At $\theta_1 = \theta_c$, (7.26) reduces to

$$\frac{k_0 n_1 h \cos\theta_c - m_{\max}\pi}{2} = 0 \tag{7.29}$$

rearranging for m, the mode number

$$m_{\max} = \frac{k_0 n_1 h \cos\theta_c}{\pi}. \tag{7.30}$$

We can see that if we evaluate $m_{\max}$, and find the nearest integer that is less than the evaluated $m_{\max}$, then this will be the highest order mode number, which we can denote as $[m_{\max}]_{\mathrm{int}}$. The number of modes will actually be $[m_{\max}]_{\mathrm{int}} + 1$, since the lowest order mode (usually called the fundamental mode), has a mode number $m = 0$.

In passing we can also note a very interesting characteristic of the symmetrical waveguide, from (7.26) and (7.27), the eigenvalue equations for TE and TM polarizations. Both equations allow a solution when $m = 0$. This is because the term in the square root on the right-hand side of the equation is always positive, and hence the square root is always a real number. This in turn is because θ_1 is always greater than the critical angle, and hence $\sin^2\theta_1$ is always greater than $\sin^2\theta_c$ ($= n_2^2/n_1^2$). This implies that the lowest order mode is always allowed. This means that the fundamental mode will always

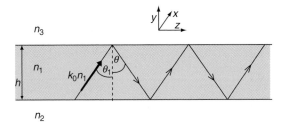

Fig. 7.5. Propagation in an asymmetric planar waveguide

propagate, and the waveguide is never "cut-off". We will see shortly that this is not the case for the asymmetric waveguide.

7.4.2 The Asymmetrical Planar Waveguide

Let us now consider the slightly more complex asymmetrical planar waveguide. This is shown in Fig. 7.5. In this case $n_2 \neq n_3$.

   We can approach a simple analysis in the same way as we did for the symmetrical planar waveguide, but the phase change on reflection as the upper and lower waveguide boundaries will not be the same. This time the eigenvalue equation for TE modes becomes

$$[k_0 n_1 h \cos\theta_1 - m\pi] = \tan^{-1}\left[\frac{\sqrt{\sin^2\theta_1 - \left(\frac{n_2}{n_1}\right)^2}}{\cos\theta_1}\right] + \tan^{-1}\left[\frac{\sqrt{\sin^2\theta_1 - \left(\frac{n_3}{n_1}\right)^2}}{\cos\theta_1}\right].$$

(7.31)

Once again this equation can be solved (numerically or graphically) to find propagation angle θ_1, for a given value of m. Note, however, that there is not always a solution to (7.31) for $m = 0$, because it is possible for one of the terms within the square roots on the right-hand side of the equation to be negative. This is because the critical angle for the waveguide as a whole will be determined by the larger of the critical angles of the two waveguide boundaries. Obviously for TIR at both waveguide boundaries, the propagating mode angle must be greater than both critical angles. However, the mode angle of the waveguide may not satisfy the condition for both critical angles to be exceeded if the guide is too thin (h is small), or if the refractive index difference between core and claddings is too small. Let us consider this in detail by solving the eigenvalue equation for TE modes, (7.31).

7.5 Solving the Eigenvalue Equation for Symmetrical and Asymmetrical Waveguides

The eigenvalue equation can be solved in a number of ways, typically numerical. However, it can be more instructive to solve the equation graphically.

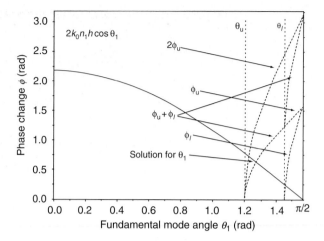

Fig. 7.6. Solution of the eigenvalue equation for $m = 0$

First, we need to choose some waveguide parameters: let $n_1 = 1.5$, $n_2 = 1.49$, $n_3 = 1.40$, $\lambda_0 = 1.3\,\mu\text{m}$ and $h = 0.3\,\mu\text{m}$.

Whilst the refractive indices are not representative of a silicon waveguide, it will enable us to make a point about modal characteristics. Figure 7.6 shows graphical solution of the TE eigenvalue equation (7.31) for $m = 0$. The three terms in the equation are plotted separately for clarity. The two phase change terms, ϕ_u and ϕ_ℓ, are added and also shown, representing the right-hand side of (7.31). Where this curve intersects with the left-hand side of the equation, is the solution for the mode angle θ_1. This is shown in Fig. 7.6. Note that this solution occurs at an angle θ_1, that lies between θ_u and θ_ℓ, the critical angles for the upper and lower interfaces, respectively (also shown in the figure by the dotted vertical lines). Thus, the conditions for TIR are not met, and the waveguide is cut-off. If we increase the waveguide thickness, h, the curves would eventually intersect at an angle greater than both θ_u and θ_ℓ. This confirms that the asymmetrical waveguide is cut-off for some conditions. Alternatively, if we consider the symmetrical guide (with $n_2 = n_3 = 1.4$), the right-hand side of the equation becomes $2\phi_u$ (also plotted). It is clear that the solution will always be at an angle greater than θ_u, and hence confirming that the symmetrical waveguide is never cut-off.

The corresponding graphical solution for a silicon waveguide is shown in Fig. 7.7. The waveguide parameters are: $n_1 = 3.5$ (silicon), $n_2 = 1.5$ (silicon dioxide), $n_3 = 1.0$ (air), $\lambda_0 = 1.3\,\mu\text{m}$ and $h = 0.15\,\mu\text{m}$.

It is interesting to note there is relatively little difference between the symmetrical and the asymmetrical silicon waveguide, because typical cladding material for silicon is silicon dioxide (SiO_2), which has a refractive index of 1.5. Hence either cladding with air or SiO_2 means that the cladding is very different to the core refractive index of approximately 3.5.

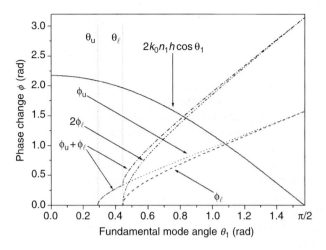

Fig. 7.7. Solution of the eigenvalue equation for $m = 0$ (silicon-on-insulator)

7.6 Monomode Conditions

It is often convenient for a waveguide to support only a single mode, for a given polarization of light. Such a waveguide is referred to as monomode. We will return to this point later in this chapter. For now, consider again the TE polarization eigenvalue equation (7.26) for a symmetrical waveguide, reproduced below

$$\tan\left[\frac{k_0 n_1 h \cos\theta_1 - m\pi}{2}\right] = \left[\frac{\sqrt{\sin^2\theta_1 - \left(\frac{n_2}{n_1}\right)^2}}{\cos\theta_1}\right]. \tag{7.32}$$

Considering the second mode, with mode number $m = 1$, the limiting condition for this mode is that the propagation angle is equal to the critical angle. Under these circumstances the square root term reduces to zero. Since the propagation angle for the second mode will be less than for the fundamental mode, for all angles greater than this critical angle, the waveguide will be monomode. Thus (7.32) reduces to

$$\tan\left[\frac{k_0 n_1 h \cos\theta_c - \pi}{2}\right] = 0$$

i.e.

$$\cos\theta_c = \frac{\pi}{k_0 n_1 h} = \frac{\lambda_0}{2 n_1 h}. \tag{7.33}$$

Hence for monomode conditions

$$\theta_c < \cos^{-1}\left(\frac{\lambda_0}{2 n_1 h}\right). \tag{7.34}$$

Similar expressions can be derived from the other eigenvalue equations used thus far.

7.7 Effective Index

Thus far we have used a simple model to look at the characteristics of a planar waveguide in order to gain an understanding of modes. We can take the model still further to understand a little more. Previously, we briefly discussed the idea of propagation constants in the y and z directions. These were defined in (7.21) and (7.22) as

$$k_z = n_1 k_0 \sin \theta_1 \tag{7.35}$$

and

$$k_y = n_1 k_0 \cos \theta_1. \tag{7.36}$$

The propagation constant in the z direction is of particular interest as it indicates the rate at which the wave propagates in the z direction. k_z is often replaced in many texts by the variable β, and the two terms are used interchangeably. We can define a parameter N, called the effective index of the mode, such that

$$N = n_1 \sin \theta_1 \tag{7.37}$$

i.e. (7.35) becomes

$$k_z = \beta = N k_0. \tag{7.38}$$

Note the similarity with (7.19). This is equivalent to thinking of the mode as propagating straight down the waveguide, without "zig-zagging" back and forth, with refractive index N. Having defined β, we can immediately consider the range of values that β can take.

The lower bound on β is determined by the critical angles of the waveguide. For the asymmetrical waveguide the smaller of the two critical angles is usually defined by the upper cladding layer, as this usually has a lower refractive index than the lower cladding layer. In the case of silicon, the lowest possible value of the refractive index of the upper cladding is 1.0, if the upper cladding was air. This means that TIR is limited by the lower cladding, which will have a larger critical angle that needs to be exceeded for TIR to occur. Hence $\theta_1 \geq \theta_\ell$. Hence the lower bound on β is given by

$$\beta \geq n_1 k_0 \sin \theta_\ell = k_0 n_2. \tag{7.39}$$

The upper bound on β is governed by the maximum value of θ, which is clearly $90°$. In this case $\beta = k = n_1 k0$. Hence

$$k_0 n_1 \geq \beta \geq k_0 n_2. \tag{7.40}$$

Using (7.38)

$$n_1 \geq N \geq n_2. \tag{7.41}$$

7.8 Electromagnetic Theory

Starting with Maxwell's equations, if we assume a loss-less, non-conducting medium, limit ourselves to propagation in the z direction, and consider one polarization at a time (TE or TM), we can derive a scalar equation describing wave propagation in our planar waveguide

$$\frac{\partial^2 E_x}{\partial y^2} + \frac{\partial^2 E_x}{\partial z^2} = \mu_m \varepsilon_m \frac{\partial^2 E_x}{\partial t^2}, \tag{7.42}$$

where ε_m is the permittivity of the waveguide, μ_m is the permeability of the waveguide, and in this case the electric field polarized in the x direction corresponds to TE polarization. This is called the scalar wave equation.

Since there is only electric field in the x direction, we can write the equation of this field as

$$E_x = E_x(y) \, e^{-j\beta z} \, e^{j\omega t}. \tag{7.43}$$

This means that there is a field directed (polarized) in the x direction, with a variation in the y direction yet to be determined, propagating in the z direction with propagation constant β, with sinusoidal ($e^{j\omega t}$) time dependence.

Solution of the wave equation provides us with expressions of the electric fields in the core and claddings. So we use the boundary conditions to find the variation in the y direction (i.e. to find $E_x(y)$). In the upper cladding, we will have

$$E_x(y) = E_u \, e^{-k_{yu}(y-\frac{h}{2})} \quad \text{for } y \geq (h/2). \tag{7.44}$$

In the core we will have

$$E_x(y) = E_c \, e^{-jk_{yc}y} \quad \text{for } -(h/2) \leq y \leq (h/2). \tag{7.45}$$

In the lower cladding we will have

$$E_x(y) = E_\ell \, e^{k_{yl}(y+\frac{h}{2})} \quad \text{for } y \leq -(h/2), \tag{7.46}$$

where

$$k_{yi}^2 = k_0^2 n_i - \beta^2.$$

Note that n and k_y are written n_i and k_{yi} because they can now represent any of the three media (core, upper cladding, or lower cladding), by letting $i = 1, 2$, or 3.

7.9 Another Look at Propagation Constants

In solving the wave equation we assumed field solutions with propagation constants in each of the three media (core and upper and lower cladding). However, we did not really discuss why the propagation constants took the form they did.

Our general solution was of the form

$$E_x = E_c\, \mathrm{e}^{-k_y y}\, \mathrm{e}^{-\mathrm{j}\beta z}\, \mathrm{e}^{\mathrm{j}\omega t}. \tag{7.47}$$

However, in the three media, core, upper cladding and lower cladding, the propagation constant k_y took different forms. In the claddings k_y was a real number, whereas in the core k_y was an imaginary number (which later resulted in a cosine function). This simply corresponds to the condition that TIR is satisfied at both boundaries. Mathematically it corresponds to whether β is greater or lesser than $k_0 n_i^2$. As we know, a term $\mathrm{e}^{\mathrm{j}\varphi}$ corresponds to a propagating sinusoidal/cosinusoidal type field, whereas $\mathrm{e}^{-\varphi}$ simply corresponds to an exponentially decaying field. Thus imaginary propagation constants in the claddings, while valid solutions to the wave equation, represent fields propagating in the y direction through the claddings, and hence *not* to a totally internally reflected field. Hence all solutions other than totally internally reflected (guided) waves were ignored. This does, however, provide us with even more insight into the modal solutions of the planar waveguide. Clearly, the field penetrates the cladding to a degree determined by the decay constant (previously called propagation constant) in the cladding. Hence as the wave propagates, part of the field is propagating in the cladding.

7.10 Mode Profiles

Now that we have field solutions for modes in the planar waveguide, we can plot the field distribution, $E_x(y)$ or the intensity distribution, $|E_x(y)|^2$. In order to demonstrate more than a single mode, a larger waveguide has been defined than previously, with the following parameters: $n_1 = 3.5$, $n_2 = 1.5$, $n_3 = 1.0$, $\lambda_0 = 1.3\,\mu\mathrm{m}$ and $h = 1.0\,\mu\mathrm{m}$.

For even functions, the field in the core is a cosine function, and in the claddings, an exponential decay. Therefore, solutions for $m = 0$, and $m = 2$ are even functions, and are plotted in Figs. 7.8 and 7.9, respectively. The field

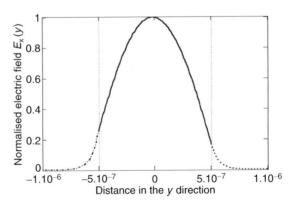

Fig. 7.8. Electric field profile of the fundamental mode ($m = 0$)

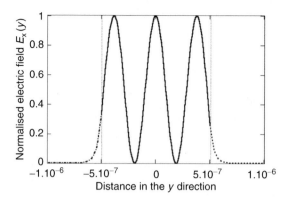

Fig. 7.9. Intensity profile of the second even mode ($m = 2$)

distribution is plotted for $m = 0$ (Fig. 7.8), and the intensity distribution is plotted for $m = 2$ (Fig. 7.9).

7.11 Silicon on Insulator Photonics

7.11.1 Introduction

Integrated optics in silicon is interesting for a combination of technological and cost reasons. The cost issues are relatively obvious, and are related to the absolute cost of silicon and silicon-on-insulator (SOI) wafers as compared to more exotic materials such as the III–V compounds or the insulator lithium niobate, to the cost of processing the wafers, and to the packing density that can be achieved. Silicon is a well understood and robust material, and the processing of silicon has been developed by the electronics industry to a level that is more than sufficient for most integrated optical applications. The minimum feature size for many applications is of the order of 1–2 µm, which in terms of microelectronics, is very old technology, although there is a trend towards miniaturization. Furthermore, as silicon microelectronics continues to advance, new and improved processing becomes available. It could be argued that an alternative technology, that of silica-based integrated optics, is lower cost, but this is a passive material, with little prospect of active devices such as sources, detectors, or optical modulators (other than thermally operated) becoming available. Consequently this technological issue is intimately linked to cost.

One significant technological issue is associated with the possibility of optical phase and amplitude modulation in silicon. Refractive index modulation is possible via several techniques, although it is now widely accepted that free carrier manipulation is the most efficient of these. This is not a fast modulation mechanism when compared to the field effect mechanisms available in other technologies, dominated by the linear electro-optic (Pockels) effect, used to advantage in, for example, lithium niobate (LiNbO$_3$) at operating speeds of several tens of GHz, but is sufficient for many communications and sensor applications. The Pockels effect is not observed in silicon due to the centro-symmetric nature of the crystal structure.

Sources and detectors (beyond 1.1 μm) in silicon are a current active research topic for many groups around the world. The topic of light sources in silicon is a particularly active research activity at present, and is discussed further elsewhere in this text. Nevertheless the possibility of accurate micromachining of silicon has meant that sources and detectors can be accurately aligned to silicon waveguides in a hybrid circuit, and silicon sources and detectors will need to be very cost effective to displace this approach, or offer very significant improvements in performance.

7.11.2 Silicon on Insulator Waveguides

So far in this chapter, when discussing optical waveguides, the discussion has been limited to a theoretical three-layer structure comprising a core and two cladding layers, the latter infinite in the x and y directions. Clearly, this is not a practical structure. However, the SOI planar waveguide approximates this theoretical waveguide. The configuration of an SOI wafer is shown in Fig. 7.10. The silicon guiding layer is typically up to a few micrometers in thickness, and the buried silicon dioxide is typically about half a micron, although individual designs vary. The purpose of the buried oxide layer is to act as the lower cladding layer, and hence prevent the field associated with the optical modes from penetrating the silicon substrate below. Therefore, as long as the oxide is thicker than the evanescent fields associated with the modes it will be satisfactory. Sometimes a surface oxide layer is also introduced as a passivation layer.

If we consider the three-layer model used previously, the waveguide is transformed from an asymmetrical waveguide into a symmetrical one by the addition of a surface oxide layer. In practice, however, the refractive indices of both air ($n = 1$) and silicon dioxide ($n = 1.5$), are so different from that of silicon ($n = 3.5$), the two configurations are very similar.

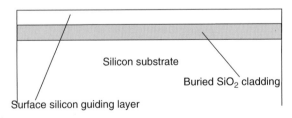

Fig. 7.10. SOI planar waveguide

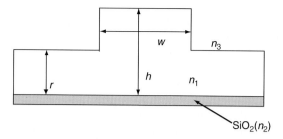

Fig. 7.11. Rib waveguide geometry

While the planar waveguide is convenient as an introduction to the subject of optical waveguides, it is of limited practical use, because light is confined only in one dimension. For many applications, two-dimensional confinement is required. This is achieved in silicon, by etching a two-dimensional waveguide, the most common being the so-called "rib" waveguide. The cross-section of a rib waveguide is shown in Fig. 7.11.

Clearly, this is just one way to achieve optical confinement in two dimensions, but based on what we have studied so far in this chapter, it is perhaps not obvious why the rib structure achieves optical confinement in the lateral direction. This is discussed further below.

Modes of Two-Dimensional Waveguides

In common with modes of optical fibres, two-dimensional rectangular waveguide modes require two subscripts to identify them. However, it is not appropriate to use the same convention as used in optical fibres, because rectangular guides are clearly not circularly symmetric. Therefore, it is natural to use Cartesian co-ordinates. Unfortunately, two slightly different conventions have developed, although they are very similar. In a rectangular waveguide there exist two families of modes, the HE mode and the EH modes. In common with the skew modes of an optical fibre these are mostly polarized in the TE or TM directions. Therefore, these modes are usually referred to as the E^x or E^y modes depending whether they are mostly polarized in the x or y directions. Two subscripts are then introduced to identify the mode. Therefore, the modes are designated $E^x_{p,q}$ or $E^y_{p,q}$, where the integers p and q represent the number of field maxima in the x and y directions, respectively. These modes are also referred to as $HE_{p,q}$ and $EH_{p,q}$ modes. Hence the fundamental modes are referred to as $E^x_{1,1}$ and $E^y_{1,1}$. This is probably the most common convention.

However, a second convention also exists in which the form of mode identification is essentially the same, but the integers start at "0" rather than at 1. This has developed from the convention used for planar waveguides previously, in which the fundamental mode is the mode for which the integer

$m = 0$. Hence in the second convention for rectangular waveguides, the fundamental modes are referred to as $E_{0,0}^x$ and $E_{0,0}^y$ (or HE$_{0,0}$ and EH$_{0,0}$). Hence some care must be taken to indicate the labelling convention used.

7.11.3 The Effective Index Method of Analysis

In this section, the effective index is used to find approximate solutions for the propagation constants of two-dimensional waveguides. This is done without resorting to any discussion of electric fields within the waveguide, and so the simplicity of the method, known as the effective index method, is very attractive.

The method is first discussed in general terms, before an example is carried out. Consider a generalized two-dimensional waveguide, as shown in Fig. 7.12a.

The approach to find the propagation constants for the waveguide shown in Fig. 7.12a is to regard the waveguide as a combination of two planar waveguides, one horizontal and one vertical. We then successively solve the planar waveguide eigenvalue equations first in one direction and then the other, taking the effective index of the first as the core refractive index for the second. Care must be taken to consider the polarization involved. If we are considering an electric field polarized in the x direction (TE polarization), then when solving the three-layer planar waveguide in the y direction, we use the TE eigenvalue equation. However, when we subsequently solve the vertical three-layer planar waveguide, we must use the TM eigenvalue equation, because, with respect to this imaginary vertical waveguide, the field is polarized in the TM direction. Decomposition of the two-dimensional waveguide into two planar waveguides is depicted in Fig. 7.12b.

The situation is complicated when solving a rib waveguide structure, because the refractive on either side of the core is not constant over the height of the core. Therefore, when solving the first part of the decomposed two-dimensional waveguide, we need to find the effective index, not only in the

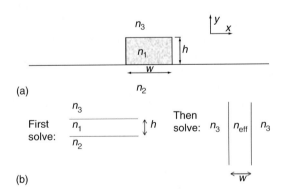

Fig. 7.12. (a) Generalized two-dimensional waveguide and (b) decomposition into two imaginary planar waveguides

core, but also in the slab regions at either side of the core in the x direction, prior to solving the second part of the decomposition. This is best illustrated by an example.

Example

Find the effective index N_{wg} of the fundamental mode of the rib waveguide in Fig. 7.11, for the following waveguide parameters: $w = 3.5\,\mu m$, $h = 5\,\mu m$, $r = 3\,\mu m$, $n_1 = 3.5$, $n_2 = 1.5$, $n_3 = 1.0$. The operating wavelength is $1.3\,\mu m$, and TE polarization is assumed.

This is representative of an example silicon waveguide. Further details are given in [1].

Solution

First, we must decompose the rib structure into vertical and horizontal planar waveguides, as shown in Fig. 7.12b. This means we must first solve the horizontal planar waveguide, shown in Fig. 7.13. Since the polarization is TE, we use the asymmetrical TE eigenvalue equation. Solving the TE eigenvalue equation for the conditions above and $m = 0$ yields a propagation angle of $87.92°$ ($1.53456\,rad$). From earlier equation

$$\beta = n_1 k_0 \sin\theta_1. \tag{7.48}$$

Therefore, the effective index of the waveguide region is given by

$$n_{effg} = \frac{\beta}{k_0} = n_1 \sin\theta_1, \tag{7.49}$$

i.e. $n_{effg} = 3.4977$.

We now need to solve the second decomposed planar waveguide of the decomposed rib structure. However, in order to do so, we must first find the effective index of the planar regions either side of the core. To do this we must solve the asymmetrical TE eigenvalue equation again, this time for a waveguide height of $r = 3\,\mu m$. This yields a propagation angle of $86.595°$, and an effective index for the planar region of $n_{effp} = 3.4938$.

We can now solve the second decomposed planar waveguide of the decomposed rib structure, using the effective indices just calculated. This time the TM eigenvalue equation should be used, and note that now we also have a symmetrical waveguide where θ_{wg} is to be found. The solution is $\theta_{wg} = 88.285°$, which corresponds to an effective index of $N_{wg} = 3.496$. Knowing the effective index we can evaluate all of the propagation constants in the core and cladding. In particular we are interested in the z-directed propagation constant, i.e.

$$\beta = k_0 N_{wg} = 16.897 \ \text{rad}\,\mu m^{-1}. \tag{7.50}$$

7.11.4 Large Single Mode Rib Waveguides

We earlier introduced an equation for the approximate number of modes in a planar waveguide. For a waveguide with dimensions of several microns the number of modes will be large. For example, consider a waveguide of height $h = 5\,\mu m$

$$m_{max} = \frac{k_0 n_1 h \cos \theta_c}{\pi}. \tag{7.51}$$

For simplicity assume a symmetrical silicon waveguide, with $n_1 = 3.5$ and $n_2 = 1.5$, thus, for an operating wavelength of $1.3\,\mu m$,

$$[m_{max}]_{int} = \frac{2 \times 3.5 \times 5 \times 10^{-6} \cos \theta_c}{\lambda_0} = 24 \tag{7.52}$$

and hence 25 modes are supported (including the $m = 0$ mode). Alternatively we could consider how thin we need to make the waveguide to support a single mode. The condition for only a single mode to propagate is

$$\theta_c < \cos^{-1} \left(\frac{\lambda_0}{2 n_1 h} \right). \tag{7.53}$$

Rearranging (7.53) in terms of waveguide height h, and substituting for $\cos \theta_c$

$$h < \left(\frac{\lambda_0/2}{n_1 \sqrt{1 - \left(\frac{n_2}{n_1} \right)^2}} \right). \tag{7.54}$$

Therefore, for single mode operation in an SOI waveguide, at a wavelength of $1.3\,\mu m$, h needs to be less than approximately $0.2\,\mu m$. It is perhaps a little surprising to discover then, that rib waveguides are used routinely with silicon-layer thickness' of several microns, and furthermore, it is routinely claimed that these waveguides are single mode. The answer to this apparent paradox is that, if the geometry of the rib waveguide is correctly designed, higher order modes leak out of the waveguide over a very short distance, leaving only the fundamental mode propagating. This was demonstrated theoretically by Soref et al. [2], in an excellent research paper in 1991, who showed that for certain geometrical constraints, only the fundamental mode survives propagation of the order of 1–2 mm and hence the device behaves as if it is single mode, as all higher order modes leak away.

This resulted in a condition for the aspect ratio

$$\frac{a}{b} \leq 0.3 + \frac{r}{\sqrt{1 - r^2}}. \tag{7.55}$$

where a is the normalized waveguide width, b is the normalized waveguide height, and r is the normalized side slab height.

7.11.5 Refractive Index and Loss Coefficient in Optical Waveguides

So far we have regarded refractive index as a real quantity, but in general it is complex. Therefore, our references to refractive index so far have, strictly speaking, referred only to the real part of refractive index. Complex refractive index can be defined as

$$n' = n_R + jn_I. \tag{7.56}$$

We know that propagation constant can be defined as

$$k = nk_0. \tag{7.57}$$

We have previously expressed a propagating field as

$$E = E_0 \, e^{j(kz-\omega t)}. \tag{7.58}$$

Therefore, if we introduce complex refractive index to (7.61) and (7.63), we obtain

$$E = E_0 \, e^{j(k_0 n' z-\omega t)} = E_0 \cdot e^{jk_0 n_R z} \cdot e^{-k_0 n_I z} \cdot e^{-j\omega t}. \tag{7.59}$$

Therefore, there is a term $\exp(-k_0 n_I z)$. In the same way that the propagation constant in the z direction is redesignated β, this term is often redesignated $\exp(-\frac{1}{2}\alpha)$. The term α is called the loss coefficient. The factor of $1/2$ is included in the definition above because, by convention, α is an intensity loss coefficient. Therefore, we can write

$$I = I_0 \, e^{-\alpha z}. \tag{7.60}$$

Clearly (7.60) describes how the intensity decays with propagation distance z, through a material.

Having defined the loss coefficient, it is worth considering what sort of propagation loss we can tolerate for an integrated optical waveguide. While there is no absolute threshold that makes a particular material technology acceptable, the widely accepted benchmark for loss is of the order of $1\,\mathrm{dB\,cm^{-1}}$. This is because an integrated optical circuit is typically a few cm in length. Since an optical loss of $3\,\mathrm{dB}$ corresponds to a halving of the optical power, it is clear that a loss of much more than $1\,\mathrm{dB\,cm^{-1}}$ will rapidly result in a very poor signal to noise ratio at the detector. Add to this additional losses due to coupling to or from the optical circuit, or losses within the circuit not associated with propagation loss, and the situation is exacerbated. Losses for SOI waveguides are typically in the range 0.1–$0.5\,\mathrm{dB\,cm^{-1}}$.

7.11.6 The Contributions to Loss in an Optical Waveguide

Losses in an optical waveguide originate from three sources: scattering, absorption and radiation. Within each category are further subdivisions of the contribution to the loss, although the relative contribution of each of these effects is dependent upon the waveguide design, and the quality of the material in which the waveguide is fabricated. Each of the contributions to loss will be discussed in turn.

Scattering

Scattering in an optical waveguide can result from two sources: volume scattering and interface scattering. Volume scattering is caused by imperfections in the bulk waveguide material, such as voids, contaminant atoms or crystalline defects. Alternatively interface scattering is due to roughness at the interface between the core and the claddings of the waveguide. Usually, in a well-established waveguide technology, volume scattering is negligible, because the material has been improved to a sufficient level prior to fabrication of the waveguide. For new or experimental material systems, however, volume scattering should always be considered. Interface scattering, however, may not be negligible, even for a relatively well-developed material system, because losses can be significant, even for relatively smooth interfaces.

While it is now well established that silicon waveguides are low loss propagation media, one could reasonably be concerned that volume scattering could be a contributor to optical loss, since in several of the fabrication techniques used to produce SOI wafers, the potential exists for the introduction of defects, notably via ion implantation. It has been shown that the contribution to volume scattering is related to the number of defects, their size with respect to the wavelength of propagation, and the correlation length along the waveguide. In bulk media, Rayleigh scattering is the dominant loss mechanism (see for example [3]), which exhibits a λ^{-4} dependence. However, for confined waves the wavelength dependence is related to the axial correlation length of the defects [4]. For correlation lengths shorter than or of the order of the wavelength, the scattering loss exhibits a λ^{-3} dependence, because the reduction of confinement for longer wavelengths partially counters the λ^{-4} relation. For long correlation lengths compared to the wavelength, radiation losses dominate and a λ^{-1} dependence is observed.

Interface scattering has been studied by a number of authors, who have published a range of expressions for approximating the scattering from the surface or interface of an optical waveguide. However, since many of these models are very complex, an approximate technique will be introduced here, which is attractive due to its simplicity. The theory of this technique was first introduced by Tien in 1971 [5], and is based upon the specular reflection of power from a surface. This condition holds for long correlation lengths, which is a reasonable assumption in most cases. If the incident beam has power P_i, the specular reflected power, P_r, from a surface is given by [6]

$$P_r = P_i \exp\left[-\left(\frac{4\pi\sigma n_1}{\lambda_0}\cos\theta_1\right)^2\right], \tag{7.61}$$

where σ is the variance of the surface roughness (or rms roughness), θ_1 is the propagation angle within the waveguide, and n_1 is the refractive index of the core.

By considering the total power flow over a given distance, together with the loss at both waveguide interfaces based upon (7.61), Tien produced the following expression for the loss coefficient due to interface scattering:

$$\alpha_s = \frac{\cos^3 \theta}{2 \sin \theta} \left(\frac{4 \pi n_1 (\sigma_u^2 + \sigma_\ell^2)^{\frac{1}{2}}}{\lambda_0} \right)^2 \left(\frac{1}{h + \frac{1}{k_{yu}} + \frac{1}{k_{yl}}} \right), \qquad (7.62)$$

where σ_u is the rms roughness for the upper waveguide interface, σ_ℓ is the rms roughness for the lower waveguide interface, k_{yu} is the y-directed decay constant in the upper cladding, k_{yl} is the y-directed decay constant in the lower cladding, and h is the waveguide thickness.

The approximate degree of interface scattering can be evaluated from (7.62). This is demonstrated by the example in [1].

Example

Reed and Knights [1] used $n_1 = 3.5$, $n_2 = 1.5$, $n_3 = 1.0$, $h = 1.0\,\mu$m and the operating wavelength $\lambda_0 = 1.3\,\mu$m. They compared the scattering loss of two different modes of the waveguide, the TE$_0$ and the TE$_2$ modes, for both σ_u and $\sigma_\ell = 1$ nm, evaluating the scattering loss for each mode, for each nm of rms roughness at each interface. For the TE$_0$ mode

$$\alpha_s = \frac{\cos^3 \theta}{2 \sin \theta} \left(\frac{4 \pi n_1 (\sigma_u^2 + \sigma_\ell^2)^{\frac{1}{2}}}{\lambda_0} \right)^2 \left(\frac{1}{h + \frac{1}{k_{yu}} + \frac{1}{k_{yl}}} \right) = 0.04 \ \text{cm}^{-1}. \qquad (7.63)$$

Expressed in dB, this is equivalent to a loss of $0.18\,\text{dB cm}^{-1}$.

For the TE$_2$ mode

$$\alpha_s = \frac{\cos^3 \theta}{2 \sin \theta} \left(\frac{4 \pi n_1 (\sigma_u^2 + \sigma_\ell^2)^{\frac{1}{2}}}{\lambda_0} \right)^2 \left(\frac{1}{h + \frac{1}{k_{yu}} + \frac{1}{k_{yl}}} \right) = 1.33 \ \text{cm}^{-1}. \qquad (7.64)$$

Expressed in dB, this is equivalent to a loss of $5.79\,\text{dB cm}^{-1}$. It is clear from this example that the higher order modes will suffer more loss due to interface scattering than the fundamental mode. In this example, the TE$_2$ mode experiences more than 30 times more loss than the fundamental mode, resulting from interface scattering. This is due to both the differences in optical confinement, and to more reflections per unit length in the direction of propagation for the higher order modes.

Absorption

The two main potential sources of absorption loss for semiconductor waveguides are bandedge absorption (or interband absorption), and free carrier absorption. Interband absorption occurs when photons with energy greater than the bandgap, are absorbed to excite electrons from the valence band to the conduction band. Therefore to avoid interband absorption, a wavelength that is longer than the absorption edge wavelength of the waveguide material must be used. Silicon is an excellent example material to demonstrate this point. The bandedge wavelength of silicon is approximately $1.1\,\mu$m, above which silicon is used as a waveguide material. For wavelengths shorter than $1.1\,\mu$m, silicon absorbs very strongly, and is one of the most common materials used for photodetectors in the visible wavelength range, and at very short infrared wavelengths. Therefore, semiconductor waveguides should suffer negligible bandedge absorption, provided a suitable wavelength of operation is chosen.

Free carrier absorption, however, may be significant in semiconductor waveguides. The concentration of free carriers will affect both the real and imaginary refractive indices. For devices fabricated in silicon, modulation of the free carrier density is used to deliberately modulate refractive index, an effect that will be discussed in more detail later in this chapter when optical modulators are considered. Changes in absorption in semiconductors can be described by the well-known Drude–Lorenz equation [7]

$$\Delta\alpha = \frac{e^3\lambda_0^2}{4\pi^2 c^3 \varepsilon_0 n} \left(\frac{N_e}{\mu_e(m_{ce}^*)^2} + \frac{N_h}{\mu_h(m_{ch}^*)^2} \right), \qquad (7.65)$$

where e is the electronic charge, c is the velocity of light in vacuum, μ_e is the electron mobility, μ_h is the hole mobility, m_{ce}^* is the effective mass of electrons, m_{ch}^* is the effective mass of holes, N_e is the free electron concentration, N_h is the free hole concentration, ε_0 is the permittivity of free space and λ_0 is the free space wavelength.

Some of the parameters in (7.65) are interdependent, so care must be taken while evaluating the effect of free carrier absorption. However, in order to demonstrate the significance of free carrier absorption let us consider the additional absorption introduced by free carriers. Soref and Bennet evaluated (7.65) for values of N in the range 10^{18}–10^{20} cm^{-3}, and showed that an injected hole and electron concentration of 10^{18} cm^{-3} introduces a total additional loss of approximately 2.5 cm^{-1}. This corresponds to a loss in dB of 10.86 dB cm^{-1}, indicating the dramatic effect doping of the semiconductor can have on the loss of a waveguide.

Radiation

Radiation loss from a straight optical waveguide should ideally be negligible. This type of loss implies coupling from the waveguide into the surrounding media, typically the upper or lower cladding, or for a rib waveguide, into the planar region adjacent to the guide. If the waveguide is well designed this loss will not normally be significant, although unwanted perturbations in the waveguide, due to, for example, a slightly damaged fabrication mask, may cause scattering of light from one mode to another. The second mode, may in turn result in some radiative loss if that mode is leaky. Another situation that may result in some radiative loss, is curvature of the waveguide, as this will change the angle of incidence at the waveguide wall, which may in turn result in some radiative loss.

For a multilayer waveguide structure such as the SOI rib waveguide structure depicted in Fig. 7.11, the possibility of radiative loss exists if the lower waveguide cladding is finite. In the case of SOI, the buried oxide layer must be sufficiently thick to prevent optical modes from penetrating the oxide layer and coupling to the silicon substrate. Clearly, the required thickness will vary from mode to mode, as we have seen that each mode penetrates the cladding to a different depth. Furthermore, the penetration depth also varies with the waveguide dimensions with respect to the wavelength of operation. Hence it is important to ensure that a sufficiently thick buried oxide is utilized.

7.11.7 Coupling to the Optical Circuit

Coupling of light to an integrated optical circuit is conceptually trivial, but in practice is a non-trivial problem. This is because of the small size of optical waveguides micrometres in either cross-sectional dimension. A variety of techniques exist for performing the coupling task, the most common being end-fire coupling, butt coupling, prism coupling and grating coupling. End fire coupling and butt coupling are very similar, involving simply shining light onto the end of the waveguide. The distinction that is usually made between these two methods is that butt coupling involves simply "butting" the two devices or waveguides up to one another, such that the mode field of the "transmitting" device, falls onto the endface of the second device, whereas end-fire coupling incorporates a lens to focus the input beam onto the end-face of the "receiving" device. Therefore, light is introduced into the end of the waveguide, and can potentially excite all modes of the waveguide. Prism coupling and grating coupling, however, are distinctly different approaches, because they introduce an input beam through the surface of a waveguide, at a specific angle. This enables phase matching to a particular propagation constant within the waveguide, thereby enabling excitation of a specific mode. The principles of these four coupling techniques are shown schematically in Fig. 7.13.

For the purposes of semiconductor waveguide evaluation, prism coupling is not particularly useful, because the prism should typically have a higher refractive index that the waveguide, which seriously restricts the possibilities, particularly for silicon, that has a high refractive index of 3.5. The remaining three techniques can be useful for silicon-based technology, and will be discussed in turn.

Grating Couplers

Grating couplers are useful because they provide a means of coupling to individual modes, and because they are useful for coupling to waveguide layers of a wide range of thickness. However, because the input beam must be introduced at a specific angle, grating couplers are not sufficiently robust for commercial devices, but are a valuable development tool.

In order to couple light into a waveguide mode, as depicted in Fig. 7.13a or b, it is necessary for the components of the phase velocities in the direction

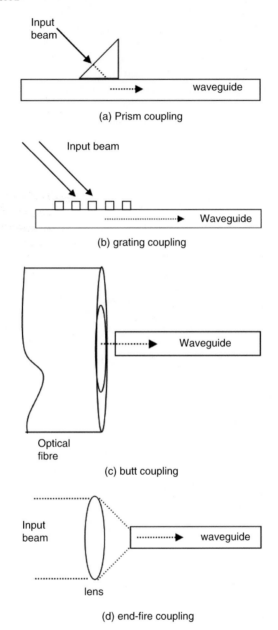

(a) Prism coupling

(b) grating coupling

(c) butt coupling

(d) end-fire coupling

Fig. 7.13. Four techniques for coupling light to optical waveguides

of propagation (z direction) to be the same. This is referred to as the phase-match condition, and in this case it means that the propagation constants in the z direction must be the same. Consider first a beam (or ray) incident upon the surface of the waveguide at an angle θ_a (Fig 7.14). The ray will

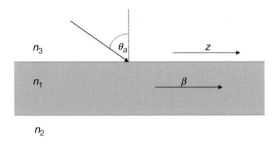

Fig. 7.14. Light incident upon the surface of a waveguide

propagate in medium n_3 with a propagation constant $k_0 n_3$, in the direction of propagation. Therefore, the z-directed propagation constant in medium n_3 will be

$$k_z = k_0 n_3 \sin \theta_a. \tag{7.66}$$

Therefore, the phase-match condition will be

$$\beta = k_z = k_0 n_3 \sin \theta_a, \tag{7.67}$$

where β is the waveguide propagation constant.

However, we know that $\beta \geq k_0 n_3$. Therefore the condition of (7.67) can never be met, since $\sin \theta_a$ will be less than unity. This is why a prism or a grating is required to couple light into the waveguide, because both can satisfy the phase-match condition if correctly designed.

A grating is a periodic structure. If it is to used as an input or output coupler, it is usual to fabricate the grating on the waveguide surface. The periodic nature of the grating causes a periodic modulation of the effective index of the waveguide. For an optical mode with propagation constant β_W when the grating is not present, the modulation results in a series of possible propagation constants, β_p given by

$$\beta_p = \beta_W + \frac{2p\pi}{\Lambda}, \tag{7.68}$$

where Λ is the period of the grating and $p = \pm 1, \pm 2, \pm 3, \ldots$

These modes are equivalent to the different diffraction orders of a diffraction grating. Clearly, the propagation constants corresponding to positive values of p cannot exist in the waveguide because the propagation constant β_p will still be less than $k_0 n_3$. Therefore, only the negative values of p can result in a phase match. It is usual to fabricate the grating such that only the value $p = -1$ results in a phase match with a waveguide mode. Therefore the waveguide propagation constant becomes

$$\beta_p = \beta_W - \frac{2\pi}{\Lambda}. \tag{7.69}$$

Therefore, the phase-match condition becomes

$$\beta_W - \frac{2\pi}{\Lambda} = k_0 n_3 \sin \theta_a \tag{7.70}$$

and

$$\Lambda = \frac{\lambda}{N - n_3 \sin \theta_a} \tag{7.71}$$

Butt Coupling and End-Fire Coupling

The parameters that affect the efficiency of butt coupling and end-fire coupling are similar, so these techniques will be discussed together. When a light beam in incident upon the endface of an optical waveguide, the efficiency with which the light is coupled into the waveguide is a function of (i) how well the fields of the excitation and the waveguide modes match; (ii) of the degree of reflection from the waveguide facet; (iii) the quality of the waveguide endface; (iv) and the spatial misalignment of the excitation and waveguide fields. There can also be a numerical aperture mismatch in which the input angles of the optical waveguide are not well matched to the range of excitation angles, but this latter term is neglected here.

Overlap of Excitation and Waveguide Fields

Matching of the excitation fields is usually evaluated by carrying out the overlap integral between the excitation field and the waveguide field. In order to evaluate this effect fully, all modes of the waveguide should be included in the waveguide field. However, in practice, many applications require only single mode operation, and consequently the overlap integral is evaluated between the input field and the fundamental mode of the waveguide. The overlap integral Γ of two fields E and ε, is given by

$$\Gamma = \frac{\int_{-\infty}^{\infty} dy \int_{-\infty}^{\infty} E \cdot \varepsilon \cdot dx}{\left[\int_{-\infty}^{\infty} dy \int_{-\infty}^{\infty} E^2 \, dx \cdot \int_{-\infty}^{\infty} dy \int_{-\infty}^{\infty} \varepsilon^2 dx \right]^{\frac{1}{2}}}, \tag{7.72}$$

where the denominator is simply a normalising factor. The factor Γ lies between 0 and 1, and therefore, represents the range between 0 coupling and 100% coupling due to field overlap.

Reflection from the Waveguide Facet

Reflection from the waveguide endface is determined by the refractive indices of the media involved in coupling from one medium to another, and is described by the Fresnel equations introduced earlier. The reflection coefficient r_{TE} for TE polarization is

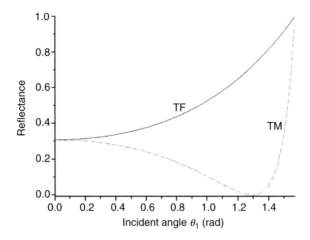

Fig. 7.15. Reflection at an air/silicon interface

$$r_{\mathrm{TE}} = \frac{-\sin(\theta_1 - \theta_2)}{\sin(\theta_1 + \theta_2)}. \tag{7.73}$$

Hence the reflectivity R is given by

$$R_{\mathrm{TE}} = r_{\mathrm{TE}}^2 = \frac{\sin^2(\theta_1 - \theta_2)}{\sin^2(\theta_1 + \theta_2)}. \tag{7.74}$$

Similarly, R_{TM} can be found to be

$$R_{\mathrm{TM}} = r_{\mathrm{TM}}^2 = \frac{\tan^2(\theta_1 - \theta_2)}{\tan^2(\theta_1 + \theta_2)}. \tag{7.75}$$

The two functions are plotted in Fig. 7.15. for an air/silicon interface (i.e. $n_1 = 1.0$, $n_2 = 3.5$). Note that at normal incidence ($\theta_1=0$), the reflection of both TE and TM polarizations is the same. Furthermore, end-fire coupling introduces light at near normal incidence. Consequently the approximation is usually made that the Fresnel reflection at the waveguide facets is due to normal incidence. In this case, (7.74) and (7.75) both reduce to

$$R = \left| \frac{n_1 - n_2}{n_1 + n_2} \right|^2. \tag{7.76}$$

For a silicon/air interface this reflection is approximately 31%, which introduces an additional loss of 1.6 dB. A loss of 1.6 dB for each facet of the waveguide is considerable, and is reduced in commercial devices by the use of anti-reflection coatings.

The Quality of the Waveguide Endface

The quality of the waveguide endface is very dependent upon the preparation technique. Three main options are available for endface preparation of semiconductor waveguides: cleaving, polishing and etching (see [1]).

7.11.8 Measurement of Propagation Loss in Integrated Optical Waveguides

Having considered the contributions to both loss and coupling efficiency, it is worth to consider briefly the common methods of measuring the loss of optical waveguides. Several techniques are available for waveguide loss measurement, and the method implemented will depend upon the information required from the experiment. For example, if the loss associated with particular modes of a waveguide is required, then clearly a coupling technique that is mode selective is required, such as prism coupling or grating coupling. Alternatively, if the total loss of the waveguide is required, end-fire coupling or butt coupling is attractive due to the simplicity with which either technique can be implemented.

Insertion Loss and Propagation Loss

When an optical waveguide is measured, the objective of the measurement must be made clear. There is often confusion between insertion loss and propagation loss. The insertion loss of a waveguide or device, is the total loss associated with introducing that element into a system, and therefore includes both the inherent loss of the waveguide itself, and the coupling losses associated with exciting the device. This measurement is typically associated with system design or specification, since the total loss of introducing the device clearly affects the system performance.

Alternatively, the propagation loss is the loss associated with propagation in the waveguide or device, excluding coupling losses. To optimize the performance of a given device during development or research, it is usually the propagation loss that is of interest, since the contributions to the loss due to waveguide design or material properties, are associated with the propagation loss. It is sometimes necessary to derive one of these parameters from the other.

There are three main experimental techniques associated with waveguide measurement. They are (i) the cut-back method, (ii) the Fabry–Perot resonance method and (iii) scattered light measurement. Each method are described.

The Cut-Back Method

The cut-back method is conceptually the simplest method of measuring an optical waveguide, and is associated with either end-fire coupling or butt coupling. A waveguide of length L_1 is excited by one of the coupling methods mentioned, and the output power from the waveguide, I_1, and the input power to the waveguide, I_0 are recorded. The waveguide is then shortened to another length, L_2, and the measurement repeated to determine I_2, while the input power I_0 is kept constant. The propagation loss of the length of waveguide

$(L_1 - L_2)$ is therefore related to the difference in the output power from each measurement. We can express this as

$$\frac{I_1}{I_2} = \exp(-\alpha(L_1 - L_2)), \tag{7.77}$$

i.e.

$$\alpha = \left(\frac{1}{L_1 - L_2}\right) \ln \left[\frac{I_2}{I_1}\right]. \tag{7.78}$$

Note that this expression has been produced only from two data points, and has assumed that the input coupling, the condition of the waveguide endfaces, and the input power all remain constant. The accuracy of the technique can be improved by taking multiple measurements and plotting a graph of the optical loss against waveguide length. For more details, see [1].

A variation of the cut-back method, although inherently less accurate is to carry out an insertion loss measurement for a single waveguide length, and calculate the coupling loss by evaluating the reflection from each waveguide endface, together with the mode field mismatch. Whilst this technique is usually a little less accurate than the cut-back method, it has the enormous advantage of being non-destructive.

The Fabry–Perot Resonance Method

An optical waveguide with polished endfaces (facets), is similar in structure to the cavity of a laser. Light propagates along the waveguide, and may be reflected at either facet. Therefore, the waveguide structure may be regarded as a resonant cavity with the waves undergoing multiple reflections as they pass along the waveguide and back. Such a cavity is called a Fabry–Perot cavity. The optical intensity transmitted through such a cavity, I_t, is related to the incident light intensity, I_0, by the well-known equation

$$\frac{I_t}{I_0} = \frac{(1 - R)^2 e^{-\alpha L}}{(1 - R e^{-\alpha L})^2 + 4R e^{-\alpha L} \sin^2\left(\frac{\phi}{2}\right)}. \tag{7.79}$$

where R is the facet reflectivity, L is the waveguide length, α is the loss coefficient, and ϕ is the phase difference between successive waves in the cavity. Equation (7.79) has a maximum value when $\phi = 0$ (or multiples of 2π), and a minimum value when $\phi = \pi$. i.e.

$$\frac{I_{\max}}{I_0} = \frac{(1 - R)^2 e^{-\alpha L}}{(1 - R e^{-\alpha L})^2}. \tag{7.80}$$

$$\frac{I_{\min}}{I_0} = \frac{(1 - R)^2 e^{-\alpha L}}{(1 - R e^{-\alpha L})^2 + 4R e^{-\alpha L}} = \frac{(1 - R)^2 e^{-\alpha L}}{(1 + R e^{-\alpha L})^2}. \tag{7.81}$$

Therefore, we can evaluate the ratio of the maximum to minimum intensity as, ζ

$$\zeta = \frac{I_{\max}}{I_{\min}} = \frac{(1 + R\,\mathrm{e}^{-\alpha L})^2}{(1 - R\,\mathrm{e}^{-\alpha L})^2}. \tag{7.82}$$

We can rearrange (7.82) as

$$\alpha = -\frac{1}{L} \ln\left[\frac{1}{R}\frac{\sqrt{\zeta}-1}{\sqrt{\zeta}+1}\right]. \tag{7.83}$$

Therefore, if we know the reflectivity, R, and if we can measure the ratio of the maximum intensity to minimum intensity, ζ, the loss coefficient can be evaluated.

The transfer function of the Fabry–Perot cavity is described by (7.79), and is periodic when the phase, ϕ, passes through multiples of 2π. Therefore if we can experimentally sweep the Fabry–Perot cavity through a few cycles of 2π, ζ can be measured. Such cycling can be achieved by varying the temperature of the waveguide and hence the length and refractive index, or alternatively, by varying the wavelength of the light source.

Scattered Light Measurement

The measurement of scattered light from the surface of a waveguide can also be used to determine the loss. The assumption underlying this method is that, the amount of light scattered is proportional to the propagating light. Therefore, if scattered light is measured as a function of waveguide length, the rate of decay of scattered light will mimic the rate of decay of light in the waveguide. Optical fibres can be used to collect scattered light from the surface of a waveguide, and can be scanned along the surface. Alternatively, an image of the entire surface can be made, and the decay of scattered light determined accordingly. However, it is clear that light is only scattered significantly if the loss of the waveguide is high, and relatively high power is propagating in the waveguide. In many situations, neither is desirable, and hence this method tends to be used for initial studies of waveguide materials, when losses are high, but is discarded in favour of other techniques when the waveguide has been optimized.

7.11.9 Optical Modulation Mechanisms in Silicon

One of the requirements of an integrated optical technology, particularly one related to communications, is the ability to perform optical modulation. This implies a change in the optical field due to some applied signal, typically, although not exclusively an electrical signal. The change in the optical field is usually derived from a change in refractive index of the material involved, with the applied field, although other parametric changes are possible. It is now widely accepted that the most efficient means of implementing optical modulation in silicon via an electrical signal, is to use carrier injection or depletion. This will be discussed later. First, however, let us consider other electrical modulation techniques to discover why they are not useful in silicon, even though they are used in other integrated optical technologies. The primary candidates for electrically derived modulation are the Pockels effect, the Kerr effect and the Franz–Keldysh effect. Soref and Bennett [8] have examined electric field effects in silicon, and have shown that none of these effects are efficient, and that the plasma dispersion effect or the thermo-optic effect are more useful (see [1]).

The Plasma Dispersion Effect

We previously noted that the concentration of free charges in silicon contributes to the loss via absorption. We also observed that the imaginary part of the refractive index is determined by the absorption (or loss) coefficient. Therefore, it is clear that changing the concentration of free charges can change the refractive index of the material. We saw that the Drude–Lorenz equation relating the concentration of electrons and holes to the absorption was

$$\Delta\alpha = \frac{e^3\lambda_0^2}{4\pi^2c^3\varepsilon_0 n}\left(\frac{N_e}{\mu_e(m_{ce}^*)^2} + \frac{N_h}{\mu_h(m_{ch}^*)^2}\right). \tag{7.84}$$

The corresponding equation relating the carrier concentrations, N, to change in refractive index, Δn, is

$$\Delta n = \frac{-e^2\lambda_0^2}{8\pi^2c^2\varepsilon_0 n}\left(\frac{N_e}{m_{ce}^*} + \frac{N_h}{m_{ch}^*}\right). \tag{7.85}$$

Soref and Bennett [8] studied results in the scientific literature to evaluate the change in refractive index, Δn, to experimentally produced absorption curves for a wide range of electron and hole densities, over a wide range of wavelengths. In particular they focused on the communications wavelengths of 1.3 and 1.55 μm. Interestingly their results were in good agreement with the classical Drude–Lorenz model, only for electrons. For holes they noted a $(\Delta N)^{0.8}$ dependence. They also quantified the changes that they had identified from the literature, for both changes in refractive index and in absorption [8]. They produced the following extremely useful expressions, which are now used almost universally to evaluate changes due to injection or depletion of carriers in silicon

At $\lambda_w = 1.55\,\mu m$

$$\Delta n = \Delta n_e + \Delta n_h = -[8.8 \times 10^{-22}\Delta N_e + 8.5 \times 10^{-18}(\Delta N_h)^{0.8}], \tag{7.86}$$

$$\Delta\alpha = \Delta\alpha_e + \Delta\alpha_h = 8.5 \times 10^{-18}\Delta N_e + 6.0 \times 10^{-18}\Delta N_h, \tag{7.87}$$

where Δn_e is the change in refractive index resulting from change in free electron carrier concentrations; Δn_h the change in refractive index resulting from change in free hole carrier concentrations; $\Delta\alpha_e$ is the change in absorption resulting from change in free electron carrier concentrations and $\Delta\alpha_h$ is the change in absorption resulting from change in free hole carrier concentrations.

The Thermo-Optic Effect

In addition to the electric field effects and injection of free carriers into silicon, one other modulation technique has proved viable for optical modulation devices in silicon. It is the thermo-optic effect, in which the refractive index of silicon is varied by applying heat to the material. The thermo-optic coefficient in silicon is [9]

$$\frac{dn}{dT} = 1.86 \times 10^{-4}\ \mathrm{K}^{-1}. \tag{7.88}$$

Therefore if the waveguide material can be raised in temperature by approximately 6°C in a controllable manner, a refractive index change of 1.1×10^{-3} results. There are of course issues about controlling the temperature rise to the locality of the waveguide, and of efficiency of the mechanism used to deliver the thermal energy. However, experimental results of [9] suggest that a 500 µm device length can deliver a phase shift of π radians for an applied power of 10 mW, if the waveguide is thermally isolated from the substrate. This corresponds to a thermal change of approximately 7°C, and hence a refractive index change of approximately 1.3×10^{-3} over the length of the device.

It is worth noting that the refractive index change is positive with applied thermal energy, whereas the injection of free carriers results in a decrease in refractive index. Therefore the two effects could compete in a poorly designed modulator.

7.12 Devices

7.12.1 The Mach Zehnder Interferometer

Interferometers are central to many optical circuits, but one of the most frequently used interferometers is the famous Mach Zehnder interferometer. The device is shown schematically in Fig. 7.16. The Mach Zehnder interferometer is a common device in optical circuits, being the basis of several other devices such as modulators, switches and filters. Let us briefly review the operation of the interferometer.

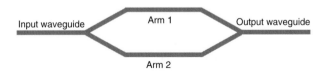

Fig. 7.16. Schematic of a waveguide Mach Zehnder interferometer

First consider an input wave, at the input waveguide, and let the wave be of TE polarization. Assuming the waveguide splitter (Y-junction) at the input of the interferometer divides the wave evenly, the intensity in arm 1 and arm 2 of the interferometer will be the same. We can represent the electric fields of the propagating modes in arm 1 and arm 2 of the interferometer as E_1 and E_2, respectively, where

$$E_1 = E_0 \sin(\omega t - \beta_1 z) \tag{7.89}$$

and

$$E_2 = E_0 \sin(\omega t - \beta_2 z) \tag{7.90}$$

For the moment, we have considered the two fields to have the same amplitude, but different propagation constants. The two fields propagate along their respective arms of the interferometer and recombine at the output waveguide. When the input Y-junction divides the input field, the two fields formed in arm 1 and arm 2 will be in phase. However, when the fields recombine, they may no longer be in phase, either due to different propagation constants in the arms, or due to different optical path lengths in the arms. The intensity at the output waveguide, S_T will be

$$S_T = [(E_1 + E_2) \times (H_1 + H_2)] = S_0(E_1 + E_2)^2. \tag{7.91}$$

Therefore we need to evaluate the term $(E_1 + E_2)^2$. Let us assume different path lengths for the two waves, of L_1 and L_2. Expanding and substituting for E_1 and E_2 gives

$$S_T = S_0 \left\{ E_0^2 \sin^2(\omega t - \beta_1 L_1) + E_0^2 \sin^2(\omega t - \beta_2 L_2) \right. \\ \left. + 2E_0^2 \sin(\omega t - \beta_1 L_1) \sin(\omega t - \beta_2 L_2) \right\} \tag{7.92}$$

Using trigonometric identities, (7.92) can be rewritten as

$$S_T = S_0 \left\{ E_0^2 \left(\frac{1}{2}[1 - \cos(2\omega t - 2\beta_1 L_1)] \right) + E_0^2 \left(\frac{1}{2}[1 - \cos(2\omega t - 2\beta_2 L_2)] \right) \right. \\ \left. + E_0^2[\cos(\beta_2 L_2 - \beta_1 L_1) - \cos(2\omega t - \beta_2 L_2 - \beta_1 L_1)] \right\}. \tag{7.93}$$

Since optical frequencies are very high, only the time average of these waves can be observed. Hence all terms in (7.93), must be replaced with their time average equivalent, yielding (7.94)

$$\begin{aligned} S_T &= S_0 \left\{ \frac{E_0^2}{2} + \frac{E_0^2}{2} + E_0^2[\cos(\beta_2 L_2 - \beta_1 L_1)] \right\} \\ &= S_0 \{ E_0^2[1 + \cos(\beta_2 L_2 - \beta_1 L_1)] \}. \end{aligned} \tag{7.94}$$

Equation (7.94) is the well-known transfer function of the interferometer, which is plotted in Fig. 7.17, normalized to a maximum amplitude of 1. The term $(\beta_2 L_2 - \beta_1 L_1)$, represents the phase difference between the waves from

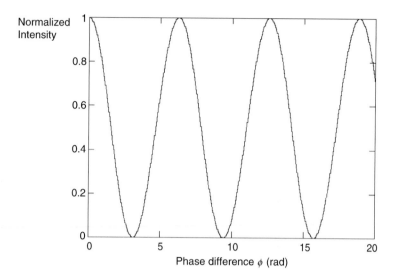

Fig. 7.17. The normalized transfer function of a Mach Zehnder interferometer

each arm of the interferometer. Clearly, if the two arms of the interferometer have identical waveguides, and hence identical propagation constants, the transfer function will have maxima when the path length difference, $|L_2 - L_1|$, results in a phase difference of a multiple of 2π radians. Similarly, the transfer function will have mimima when the phase difference is a multiple of π rad.

Thus, the intensity at the output of the Mach Zehnder interferometer can be manipulated to be a maximum or a minimum, by varying the relative phase of the two arms of the interferometer. This could be accomplished by inserting an optical phase modulator into one of the interferometer arms. In the case of the silicon modulator discussed in Sect. 7.11, the unwanted additional absorption due to free carriers will result in an imbalance in the intensities in each of the interferometer arms, and consequently, in imperfect interference in the output waveguide, and in particular in an imperfect "null" in the transfer function. If such an intensity modulator is to be used over less than 2π rad, one arm can be biased in terms of loss, to compensate.

7.12.2 The Waveguide Bend

The majority of the waveguides discussed so far in this chapter have been depicted as simple straight structures, uniform in the z direction (except for the waveguides forming the Mach Zehnder interferometer). However, to make the optical circuits practical we need to be able to send light to various parts of the circuit. This means that we must be able to form bends in the waveguide. Alternatively, one could imagine a series of straight waveguides joined together, but the abrupt junction of such waveguides would result in scattering centres, and hence losses would result at each abrupt change.

A curved waveguide allows a gradual transition from one direction to another that can have negligible loss.

At first sight, it may seem extravagant to regard a simple curve in a waveguide as a separate device, and hence perhaps it is unexpected to find bends covered in this chapter. However, like many other optical devices, the bend requires careful design in order that it is not lossy. For this reason it is included in this chapter, so that we can use the knowledge we have gained in earlier chapters to understand the design issues for the waveguide bend.

Consider Fig. 7.18, which shows an optical fibre bend top view. An illustration of the lateral optical field is also shown, including the evanescent fields that extend into the cladding. It is convenient to consider the optical fibre, as the symmetrical nature of the fibre simplifies the explanation of loss. We will consider a rib waveguide subsequently.

Consider the mode shown in Fig. 7.19, travelling around the fibre bend. Because the arc of the bend at the outside of the bend is longer than the arc at the inside of the bend, light at the outer cladding must propagate more quickly than light at the inner cladding, in order to maintain the phase relationship across the mode.

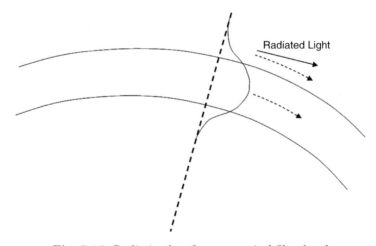

Fig. 7.18. Radiation loss from an optical fibre bend

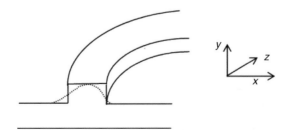

Fig. 7.19. Waveguides bend showing distorted lateral mode shape

As the evanescent tail extends into the outer cladding, eventually a distance is reached where light in the outer cladding will need to exceed the velocity of unguided light in the same material, in order to maintain the mode. This is of course, impossible, hence light is radiated and lost from the mode. A similar situation occurs in a waveguide mode, although the geometrical shape of the waveguide is more complex.

Marcatili and Miller [10], carried out an analysis of such a waveguide bend to determine the loss coefficient. While their analysis was not for a rib waveguide, the use of appropriately calculated propagation constants and decay constants inside and outside the rib gives reasonable results for a rib waveguide. They used an assumption that waveguide radiation from a bend was similar to emission of photons from an abruptly ended waveguide. This enabled them to determine a criteria for defining when light was lost from the guide, and hence define a distance from the side of the bend when light could be considered sufficiently far to no longer be part of the propagating mode. This is a consequence of having to consider the distortion of the mode travelling around a bend, as shown in Fig. 7.19, and to decide when light is sufficiently far from the waveguide to be considered lost. However, their analysis was based upon modes defined within the straight waveguide. Nonetheless, their result is widely used, partly due to the mathematical convenience. They showed that the loss coefficient from the bend was of the form

$$\alpha_{\text{bend}} = C_1 \exp(-C_2 R), \tag{7.95}$$

where R is the bend radius, and C_1 and C_2 are related to the waveguide and mode properties. The constants C_1 and C_2 are given by

$$C_1 = \frac{\lambda_0 \cos^2\left(k_{xg}\frac{w}{2}\right)\exp(k_{xs}w)}{w^2 k_{xs} n_{\text{effp}}\left[\frac{w}{2} + \frac{1}{2k_{xg}}\sin(wk_{xg}) + \frac{1}{k_{xs}}\cos^2\left(k_{xg}\frac{w}{2}\right)\right]} \tag{7.96}$$

and

$$C_2 = 2k_{xs}\left(\frac{\lambda_0\beta}{2\pi n_{\text{effp}}} - 1\right), \tag{7.97}$$

where β is the z-directed propagation constant, k_{xg} is the x-directed propagation constant in the waveguide, k_{xs} is the x-directed decay constant representing the evanescent field, w is the waveguide width, n_{effp} is the effective index outside the rib, and λ_0 is the free space wavelength.

Equation (7.95) shows that the loss coefficient is critically dependant upon the radius of curvature of the bend. Consequently the radius of curvature must be as large as possible to minimize loss. However, for most applications a small device footprint is desirable, implying that the radius should be as small as possible. Furthermore, the constants C_1 and C_2 are critically dependant upon the x-directed loss coefficient, k_{xs}. In both constants, reducing the value of k_{xs} reduces the loss. Hence this tells us that increasing the confinement of the waveguide will reduce the loss from the bend. An alternative way to look at this statement is that tightly confining waveguides can tolerate tighter bends for a given bend loss. Thus, the SOI technology is potentially a good candidate for small bend dimensions, as it is a highly confining technology due to the large refractive index of silicon. Of course, the degree of lateral confinement will vary with the etch depth of the rib waveguide, and hence careful design of the rib associated with a bend is very important. To demonstrate the losses involved, as predicted by this model, let us consider both small and large SOI waveguide structures. Consider the rib waveguide of Fig. 7.20.

First, let us consider a waveguide with the following parameters: $h = 5\,\mu m$, $r = 3\,\mu m$ and $w = 3.5\,\mu m$. For these conditions, and assuming TE polarization at a wavelength of $\lambda = 1.55\,\mu m$, we can evaluate the bend radius from (7.95), for a bend loss of $0.1\,dB\,cm^{-1}$. The resulting bend radius is approximately $7.25\,mm$, rather large. It is worth noting that this figure agrees reasonably well with the sort of figures achieved experimentally giving some confidence in the model. See for example, the results of Rickman and Reed [11] for large waveguides with bend radii of the order of 7–10 mm. However, it should also be remembered that this is a simple model of a waveguide bend, and hence is rather approximate. It is helpful to consider the variation of bend loss with radius of the bend. This is plotted in Fig. 7.21, for the waveguide defined earlier.

Before moving to a smaller waveguide, it is interesting to consider the effect of etching the waveguide a little more deeply, and hence better confining the mode. Let the conditions now be: $h = 5\,\mu m$, $r = 2.5\,\mu m$, and $w = 3.5\,\mu m$. Therefore the waveguide sidewalls have been etched by a further half of one micrometer. The radius of a waveguide with a loss of $0.1\,dB\,cm^{-1}$ now falls to only 3 mm, significantly smaller.

Let us now consider a much smaller waveguide. Let the waveguide parameters be: $h = 2\,\mu m$, $r = 1.2\,\mu m$ and $w = 1.4\,\mu m$. Proportionally, these are the same dimensions as for the large rib discussed earlier. However, the bend radius of a waveguide with a loss of $0.1\,dB\,cm^{-1}$ now falls to 0.73 mm. We can plot the variation in loss as a function of bend radius, as we did for a larger rib. This is shown in Fig. 7.22. In a similar fashion to the earlier example, let us now etch a little deeper into a small waveguide, so that the parameters

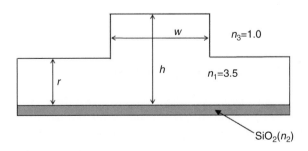

Fig. 7.20. Rib waveguide geometry

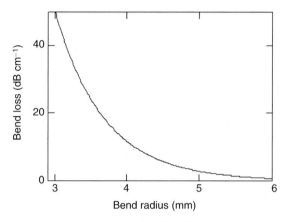

Fig. 7.21. Bend loss variation with bend radius for a large rib waveguide

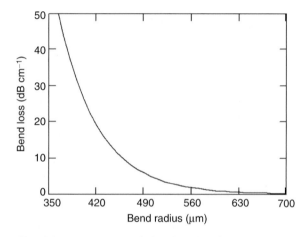

Fig. 7.22. Bend loss variation with bend radius for a small rib waveguide

become: $h = 2\,\mu m$, $r = 1.0\,\mu m$ and $w = 1.4\,\mu m$. In this case, the waveguide now can tolerate a bend radius of only $305\,\mu m$. Clearly, the smaller the waveguides become the smaller the bend radius can be, and hence save on valuable real-estate area on the silicon wafer. This is just one of the reasons that there is currently a trend towards smaller photonic circuit dimensions.

7.12.3 The Waveguide to Waveguide Coupler

In Sects. 7.2 and 7.1 of this chapter, we saw that power propagating along a waveguide, includes some power travelling outside the waveguide in the cladding. This was characterized in our planar waveguide by the penetration depth of the cladding. The field that extends beyond the waveguide is referred to as the evanescent field. We can use this field to couple light from one

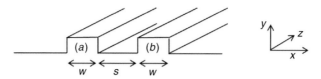

Fig. 7.23. Two waveguides separated by a small distance s

waveguide to another if the waveguides are sufficiently close that the evanescent fields overlap, in a device called a waveguide to waveguide coupler, or an evanescent coupler. Zappe [12] presented a useful simplified analysis of such a device, and we follow his approach here.

Consider the two identical waveguides, (a) and (b), shown in Fig. 7.23. The waveguides have width w, and are separated by a small distance s. Let the field in waveguide (a) be described by the equation

$$E_a = a_0(x, y) e^{j\beta z} e^{jwt}. \tag{7.98}$$

Similarly the field in waveguide (b) will be

$$E_b = b_0(x, y) e^{j\beta z} e^{jwt}. \tag{7.99}$$

The way the fields interact can be described by coupled mode theory (e.g. [13]). Simplified versions of the coupling equations that relate the amplitudes within each waveguide are

$$\frac{da_0}{dz} = \kappa b_0 \tag{7.100}$$

and

$$\frac{db_0}{dz} = -\kappa a_0, \tag{7.101}$$

where κ is a coupling coefficient. In this simplified equations we have assumed that the waveguides are identical and hence a phase match exists between the modes in each waveguide, no attenuation, and co-directional coupling. If we now assume that one of the waveguides is excited at $z = 0$, and the other is not, then we have

$$a_0(z = 0) = 0 \tag{7.102}$$

and

$$b_0(z = 0) = c_0. \tag{7.103}$$

The solutions of these equations take the form

$$a_0(z) = c_0 \sin(\kappa z) \tag{7.104}$$

and

$$b_0(z) = c_0 \cos(\kappa z). \tag{7.105}$$

Therefore, it is clear that a field in one waveguide gives rise to a field in the other, over some propagation distance z, and further that the transfer of power from one guide to the other is periodic, with a period given by the coupling length, referred to as L_π. We can see that complete transfer of power from one guide to the other takes place when $L_\pi = m\pi/2\kappa$, for integer values of m. Therefore, we can also imagine coupling lengths that are a specific fraction of the coupling length to couple a predetermined proportion of the power from one waveguide to the other. A particularly common example is the coupler with coupling length $L_c = L_\pi/2 = m\pi/4\kappa$, which couples half of the energy from one waveguide to the other. This structure is known as the 3 dB coupler for obvious reasons.

The fact that power transfers from one guide to another in a periodic manner means that over multiple coupling length the power will transfer back and forth between the waveguides. This means that if a single coupling length is very short, it may be more convenient to make devices with multiple coupling lengths, although care must be taken to ensure that fabrication tolerances are not exacerbated in this case.

The coupling efficiency is determined by the coupling coefficient, κ, which can be expressed as [14]

$$\kappa = \frac{2k_{xc}^2 k_{xs} \exp(-k_{xs}s)}{\beta w(k_{xs}^2 + k_{xc}^2)}, \qquad (7.106)$$

where k_{xc} is the x-directed propagation constant in the core, k_{xs} is the x-directed decay constant (between the waveguides), w is the waveguide width and s is the waveguide separation. The exponential term in (7.106) is clearly a strong influence on the value of κ. This shows that the waveguide spacing and decay constant are very important parameters to the coupler. Clearly this is expected, as the degree to which the evanescent fields overlap is dependant upon these parameters.

Equation (7.106) shows that the coupling coefficient is not only a function of modal confinement, but also of the propagation constant. This implies that coupling efficiency will vary with wavelength, and hence by cascading a series of couplers of slightly different design, it is possible to achieve a specific wavelength dependence.

Further details can be found in [1].

References

1. G.T. Reed, A.P. Knights, Silicon Photonics: An Introduction, Wiley, UK, January 2004, ISBN 0-470-87034-6
2. R.A. Soref, J. Schmidtchen, K. Petermann, Large single mode rib waveguides in GeSi–Si and Si-on-SiO$_2$, Journal of Quantum Electronics, *27*, 1971–1974, 1991
3. J.M. Senior, Optical Fiber Communications, Principles and Practice, 2nd Edition, Prentice-Hall, London, UK, 1992, ISBN 0-13-635426-2
4. J. Nayyer, Y. Suematsu, H. Tokiwa, Mode coupling and radiation loss of clad type optical waveguides due to the index inhomogeneities of the core material, Optical and Quantum Electronics, *7*, 481–492, 1975

5. P.K. Tien, Light waves in thin films and integrated optics, Applied Optics, *10*, 2395–2413, 1971

6. R.G. Hunsperger, Integrated Optics: Theory and Technology, 3rd Edition, Springer, Berlin Heidelberg New York, 1991, ISBN 0-387-53305-2

7. R.A. Soref, J.P. Lorenzo, All-silicon active and passive guided wave components for $\lambda = 1.3$ and $1.6\,\mu$m', IEEE Journal of Quantum Electronics, *QE-22*, 873–879, 1986

8. R.A. Soref, B.R. Bennett, Electrooptical Effects in Silicon, IEEE Journal of Quantum Electronics, *QE-23*, 123–129, 1987

9. S.A. Clark, B. Culshaw, E.J.C. Dawney, I.E. Day, Thermo-optic phase modulators in SIMOX material, Proceedings of the SPIE, *3936*, 16–24, 2000

10. E.A.J. Marcatili, S.E. Miller, Improved relationships describing directional control in electromagnetic wave guidance, Bell System Technical Journal, *48*, 2161–2188, 1969

11. A.G. Rickman, G.T. Reed, Silicon on insulator optical rib waveguides: loss, mode characteristics, bends and y-junctions, IEE Proceedings of the Optoelectronics, *141*, 391–393, 1994

12. H.A. Zappe, Introduction to Semiconductor Integrated Optics (Chapter 8), Artech House, 1995, ISBN 0-89006-789-9

13. R.G. Hunsperger, Integrated Optics: Theory and Technology, 3rd Edition, Springer, Berlin Heidelberg New York, 1991, ISBN 0-387-53305-2

14. S. Somekh, E. Garmire, A. Yariv, H.L Garvin, R.G Hunsperger, Channel optical waveguide directional couplers, Applied Physics Letters, *22*, 46–47, 1973

Submicron Silicon Strip Waveguides

D. Van Thourhout, W. Bogaerts, and P. Dunon

Summary. Submicron silicon strip waveguides have recently attracted a lot of attention and are currently studied by several research groups. The use of submicron silicon strip waveguides allows realizing extremely compact photonic circuits, using standard CMOS-processing methods. Therefore, they are a promising candidate for fabricating future photonic ICs using a cost-effective and high-yield process. We review the state-of-the-art of submicron silicon strip waveguides, starting from its basic properties including the single mode condition, scattering losses, bend losses, and polarization behavior. Subsequently we review the fabrication methods employed for realizing these circuits and discuss the compatibility with advanced CMOS-fabrication methods. Then basic waveguide structures such as splitters and fiber couplers are studied. Finally some examples of more complex devices, including cascaded Mach-Zehnder interferometers and arrayed waveguide gratings are shown.

8.1 Introduction

While state-of-the-art electronic circuits have unit dimensions between a few tens of nanometers to a few micrometers for their basic functional blocks, the dimensions of photonic devices are typically much larger, varying from a few micrometer (e.g., for the core of a single mode fiber) to several centimeter (e.g., for arrayed waveguide grating routers). This huge size difference makes it not interesting in most cases to integrate electronics and photonics on a single substrate. To make optoelectronic integration more attractive there is a need for considerably more compact passive optical interconnect structures (waveguides, bends, splitters, combiners, etc.), more compact wavelength selective devices (high Q-resonators, highly dispersive elements), and more compact nonlinear functions. Realizing this requires the use of ultra-high optical index contrast waveguides. In the past, so-called super-high-Δ planar lightwave circuits (PLC) have been presented, with a 1.5% index contrast between the silica cladding and the germanium-doped waveguide core [1]. However, the minimum acceptable bending radius in this material system is still 2 mm and therefore, incompatible with tight optoelectronic integration. III–V based waveguides allow a bending radius from 500 μm to a few tens of μm and very compact devices have been presented for so-called "deeply etched" waveguides, which were defined by etching completely through the high-index guiding layer and therefore show a very high lateral index contrast. However, when using a very small bending radius, light leaks out of the bend through the substrate because of the low vertical index contrast. To avoid this, also the vertical-index contrast has to be increased. Such a high-index contrast has been demonstrated in InP-based membrane type devices, GaAs/AlO$_x$ based devices, and SOI-based devices, all of which consist of a thin high refractive index semiconductor core layer between low index cladding layers (dielectric or air) and allow for the realization of ultracompact waveguide elements.

An additional problem of photonic devices, however, is the fact that they, for several reasons, often suffer from a low yield compared to their electronic counterparts [2], which makes optoelectronic integration not very appealing. Therefore, it is important to adopt, as much as possible, fabrication methods and tools used in the silicon-based electronics industry. From this point of view, SOI-based devices show the most potential for integration with electronic devices and therefore form one of the most promising solutions for dense on-chip optical interconnects. These waveguides are typically fabricated starting from an SOI wafer, consisting of a silicon substrate, an SiO$_2$ box layer (thickness t_{box}), and a silicon guiding layer (thickness h). The waveguides are formed in this guiding layer. In some cases a SiO$_2$ cladding layer is deposited following the etch process. Figure 8.1 shows the basic waveguide structure. The most relevant parameters are the waveguide width w and the height h. Due to the extremely high-index contrast, these will typically be limited to a few hundreds of nanometer for single mode waveguides.

Table 8.1 gives a review of publications discussing submicron silicon strip waveguides, showing a strongly enhanced activity in this domain during the last few years. Two main categories of waveguide structures can be distinguished. Some groups are aiming to reach polarization independence by choosing the waveguide width equal to its height [14, 19, 23, 26]. Most groups, however, opt for an asymmetric guide with a core height varying between 200

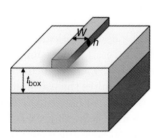 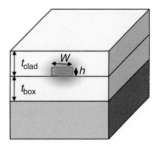

Fig. 8.1. Basic Silicon strip waveguide structure with air top cladding (*left*) and silica top cladding (*right*)

Table 8.1. Publication overview submicron SOI wires

aff.	date	h (nm)	w (nm)	loss (db cm^{-1})	box (um)	top clad	fab.	ref.
IMEC	Apr 2004	220	500	2.4	1	no	DUV	[3–7]
IBM	Apr 2004	220	445	3.6	2	no	EBeam	[8,9]
Cornell	Aug 2003	270	470	5.0	3	no	EBeam	[10,11]
NTT	Feb 2005	300	300	7.8	3	yes	EBeam	[12–14]
		200	400	2.8				
Yokohama	Dec 2002	320	400	105.0	1	no	EBeam	[15–19]
MIT	Dec 2001	200	500	32.0	1	yes	G-line	[20,21]
		50	200	0.8			+oxidation	
LETI / LPM	Apr 2005	300	300	15.0	1	yes	DUV	[22,23]
		200	500	5.0				
Columbia	Oct 2003	260	600	110.0	1	yes	EBeam	[24,25]
NEC	Oct 2004	300	300	19	1	yes	EBeam	[26]

and 340 nm and a width varying from 400 to 600 nm. The latter is in most cases determined by the single mode condition (see later). Reported losses for single mode waveguides vary from $+100\,\text{dB cm}^{-1}$ to less than $3\,\text{dB cm}^{-1}$. In [20] a $0.8\,\text{dB cm}^{-1}$ loss was reported. However, the dimensions of this waveguide were too small to strongly confine the light and this wire can no longer be considered as a high contrast waveguide.

In Sect. 8.2 we will investigate the basic properties of submicron silicon strip waveguides. Unless otherwise noted all calculations were performed using $h = 220\,\text{nm}$ and $w = 500\,\text{nm}$ at a wavelength of 1550 nm and for TE-polarization. A full-vectorial mode-solver based on the film-matching method was used.

8.2 Basic Properties

8.2.1 Refractive Index, Group Index, Single Mode Condition

The refractive indices of the silicon core layer and the SiO_2 cladding at a wavelength of 1550 nm are, respectively, given by 3.45 and 1.46. Figure 8.2 shows the effective index for the lowest order TE-like and TM-like modes for a waveguide with a 220 nm high core as function of the waveguide width, respectively, without and with a SiO_2 overcladding. For the air cladded waveguide we see an anticrossing between the dispersion curves for the first-order TE-like mode and the zeroth-order TM-like mode at $w = 680\,\text{nm}$ and at this point the modes have a hybrid character. For a width $w = 600\,\text{nm}$ the first-order TE-like mode is cutoff and for smaller widths the waveguides are single mode. Adding a SiO_2 top-cladding slightly decreases the refractive index contrast. The single mode width is decreased to 480 nm.

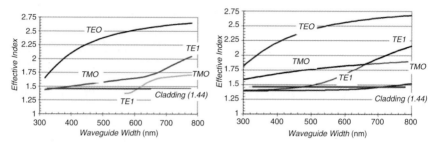

Fig. 8.2. Effective index of zeroth- and first-order mode for a waveguide with a 220 nm high core as a function of the waveguide width, respectively, without (*left*) and with (*right*) an SiO$_2$ overcladding

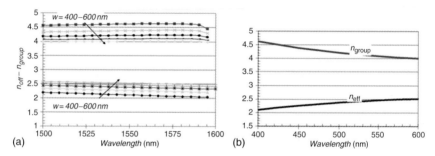

Fig. 8.3. Effective index and group index as a function of wavelength (**a**) and as a function of waveguide width (**b**)

The wavelength dependence of the silicon refractive index can be found in [27]. Since we are operating relatively far away from the bandedge the material dispersion is limited and negligible compared to the waveguide dispersion. Figure 8.3a shows the effective index and the group index ($n_{\mathrm{g}} = n_{\mathrm{eff}} - \lambda dn_{\mathrm{eff}}/d\lambda$) as a function of wavelength. Because of the high waveguide dispersion, the group index is considerably higher than the effective index, which may be relevant when determining the latency of an optical link. Also important filter characteristics, such as the free spectral range of ring resonators and arrayed waveguide gratings (AWGs) are determined by the group index. Figure 8.3b gives the variation of the group index as a function of the waveguide width and shows an increase in the group index for smaller waveguides.

8.2.2 Losses

Published losses for single mode waveguides vary from a few db cm^{-1} to several hundreds of db cm^{-1} (see Table 8.1). The most relevant loss factors are fundamental material absorption (e.g., due to free carrier absorption), substrate leakage (see Sect. 8.2.3), and scattering at interface roughness. The first two factors can be reduced to negligible values by, respectively, choosing sufficiently high resistivity wafers and a sufficiently high box layer thickness (see later). Reducing scattering losses requires careful process optimization to reduce sidewall roughness. Several authors have developed models to evaluate the influence of roughness on the propagation loss. Most authors agree that the scattering loss increases with index contrast and field strength but sources disagree on the exact relationship. Using the simple model proposed by Tien [28], the propagation loss is calculated as

$$\alpha_s \propto \frac{\sigma^2 E_s^2}{\int E^2 dx} \Delta n^2. \tag{8.1}$$

With σ the interface roughness, E_s the electric field at the interface, and Δn the refractive index contrast. More advanced models also include the coherence length L_c of the roughness [29]. In [30] the loss due to scattering was calculated for square silicon wire waveguides as a function of their dimensions and for several values of roughness amplitude and correlation length. The calculated losses agree well with the values reported in literature. An example is given in Fig. 8.4. Due to the increasing intensity of the light at the sidewall interface, the losses are increasing considerably with decreasing waveguide width. However, at a certain critical width ($\sim$260 nm in Fig. 8.4), the light is no longer strongly confined in the waveguide core and the optical mode width starts increasing again, leading to a reduced loss. However, the reduced confinement also will result in an increased minimal acceptable bend radius and increased substrate leakage losses (see later).

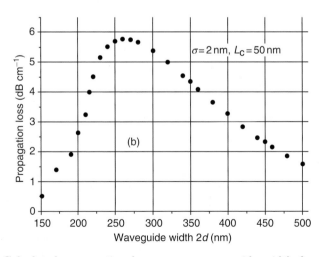

Fig. 8.4. Calculated propagation losses versus waveguide width for $\sigma = 2$ nm, $L_c = 5$ nm (from [30])

In literature, values varying from $\sigma = 11\,\text{nm}$, $L_c = 70\,\text{nm}$ [19] to $\sigma = 2\,\text{nm}$, $L_c = 50\,\text{nm}$ [9] have been reported for the roughness standard deviation and coherence length, resulting in a corresponding reduction in losses from over $100\,\text{db}\,\text{cm}^{-1}$ to below $4\,\text{db}\,\text{cm}^{-1}$ for single mode waveguides. Figure 8.5 shows the measured waveguide loss as function of waveguide width for waveguides fabricated in our facilities (crosses, [3]), demonstrating the expected increase in loss for smaller width waveguides. The data point indicated by the black square is taken from [9], the data point indicated by the black diamond is taken from [14].

Figure 8.6 shows the loss for a waveguide with height $h = 220\,\text{nm}$, width $w = 450\,\text{nm}$ and a box layer thickness of $2\,\mu\text{m}$, as function of the wavelength (taken from [9]). The minimal loss is observed around 1450–1500 nm. For higher and lower wavelength values, the loss increases considerably. According to the authors of [9], for higher wavelengths this may be explained by the increase in mode field diameter and the corresponding increase in interaction with the sidewall roughness and substrate leakage. For smaller wavelengths, the roughness amplitude increases relatively with respect to the wavelength, also leading to higher losses. Similar results were presented in [23]. In [14], the response over a similar wavelength range was almost completely flat, however. Figure 8.6 also shows the strong polarization dependence of the loss. Based on a dipole scattering mode, Bogaerts [6] found that for typical etching induced sidewall roughness consisting of vertical striations TM-polarized light is scattered much more efficiently than TM-polarized light and therefore the TM-losses should be higher. At 1550 nm, this was also experimentally observed [4, 9]. However, Vlasov [9] noted that, due to

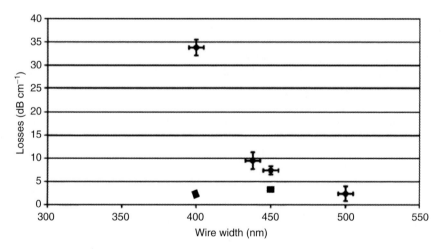

Fig. 8.5. Measured waveguide loss as function of width as reported by Bogaerts [6]. The square and the diamond denotes the loss reported by, respectively, Vlasov [9] and Tsuchizawa [14]

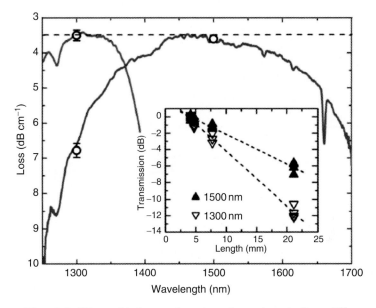

Fig. 8.6. Waveguide loss as function of wavelength (from [9])

the boundary condition imposed on discontinuity in the electric field the TE-polarized mode has a higher field strength at the sidewall and therefore should have a higher loss.

As explained earlier, process optimization has led to a propagation loss reduction from over $100\,\mathrm{db\,cm^{-1}}$ to $\sim 2\,\mathrm{db\,cm^{-1}}$ for single mode silicon strip waveguides. Despite this spectacular improvement, such losses are still much higher than those reported for silica-based waveguides [1], where, depending on the refractive index contrast, losses from $0.01\,\mathrm{db\,cm^{-1}}$ (minimum bend radius $25\,\mathrm{mm}$) to $0.07\,\mathrm{db\,cm^{-1}}$ (minimum bend radius $2\,\mathrm{mm}$) are reported. However, while the dimensions of typical PLC-based devices such as AWGs are in the order of centimeters, dimensions of silicon wire based devices have dimensions below $100\,\mu\mathrm{m}$ (see next sections), leading to total propagation losses per device that are comparable for both optical waveguide platforms.

8.2.3 Box Layer Thickness

Another important parameter is the thickness t_{box} of the SiO_2 box layer. As can be seen in Table 8.1, values between 1 and $3\,\mu\mathrm{m}$ are typically employed. Making t_{box} thicker is technologically more challenging but smaller values may cause leakage to the substrate. Figure 8.7 shows the calculated substrate leakage as a function of the oxide buffer thickness for different waveguide

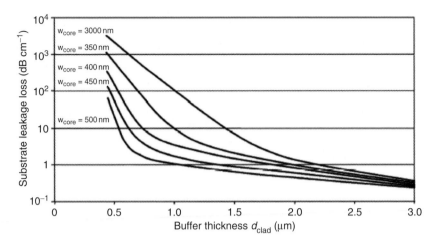

Fig. 8.7. Calculated substrate leakage (from [6])

widths [6]. For reducing the substrate leakage loss of single mode waveguides below $1\,db\,cm^{-1}$, a buffer thickness of at least $1\,\mu m$ has to be chosen. Also note that the substrate leakage loss increases rapidly with decreasing waveguide width. For TM-polarization, the loss is higher because the field is less confined in the waveguide core.

8.2.4 Polarization

Figure 8.2 shows that waveguides with a rectangular cross-section exhibit very different modal properties for TE- and TM-polarized light. As mentioned in Sect. 8.2.3, also the loss is very polarization sensitive (e.g. see Fig. 8.6). To overcome this problem, several authors have proposed using waveguides with a square cross-section (typically $300\,nm \times 300\,nm$) [14, 19, 23, 26], which results in equal properties for TE-like and TM-like modes. Since the sidewall roughness is typically present at the vertical interfaces, the polarization dependent loss will remain important, however. An additional problem to be resolved is the polarization crosstalk in bends with a small radius ($<5\,\mu m$). This issue was discussed in detail in [19]. Using FDTD-simulations the maximum polarization crosstalk induced in a $1\,\mu m$ radius bend was calculated to be $-25\,dB$, which is acceptable for practical implementations. The actual measured polarization crosstalk was much larger, however, up to $-10\,dB$. The authors explained this by a $\sim5°$ tilt of the sidewall of the fabricated waveguide. Also the scattering loss in the bend may play a role. For on-chip optical interconnect, the polarization issue is less important. For other applications polarization diversity approaches can be used [41] (see the section "Grating Couplers").

8.2.5 Temperature Dependence

Since the temperature over a CMOS chip may vary considerably, the influence of operating temperature on the waveguide properties has to be investigated, in particular when optical filters and resonators are used. The change of the refractive index of silicon is around $1.79 \times 10^{-4}\,\mathrm{K^{-1}}$ at 1550 nm [32]. The refractive index change of SiO_2 depends on the fabrication method but is an order of magnitude smaller than that of silicon. From these values the refractive index change of the propagation index and the related shift in filter peak wavelength value can easily be calculated. For most classical filter structures such as Mach-Zehnder interferometers (MZI), ring resonators, and phased-arrays the wavelength shift depends in the same way on the change in refractive index and is given by $d\lambda/\lambda = dn_g/n_g$. Based on the values given above, we calculated a wavelength shift of $140\,\mathrm{pm\,K^{-1}}$. Measured temperature dependence of the peak wavelength for ring resonators and cascaded MZI-structures shows a good linearity but a reduced shift of $80\,\mathrm{pm\,K^{-1}}$ and $89\,\mathrm{pm\,K^{-1}}$, respectively (Fig. 8.8). We believe the discrepancy between calculations and measured results may be caused by stress in the SOI-structure.

The strong temperature dependence may be a problem for stabilizing wavelength dependent structures but can also be used for making compact switches [25]. This is discussed later in this chapter.

8.2.6 Bend Radius

Obviously, one of the most attractive features offered by silicon wire waveguides is the possibility for realizing an extremely short bending radius. Both theoretical (FDTD) and experimental results have been reported by several authors. The lowest experimental losses were reported by Vlasov [9],

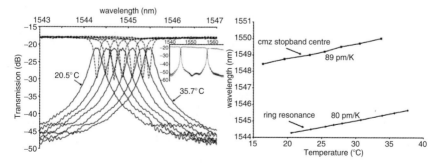

Fig. 8.8. Laser-to-detector transmission spectra to pass and drop ports of a racetrack resonator with 5 μm radius while temperature was increased from 20.5 to 35.7°C (*left*). Measured temperature dependence of passband wavelength for ring resonator filter and cascaded MZI (*right*)

who reports losses of 0.086, 0.013, and 0 dB for a 90° turn with bend radius of, respectively, 1, 2, and 5 µm (uncertainty +/− 0.005 dB per turn) for waveguides with a rectangular cross-section, using TE-polarized light at a wavelength of 1500 nm. For TM-polarization, the losses at a wavelength of 1500 nm increase. In [14], 0.15 and 0.6 dB per turn were measured for waveguides with, respectively, a rectangular and a square cross-section and a 2 µm bend radius. Also in this case the loss for 5 µm bends is negligible. Similar results were found in [23]. FDTD-simulations in [19] show very good agreement with the experimental results of [9]. The authors of [19] also investigated the influence of the bend pattern on the polarization crosstalk and the losses and mention an increased crosstalk for S-shaped bends compared to U-shaped bends.

Several authors [24, 33, 34] have proposed and/or demonstrated corner-mirror and resonator-based bend structures with the goal of further decreasing the bend loss and size. However, in view of the excellent results reported earlier for standard bend structures and the increased sensitivity to wavelength, polarization, and side wall angle for the resonant structures, it is very unlikely that they will have any practical use.

8.2.7 Waveguide Pitch

To increase the total data density, the minimum waveguide pitch is an important parameter. The minimum waveguide pitch is defined here as the minimum center-to-center spacing that is allowed without causing crosstalk between the channels and is dependent on the distance over which the channels are running alongside each other and the acceptable crosstalk level.[1] Figure 8.9 shows the minimum allowable center-to-center spacing as function of waveguide height and width (for −20 db cm^{-1} crosstalk level). For wider waveguides the minimum pitch obviously increases. Since for smaller widths the mode gets less confined in the core, however, also for decreasing width the minimum pitch increases and as a consequence there is an optimum for the waveguide width. Due to the higher confinement, waveguides with an increased core layer height h allow for a smaller pitch.

The minimum acceptable waveguide pitch also imposes an upper boundary for the data density per square centimeter. Further increasing the data density would require further increasing the refractive index contrast. In principle this can be done by decreasing the cladding refractive index (using air-cladding or low-k dielectrics) or increasing the core refractive index. At this moment, none of these seems viable, however. Alternatively, new waveguide types such as plasmon waveguides [35] may be considered. In principle, such waveguides can confine the electromagnetic radiation in a smaller volume. However, they intrinsically show extremely high propagation losses. Note that photonic crystal

---

[1] The acceptable crosstalk level will depend on the application and the difference in power levels between different channels

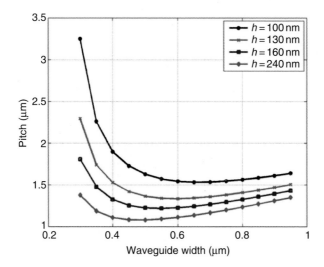

Fig. 8.9. Minimum allowable center-to-center spacing for a maximum crosstalk of $20\,\mathrm{db\,cm}^{-1}$

waveguides cannot bring a solution here since, although the optical mode itself may be more confined, they typically require several rows of holes to reach this effect.

8.3 Fabrication

8.3.1 Introduction

Table 8.2 gives a review of the fabrication methods used by several groups. Most research groups are using Ebeam lithography for defining the silicon stripe waveguides. Ebeam lithography allows for very high accuracy and is an excellent research tool. For obtaining low loss waveguides and avoiding roughness in the resist pattern, a stringent optimization of the writing process (spot size, overlap) and the resist process is required [12, 14, 16].

For large scale applications, however, such as on-chip optical interconnect, a mask-based lithography approach has to be developed. In [4] and [23] deep UV lithography is used (248 nm illumination wavelength). In [20] a G-line stepper (436 nm illumination wavelength) is used, which is in principle sufficient for defining 500 nm lines but will not be useful when finer features such as small gaps in directional couplers or star couplers have to be defined. Also mixed approaches combining mask based lithography for the wider lines and Ebeam lithography for the finer features have been demonstrated [22]. Sect. 8.3.2 will describe in detail the DUV-based fabrication process.

For transferring the resist pattern in the silicon core layer, different etching processes have been employed. In most cases a SiO_2-mask or a metal hard

Table 8.2. Fabrication approaches

aff.	loss (db cm^{-1})	box (um)	top clad	fab.	mask	etch method
IMEC	2.4	1	no	DUV	Resist	Cl$_2$/He/Hbr/O$_2$ (ICP)
IBM	3.6	2	no	EBeam	SiO$_2$	CF$_4$/CHF3/Ar (oxide) + HBr (silicon)
Cornell	5.0	3	no	EBeam		ICP
NTT	6.0	3	no	EBeam	SiO$_2$	SF$_6$/CF$_4$ etch, ECR-etch
Yokohama	105.0	1	no	EBeam	metal	ICP (CF$_4$ + Xe)
MIT	32.0 0.8		yes	G-line	SiO$_2$ oxidation	RIE (SF$_6$)
LETI	15.0 5.0	1	yes	DUV mixed	SiO$_2$	HBr etching
Columbia	110.0	1	yes	EBeam	Aluminium	RIE (CF$_4$:Ar)
Ebeam	19.0	1	yes	EBeam	SiO$_2$	ICP

mask is used. In [4] a resist mask is used, however. For etching the silicon an ICP-based or ECR-based etching procedure is preferred for making smooth sidewalls with little damage. In Sect. 8.3.2, the DUV process developed at IMEC is described in more detail.

8.3.2 CMOS-Compatible Deep-UV Lithography Based Fabrication Process

Introduction

Using CMOS-technology for nanophotonic circuits brings the potential of low cost, volume manufacturing, and solving the yield problems suffered now by photonic devices. Therefore, we investigated the possibilities and limitations of using deep UV lithography and CMOS etching processes for fabricating nanophotonic devices.

The process flow is very similar to that of conventional projection lithography. We used commercial SOI wafers from SOITEC fabricated using the UNIBOND process [37]. The SOI wafers have a 220 nm top silicon layer and a 1 µm silica box layer. In a first step the wafer is coated with photoresist and an antireflective coating which is used to avoid standing wave patterns in the photoresist. Subsequently, a 248 or 193 nm stepper is used for illumination of the resist. The pattern is typically repeated several times over the wafer and different exposure conditions can be used for the different dies allowing to make a detailed process characterization. In a next step, the resist pattern is postbaked and developed. An additional plasma treatment is used for resist hardening before etching. The photoresist is used directly as an etch mask for the silicon etch. If needed thermal oxidation or oxide deposition processes can be added to the process flow.

Deep UV Lithography

For our experiments an ASML PAS5500/750 stepper with 248 nm illumination wavelength connected to an automated track for preprocessing and postprocessing was used. The resolution of optical projection lithography is mainly determined by the illumination wavelength and the numerical aperture (NA) of the projection system and the smallest feature that can be imaged is proportional to λ/NA. Decreasing the illumination wavelength or increasing the NA therefore enhances the resolution. The drawback of increasing the resolution by increasing the NA invariably leads to a decrease in depth of focus (DOF) for the optical system which makes the fabrication much more sensitive to variations in wafer topography. For characterizing the process, both hole and line patterns were printed with different exposure doses. Lines are defined by etching two 1–3 μm wide tracks spaced by the required width (i.e., the actual line width). The results can be seen in Fig. 8.10. The hole diameters increase with increasing exposure dose, while the line width of a photonic wire decreases. For a given feature size on the mask, a wide range of printed feature sizes can be obtained. The range of exposure energies where the structure is still within specification is called the exposure latitude. One of the difficulties of the lithographic process is to print different types of structures such as isolated lines, narrow gaps, photonic crystal structures, and gratings with the same accuracy. As can be seen from Fig. 8.10, even for isolated lines, the dose-to-target (required exposure dose for reaching the designed feature size) may differ considerably depending on the designed line width. In order to print all features correctly, a bias may have to be applied in the design phase and therefore a thorough optimization of the whole process in advance is needed.

Also optical proximity effects have to be corrected. If features with dimensions close to the illumination wavelength are defined, diffraction may cause the images of neighboring structures to overlap, resulting in either destructive or constructive interference and as a consequence the features size of (semi)isolated structures may differ from that of structures in a dense array. Figure 8.11 shows two examples of optical proximity effects. In the left picture, the width of a line is influenced by a line printed next to it and even

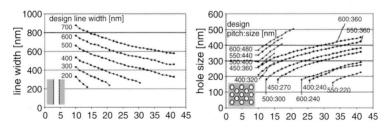

Fig. 8.10. Size of lines (*left*) and holes (*right*) as function of exposure dose (mJ cm^{-2})

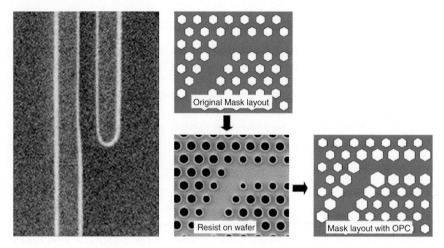

Fig. 8.11. Optical proximity effects for lines (*left*) and photonic crystals (*right*). Optical proximity corrections (OPC) on the mask can compensate for this (the corrections in the figure above are exaggerated)

the waveguide center is displaced. The right picture shows how the diameter of holes in a photonic crystal lattice is different in the bulk of the lattice and at the waveguide borders and corners. By adding a correction in the mask, OPE-effects can be compensated for. Note that, due to electron scattering, proximity effects also form a problem for Ebeam lithography. However, since the effect is coherent for mask based lithography it is more difficult to simulate.

Etching

The structures are etched on a LAM A6 platform. For the silicon etch a $Cl_2/O_2/He/HBr$ chemistry was used. The original process, including the resist thickness, was optimized for also etching the underlying silica cladding layer using a CF_4/O_2 chemistry. This, however, resulted in a very high roughness (Fig. 8.12, left), which can be overcome partly by oxidation (see section "Oxidation"). For photonic wires the deep etch is not needed, however, and the oxide etch can be omitted. To overcome a considerable bias between lithography and etch, a resist hardening plasma treatment is introduced in this case. This results in a considerable roughness reduction with a residual sidewall roughness in the order of 5 nm or less (Fig. 8.12, right) and low propagation losses (Fig. 8.5).

Oxidation

Several authors have found that a high temperature oxidation step following the waveguide etching can smoothen the sidewalls of the photonic wires [4,14, 20,21]. This was studied in detail in [20,21], where the waveguide losses before and after oxidation were determined and the oxidation kinetics were studied.

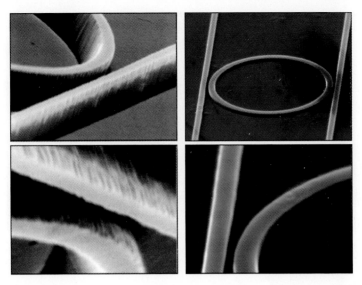

Fig. 8.12. Deeply etched structures showing high roughness and high loss ($\sim$300 db cm^{-1}) (*left*) and shallowly etched strip waveguides with reduced losses (2.4 db cm^{-1})(*right*)

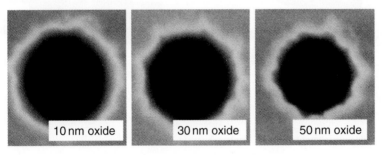

Fig. 8.13. Sidewall smoothing due to oxidation

Oxidation led to a reduction of the sidewall rms roughness from 10 to 2 nm. However, as explained above the resulting waveguide had a core thickness of only 50 nm and can no longer strongly confine the light.

Figure 8.13 shows the effect of oxidation on the sidewall roughness of a small hole. Although there is an apparent increase of the roughness at the SiO_2/air interface one may assume that the roughness which is actually relevant, namely at the Si/SiO_2 interface, is smoother due to the diffuse nature of the oxidation process.

A drawback of the oxidation process is the high temperature required ($>$1,000°C), which makes the process incompatible with back-end processing. While the oxidation can indeed reduce the losses, most of the low propagation loss results [3, 9, 14] were obtained without oxidation.

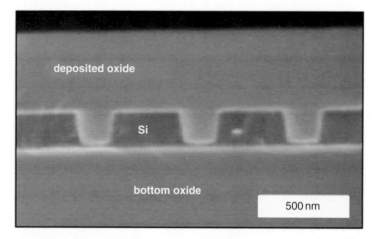

Fig. 8.14. Cross-section of photonic crystal holes with 500 nm pitch after oxide deposition

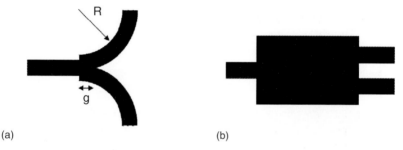

(a)   (b)

Fig. 8.15. Compact branch type coupler (**a**) and MMI-type coupler (**b**)

Oxide Deposition for Top Cladding

Due to the reduced index contrast, a silica top cladding should, for a given width, lead to a further decrease of the propagation loss. However, as can be seen from Fig. 8.2, adding a top cladding also requires considerably smaller waveguides if the single mode condition has to be fulfilled. Table 8.1 shows that several authors have indeed used a top cladding layer. Figure 8.14 shows an example of a photonic crystal structure covered with a 500 nm oxide layer deposited using a TEOS CVD process (following a 5 nm oxidation). Note that no artifacts such as voids are visible and that the deposition process creates a smooth planar top cladding.

8.3.3 CMOS-Compatibility

While deep UV-lithography is capable of printing features with the dimensions of photonic wires and photonic crystals there are some important differences between nanophotonic components and typical CMOS-structures and some issues to be taken into account when designing photonic components. These include sidewall roughness and depth of focus, as mentioned earlier. Some other potential problems are described later.

CMOS Components are Layered

In a typical process flow for the fabrication of high-end electronic devices the components are layered. Each layer contains only critical structures of a certain type (transistor gates, contact holes, interconnect wires, etc.) and a given dimension. Therefore, the process can be optimized for each layer individually. In planar nanophotonics, small alignment tolerances require that all structures are fabricated in the same lithographic step. As explained earlier, the optimal process conditions may differ strongly for different structures, however, and this requires difficult precompensation on the mask level.

Note that with typical steppers an alignment accuracy better than 100 nm between different layers can be obtained. However, for the ultracompact waveguide structures considered here this is still not sufficient and would lead to considerable transition losses. A possible solution would be to incorporate optimized transition sections, which are wider or exhibit a lower transversal index contrast and, therefore, are more tolerant to discontinuities.

Type of Structures

Also the type of structures may differ considerably between CMOS and photonics. A typical example includes super dense photonic crystal lattices. The best equivalent in CMOS is a 1:1 dense array of contact holes, in which case the hole diameter is equal to the spacing in between. In triangular photonic crystal lattices, the spacing between the holes can be reduced to virtual nothing, however.

Required Accuracy

As described in [6,38], the properties of high-contrast nanophotonic devices are extremely sensitive to deviations in the fabrication process. For wavelength dependent circuits (filters, demultiplexers, etc.) the spectral transmission shifts as a result of the waveguide thickness. In most type of circuits, the relative wavelength shift is equal to the relative change in effective index. This leads to the equation:

$$\frac{\partial \lambda}{\lambda} = -\frac{\partial k_z}{k_z} = \frac{k_x^2}{k_z^2}\frac{\partial k_x}{k_x} = -\frac{k_x^2}{k_z^2}\frac{\partial d}{d} \tag{8.2}$$

with k_z the propagation coefficient, k_x the k-vector perpendicular to the waveguide axis and d the waveguide width. Since for high contrast waveguides $k_z \approx k_x$ this means the structural dimensions need to have the same accuracy as the accuracy required for the wavelength dependent transfer of the device. In the same way, one can show that, for interference type filters such as Fabry-Pérot cavities

$$\frac{\partial \lambda}{\lambda} = \frac{\partial L}{L} \tag{8.3}$$

with L the cavity length. Again, if L is of the order of the wavelength, as needed for ultracompact resonators with large FSR and only a few modes, the tolerable thickness inaccuracy is of the same order as the tolerable resonance wavelength inaccuracy.

A lot of more advanced filter structures such as AWGs and Cascaded MZI-structures (see later) are based on multipath interference and are very sensitive to phase-errors accumulated along the way. We believe this is the main reason for the relatively high crosstalk level reported thus far for this kind of devices [7, 14, 16]. This issue is also related to another problem, namely the mask pixelization. The DUV-mask is written with a finite resolution. This may lead to coherence in the induced sidewall roughness, which enhances problems due to roughness-induced phase errors. Another problem is the limited control over the exact dimensions of structures, as e.g., has been noticed by several authors when defining ring resonator based multichannel add-drop multiplexers (e.g., see Fig. 8.25)

8.4 Devices

8.4.1 Couplers–Splitters

Except for simple point-to-point interconnections, most circuits will require splitters or couplers. This is for example the case in clock distribution networks, where the input signal has to be evenly divided toward a very large number of detectors, but also in every filter function, such as ring resonators, where couplers always form one of the basic building blocks.

Couplers and splitters have to fulfill several requirements:

1. Negligible insertion loss
2. Precise splitting and/or coupling ratio
   - For 3 dB-splitters (e.g., used in clock distribution networks or Mach-Zehnder interferometer based switches ...) an equal distribution ratio is extremely important.
   - Some types of wavelength selective filters (e.g., ring resonators, cascaded Mach-Zehnder interferometers, ladder type filters) require very accurate control over the coupling ratio, which can vary from less than 1% to 50%.
3. Wavelength independent operation
4. Tolerance to fabrication deviations

Directional coupler type devices consisting of two identical waveguides brought close together are best suited for realizing *variable coupling ratios*. They need narrow gaps between both waveguides (in some cases down to a few

tens of nanometers, mostly in the order of 200–300 nm, however) and, therefore, may be sensitive to optical proximity effects (e.g., see Fig. 8.11). They
are also not trivial to simulate because of the high index contrast. However,
they are conceptually simple and relatively broadband. Optical directional
couplers for square cross-section silicon wires, including their polarization dependent and wavelength dependent behavior were studied in detail in [26,39].
MMI-type devices with variable coupling ratio have been demonstrated in
other material systems but to our knowledge not yet employing submicron
SOI-wires.

Different designs for compact 3 dB-couplers splitting the power evenly over
two output waveguides were studied in detail in [18], both using simulations
(2D and 3D FDTD) and experimentally. The simple branch consisting of two
offset bends as sketched in Fig. 8.15a turned out to have the lowest insertion loss and to be the most manufacturable. FDTD simulations (3D) predict
a 0.2 dB excess loss for such a structure. Experimentally, an excess loss of
0.3 dB was demonstrated. By optimizing the length g and the bend radius R,
the structure can be made very tolerant to process variations. The optimized
structure was used in [17] to construct a 3-level H-tree. A 2–5 dB fluctuation
was measured over the 16 ports.

In [23], MMI-type couplers (Fig. 8.15b) were experimentally demonstrated
and a 1-to-8 distribution tree was demonstrated. The experimental imbalance
remains smaller than 0.5 dB over a 400 nm spectral range. The size of the 1×2
MMI was $2 \times 5.4\,\mu m^2$.

8.4.2 Crossings

One of the advantages of optical interconnects often mentioned is the possibility to make crosstalk free crossings, which would simplify the interconnect
problem considerably. For free space interconnects this assumption holds but
for guided wave solutions part of the light of the main channel will leak into
the intersecting channel and this will become worse with increasing optical
confinement (or in other words, increasing numerical aperture). Since SOI-
wires confine the optical mode to a very small cross-section we can expect
a large crosstalk and reflections at intersections. This problem has been investigated by Fukazawa [40] using 3D-FDTD-simulations and experimentally.
For a simple cross-section (Fig. 8.16a) the insertion loss and crosstalk were
estimated (3D-FDTD) to be 1.4 and −9.2 dB, respectively, which is totally
unacceptable for practical applications. Expanding the mode as in Fig. 8.16b
makes the angular spectrum smaller and leads to an improved performance
(<0.1 dB loss and <-30 dB crosstalk). This structure was also fabricated and
performance comparable with the simulations was demonstrated (<0.1 dB loss
and <-25 dB crosstalk). However, in practice also the intersecting waveguide
should be tapered as in Fig. 8.16c, again leading to higher losses (0.4 dB).
In [34] a resonant coupler with <0.2 dB insertion loss was presented. However, the simulations were limited to 2D FDTD and the transmission band of
the resonant coupler was limited to a few nanometer.

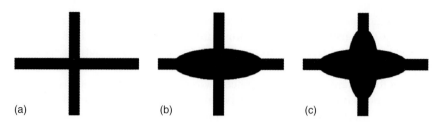

Fig. 8.16. (**a**) Standard crossing. (**b**) Crossing with single parabolic taper. (**c**) Crossing with double parabolic taper

An additional drawback of the resonant crossing and the parabolic crossing are the additional floor space required, which may be a problem in ultradense interconnect circuits. At this moment no satisfactory solution is available for realizing crossings and an interconnect layout requiring no crossings should be used, potentially using multiple wiring levels combined with vertical optical vias.

8.4.3 Fiber-Chip Couplers

Introduction

For on-chip optical interconnect, coupling between the photonic wires and an optical fiber is not necessary. For all other applications, however, the coupling of light between the fiber and the chip is one of the most essential parts of the device and determines a huge fraction of the cost of the packaged device. In some material systems, such as polymer or silica-on-silicon, the size of the basic optical waveguide mode is identical to that of the optical fiber and the coupling is straightforward. Losses below 0.1/dB per connection point have been demonstrated [1]. In a lot of cases, however, there is a large mismatch between the optical mode on and off the chip resulting in large coupling losses if no mode adapters are used. This is the case, for example, for InP-laser diodes and a fortiori, for the submicron silicon wire waveguides discussed here. In general two approaches can be followed. In the first one, the field is adapted outside of the chip, e.g., using small ball lenses or a lensed fiber. Such an approach has the advantage of not requiring additional processing for making the on-chip mode adapters but leads to high packaging costs because of the additional components required (if ball lenses are used) and because of the very high alignment accuracy required. In the second approach, the mode adapter is integrated on the chip and the alignment accuracy is relaxed, which also results in lower packaging cost. The processing becomes more complex, however, and the tradeoff between processing cost and packaging cost will determine what solution is to be preferred.

Because of the extremely large mismatch, external mode adaptation is almost impossible for silicon wires and a solution with on-chip mode adapters has to be used. A good solution has to fulfill several requirements:

- Except for a few specific cases, in general *broadband operation* is required for the fiber-chip coupler (several tens of nanometers). This requirement excludes the use of long gratings and directional coupler based devices that may have a good efficiency but work only for a narrow bandwidth range.
- *Low loss.* The loss per connection should be reduced to values comparable with current high-contrast PLC solutions, i.e., at least below 1 dB/connection and preferably below 0.5 dB/connection.
- *Low reflection.* AR-coatings can be applied but approaches avoiding this clearly have an advantage.
- Large alignment tolerance.
- *Correct fabrication tolerance.* If the coupling structure (or parts thereof) is defined in the same step as the wires themselves, very high alignment accuracy can be reached and narrow lines can be defined. Parts of the structure defined in a postprocessing step have to be more tolerant to deviations and typically will have larger minimum dimensions.
- "CMOS-compatible"-fabrication, both in terms of materials and in terms of processing (e.g., taking into account available DOF). The etching of very deep silicon structures has to be avoided.
- Size.
- Limited extra processing steps.

Different solutions have been proposed in literature and the chapter on couplers in this book describes different possible solutions. Below we are discussing two options, which were developed specifically having SOI-wires in mind and which according to us show the most promise taking into account the above requirements.

Inverse Tapers

The inverse taper approach shown in Fig. 8.17 has been demonstrated by several groups (see Table 8.3). Over a length of a few tens to a few hundreds

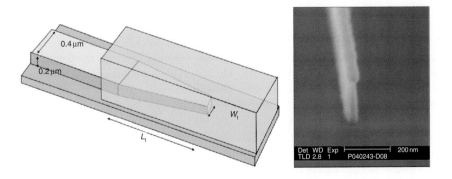

Fig. 8.17. Inverse taper based fiber-chip coupler (*left*) and tapers defined by DUV (*right*)

Table 8.3. Inverse taper review table

Aff.	h (nm)	w (nm)	L_t (um)	W_t (nm)	cladding material	cladding Size (μm^2)	loss per connection	ref.
IBM	220	445	150.0	75.0	Polymer	2×2	<0.5 dB	[9]
Cornell	270	470	40.0	100.0	SiO$_2$	2×00	<4 dB	[10]
NTT	200	400	300.0	80.0	Polymer	3×3	0.8 dB	[14]
					SiON	3×3	0.5 dB	

of micrometers the silicon waveguide is narrowed down and the optical field is pushed upward to a low-index overlay. In most cases this overlay is patterned to form a single mode waveguide, which is then used to bring the optical mode to the edge of the chip. In [10] the overlay was not patterned and the chip had to be diced directly at the end of the silicon taper. In most cases the chip is coupled to a high NA single mode fiber with a mode diameter smaller than standard single mode fiber. Using this approach low-loss broadband operation was demonstrated (<0.5 dB per connection over range 1250–1750 nm in [9, 14]). The loss is mainly determined by the taper tip width W_t and its length L_t and a compromise between acceptable loss and size has to be found.

Both polymer-based approaches [9, 14] and inorganic-based approaches [10, 14] have been demonstrated. The latter may be more reliable and can withstand high optical input powers such as those required for nonlinear optical devices, while polymers are potentially cheaper and are easier to deposit. An additional advantage of the inverse taper approach is the inherent low facet reflection.

All of the demonstrated devices until now used Ebeam lithography for defining the narrow taper tips. We recently demonstrated the fabrication of such tapers using a combination of 248 nm DUV lithography and trimming, demonstrating that these couplers are also compatible with standard CMOS processing techniques (Fig. 8.17).

Grating Couplers

The idea of using a grating for coupling light to an optical waveguide has been around for a very long time. In most cases, however, a long, weak grating is used, leading to a narrow bandwidth. Moreover, such a solution may require an additional lens because the gaussian beam exiting the grating is not adapted to the fiber mode.

We proposed an alternative approach using a short but strong grating as shown schematically in Fig. 8.18. The mode exiting this grating has the same dimensions as the single mode fiber and therefore direct butt coupling between the fiber and the chip is possible. This method has several advantages:

– *Large bandwidth.* Since a strong grating is used, the bandwidth can be several tens of nanometers.

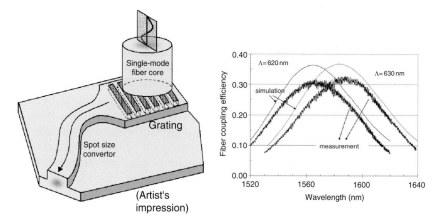

Fig. 8.18. Grating-based fiber chip coupler, only the core of the fiber is shown (*left*). Theoretical and experimental results for fabricated structure (*right*)

– Small size.
– Low reflections.
– No need for facet polishing or cleaving.
– *Wafer scale testability.* Since the devices can be accessed from the top, testing during the fabrication cycle is possible.

We designed and optimized a grating for coupling between an SOI waveguide, with a 220 nm thin Si core layer, and a single-mode optical fiber [41]. In a first stage, we have optimized the parameters of a uniform grating. For a uniform grating, the maximum coupling efficiency that can be achieved is approximately 55% and the 1-dB bandwidth for that structure is 43 nm. The parameters that were optimized are the etch depth, grating period and duty cycle, and the thickness of the buried oxide layer of the SOI. Also an index matching layer is used between the grating and the fiber. The fiber is placed not exactly vertical, but at a small angle (8°) with respect to the vertical direction, to avoid back-reflection into the fiber or waveguide. However, for the actual fabricated structure, the thickness of the BOX-layer was not optimized, leading to a reduction of the expected maximum coupling efficiency. Theoretical and experimental results for the fabricated grating are shown in Fig. 8.18. As can be seen from this picture a maximum coupling efficiency of 33% was measured. The 1 dB-bandwidth was 40 nm. For obtaining the theoretical results a 1D calculation was used. If also the lateral dimension is taken into account the discrepancy between theoretical and experimental results reduces to a few percent.

The efficiency of a uniform grating is limited by two factors. First, a uniform grating creates an approximately exponentially decaying output field along the propagation direction. The fiber-mode has a Gaussian profile. As a result, there is a mode mismatch between the two and this limits the theoretical coupling efficiency. Second, the grating does not only couple light from the waveguide upward toward the fiber, but also downward toward the substrate. The light coupled to the substrate is lost and limits the efficiency. To avoid this, a mirror (dielectric DBR or metal mirror) can be added under the grating. If the position of the mirror is correctly chosen, all the light can be coupled upward and no light is lost to the substrate.

We have optimized such a grating using a genetic algorithm. Both the width of the grating teeth and the spacing between the grating teeth is optimized. More details on the design and optimization of this structure can be found in [42]. The resulting structure is shown in Fig. 8.19, together with a field plot and the calculated transmission. The coupling efficiency is 95% and the 1-dB bandwidth is 43 nm. When regular SOI is used instead of SOI with a bottom reflector, the coupling efficiency is 63%. The regular SOI wafers are commercially available products, the fabrication of wafers with a bottom reflector requires additional processing steps. The cited values are for TE-polarization.

The 1D grating only couples TE-polarized into the circuits and TM-polarized light is lost. If a 2D grating is used, however, both polarizations couple toward different directions perpendicular with respect to each other. Therefore, the 2D grating coupler can be used as a polarization splitter [31] and for implementing a polarization diversity scheme as presented in Fig. 8.20. As discussed earlier, such a scheme could provide a solution for the extremely polarization dependent behavior of the submicron silicon wire circuits.

For making a compact transition between the grating coupler having a lateral width of 10 μm and the 500 nm wide single mode wire, compact interference type filters were designed [43] using a genetic algorithm and fabricated. A typical structure and the measured performance is shown in Fig. 8.21.

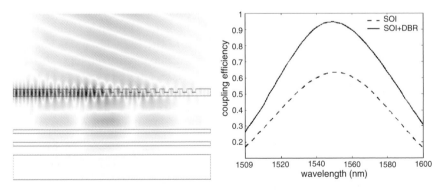

Fig. 8.19. Optimized grating structure with nonuniform grating and bottom reflector. Field plot (*left*) and calculated transmittance (*right*)

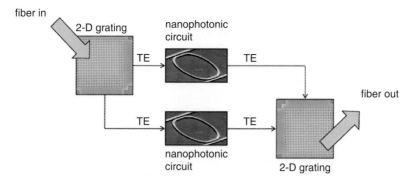

Fig. 8.20. Polarization diversity scheme using 2D-grating coupler

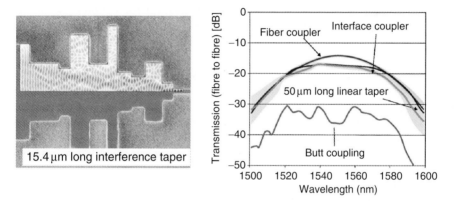

Fig. 8.21. Compact interference taper. Structure and field plot (*left*) and measured transmission (*right*)

8.4.4 Ring Resonators

Compact microring resonators can be a building block for densely integrated wavelength selective filters. Since SOI-based photonic wires enable the fabrication of bend waveguides with very short bend radius they are suited very well for fabricating ring resonators with a large free spectral range (FSR), which is important for the fabrication of useful add-drop filters. A large finesse and high extinction ratios are of utter importance too. For increasing the roll-off and for obtaining a flat-top transmission, higher-order filters can be fabricated by connecting several ring resonators in series [44]. Reliable fabrication of such devices is challenging, however, and ring resonators also form a very good test for the quality and reproducibility of the technology.

SOI-based ring resonators have been presented by several authors [11, 14, 36, 45]. We fabricated a series of devices using DUV-lithography. The most challenging part of the fabrication process is defining the narrow gap between the bus and ring waveguides in the coupler region in a reliable manner, since it is influenced strongly by optical proximity effects (e.g., see Fig. 8.11, left). The resonators have radii up to 8 μm and are laterally coupled to two waveguides for channel dropping purposes. We studied different coupling alternatives in order to achieve low crosstalk and high finesse while maintaining a large free spectral range.

The first devices fabricated are circular ring resonators coupled to two straight waveguides. The gap between ring and straight waveguide is limited to about 200 nm due to technological limitations. This makes coupling between the ring and waveguides difficult and smaller than 1%. With a 5 μm radius ring, a FSR of 17 nm is obtained but drop efficiency is very low due to the low coupling. Also, Fig. 8.22, left shows that resonances are often split. This is due to coupling to the counter-propagating mode induced by the surface roughness. As explained in [46] this effect only appears if the coupling coefficient is sufficiently low. We applied a BCB top cladding in order to enhance the coupling. This polymer has a higher refractive index (1.5) than the original air cladding, which increases the coupling coefficient slightly. The measurements in Fig. 8.22, right clearly show the resulting increase of the coupling and the absence or large reduction in the splitting of the resonances. Reduced waveguide dispersion also leads to a net increase of the FSR to 18.5 nm.

Due to low coupling and high FSR the finesse of this kind of device is high, up to 120. However, for WDM purposes coupling should be increased. Therefore, we looked at other ring resonator configurations to enhance coupling between cavity and waveguides.

One option is to include a straight coupling section, leading to a racetrack resonator (e.g., see Fig. 8.23, left). We fabricated racetracks with radii of the bend section between 1 and 6 μm. With 1 μm radius, a Q of around 2,100 is obtained but the drop efficiency is low. For larger radii, the ring losses decrease quickly. With a 2 μm radius, drop efficiency is already much larger and Q factors in the range 5,000–9,000 are obtained. We also fabricated racetracks with 5 μm radius and larger coupling section. The larger coupling leads to a high

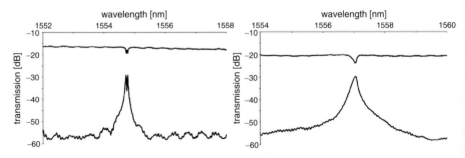

Fig. 8.22. Detail of transmission spectra of a ring resonator with 5 μm radius (*left*). Same ring with a BCB top cladding, resonance in the same wavelength range (*right*)

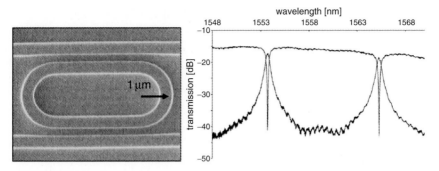

Fig. 8.23. Racetrack resonator with 1 μm radius (*left*). Transmission spectra of racetrack resonator with 5 μm radius (*right*)

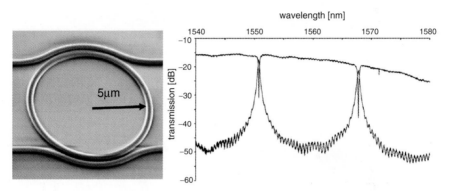

Fig. 8.24. SEM picture (*left*) and transmission spectra (*right*) of a bend-coupled ring resonator with 5 μm radius

add-drop extinction ratio of −20 dB and 50–70% drop efficiency at Q factors still larger than 3,000 (Fig. 8.23).

While racetrack resonators can clearly enhance coupling, their FSR is limited. Another option is coupling to bend waveguides (Fig. 8.24). By carefully choosing the widths of cavity and bus waveguides, a good phase matching can be obtained and coupling can be high without lowering the FSR as in the case of the racetrack resonator. We fabricated bend-coupled resonators with 5 and 8 μm radius. The amount of coupling is varied by varying the angle over which both waveguides are coupled. Coupling is large enough to obtain a relatively high extinction ratio (−10 to −15 dB) and high drop efficiency, although these first devices are still far from optimized.

We fabricated a four channel add-drop filter with racetrack resonators with different radius. By varying the radius, the resonance wavelength is changed and each cavity is tuned to another set of dropped wavelengths. One must keep in mind that with this approach the FSR also varies. Figure 8.25 shows

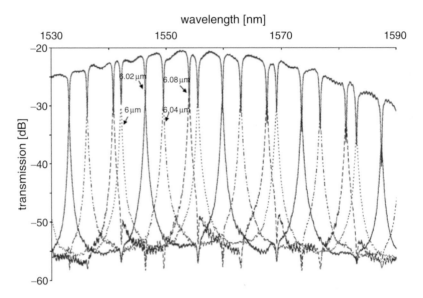

Fig. 8.25. Overlaid transmission spectra of a 1-by-4 demux with four racetracks with varying radius (indicated)

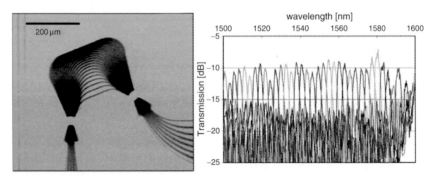

Fig. 8.26. SEM-picture of SOI-based eight-channel AWG and measured transfer curves

the overlaid transmission spectra for four resonators with 6, 6.02, 6.04, and 6.08 μm. A BCB top cladding was applied. We see that a large part of the FSR can be reached by changing the radius over only 80 nm. However, better control over the exact resonance wavelengths is needed.

8.4.5 Arrayed Waveguide Grating Devices

Silica-on-silicon AWGs are currently the most popular integrated device for multiplexing and demultiplexing multiple wavelength channels. We have designed and fabricated an 8-channel AWG (Fig. 8.26). This device has a

footprint of $380\,\mu m \times 290\,\mu m$ or about $0.1\,mm^2$. The transfer from the input
port to the 8 output waveguides is also plotted in Fig. 8.26. The small ripple
is caused by residual reflections in the grating couplers. The channel spacing is 3 nm, with a free spectral range of 24 nm. The on-chip insertion loss
was approximately 8 dB. We believe this loss is mainly caused by reflections
in the star coupler and can be reduced by adapting the transition zone between the grating arms and the star coupler. The main problem, however,
is the high crosstalk level, limited to values around $-7\,dB$, which is clearly
not sufficient for practical applications. Also in [16] an AWG fabricated using
SOI-wires was presented. The authors optimized the devices for compactness
and demonstrated a $110 \times 93\,\mu m^2$ device with a 6 nm channel spacing and a
90 nm FSR. However, also in this case, the crosstalk was limited to a few dB.

The high crosstalk may be caused by several reasons such as reflections or
overspill in the star coupler [47]. We believe the crosstalk is mainly caused by
phase errors in the grating arms, however. Due to the high refractive index
contrast even the very small roughness can lead to random phase fluctuations
over the array arms and lead to the observed crosstalk level.

8.4.6 Cascaded Mach-Zehnder Interferometers

SOI-wire based Cascaded Mach-Zehnder interferometers have been demonstrated by several authors [14, 24, 48]. As shown in Fig. 8.27, such filters consist of a series of Mach-Zehnder interferometers with constant path-length
difference but with varying coupling ratio optimized to obtain the desired
wavelength dependent transmittance. The measured transmission for the fabricated device is shown in Fig. 8.27b. A crosstalk slightly better than $-10\,dB$
was measured, limited by a nonoptimized choice of the coupling ratios and by
phase errors in the interferometers. In [14], a similar device having a $-12\,dB$
crosstalk level and a 80 nm spectral range was demonstrated. By cascading
three devices in series, the crosstalk could be increased to $-30\,dB$. Also multiple wavelength add-drops where demonstrated.

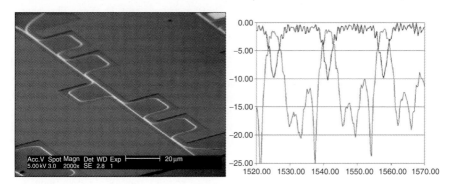

Fig. 8.27. Cascaded Mach-Zehnder interferometer. Fabricated device (*left*) and
measured performance (*right*)

8.4.7 Active Devices

The demonstration of active devices using submicron silicon wires has been limited till now and mainly modulator type devices have been demonstrated. As discussed extensively in a separate chapter in this book, the two main mechanisms for modulating the refractive index of silicon are carrier injection and thermo-optic tuning.

In [25], a compact thermo-optically tuned switch based on a Mach-Zehnder interferometer was demonstrated. The Cr–Au heaters were deposited on a $1\,\mu$m SiO_2-buffer layer. The total switching power required was $50\,mW$, the total length of the device was $\sim1.5\,mm$. A $<3.5\,\mu$s rise time was demonstrated. We demonstrated thermo-optic tuning of ring resonators and cascaded Mach-Zehnder interferometers (Fig. 8.9) [50].

In [11] FP-type modulator using deeply etching DBR-mirrors was demonstrated. Also all-optical switching in a ring resonator was demonstrated in [11]. Recently, also gain through Raman pumping has been demonstrated [49].

Acknowledgments

I would like to thank my coworkers: Wim Bogaerts, Pieter Dumon, Patrick Jaenen, Johan Wouters, Stephan Beckx, Vincent Wiaux, Dirk Taillaert, Bert Luyssaert, Gunther Roelkens, and Roel Baets.

References

1. Miya et al., IEEE Select Top. Quantum Electron. 2000, p. 38
2. L. Pavesi and D.J. Lockwood, "Silicon Photonics", Springer Berlin Heidelberg, New York, 2004
3. P. Dumon et al., "Low-loss photonic wires and ring resonators fabricated with deep UV lithography", IEEE Photon. Technol. Lett. Vol. 16 (5), 2004 pp. 1328–1330
4. W. Bogaerts, R. Baets, P. Dumon, V. Wiaux, S. Beckx, D. Taillaert, B. Luyssaert, J. Van Campenhout, P. Bienstman, and D. Van Thourhout, "Nanophotonic waveguides in silicon-on-insulator fabricated with CMOS technology", J. Lightwave Technol., Vol. 23 (1), 2005, pp. 401–412
5. P. Dumon et al., "Micro-ring resonators in silicon-on-insulator", ECIO '05, Grenoble, April 2005
6. Wim Bogaerts, "Nanophotonic waveguides and photonic crystals in silicon-on-insulator", PhD Thesis, Ghent University, April 2004
7. P. Dumon, W. Bogaerts, D. Van Thourhout, D. Taillaert, V. Wiaux, S. Beckx, J. Wouters, and R. Bockstaele, Wavelength-selective components in SOI photonic wires fabricated with deep UV lithography, Group IV Photonics, Hong Kong, p. WB5 2004
8. S.J. McNab, "Ultra-low loss photonic integrated circuit with membrane-type photonic crystal waveguides", Opt. Express Vol. 11 (22), November 2003, pp. 2927–2939

9. Y.A. Vlasov, "Losses in single-mode silicon-on-insulator strip waveguides and bends", Opt. Express Vol. 12 (8), April 2004, pp. 1622–1631

10. V.R. Almeida et al., "Nanotaper for compact mode conversion", Opt. Lett. Vol. 28, August 2003, pp. 1302–1304

11. M. Lipson, "Controlling light on a microelectronics' chip: solving the coupling, modulation and switching challenges", Group IV Photonics '04, Hong Kong, WB1

12. T. Tsuchizawa et al., "Fabrication and evaluation of submicron-square Si wire waveguides with spot size converters", Leos Annual Meeting 2002, Glasgow, paper TuU_2

13. T. Shoji et al., "Low loss mode size converter from 0.3 um square Si wire waveguides to singlemode fibres", Electron. Lett. Vol. 38 (25), December 2002, 1669–1670

14. T. Tsuchizawa et al., "Microphotonics devices based on silicon microfabrication technology", IEEE J. Select. Top. Quantum. Electron. Vol. 11 (1) 2005, pp. 232–240

15. A. Sakai et al., "Propagation characteristics of ultrahigh-δ optical waveguide on silicon-on-insulator substrate", Jpn J. Appl. Phys. Vol. 40, April 2001, pp. L383–L385

16. T. Fukazawa et al., "Very compact arrayed-waveguide grating demultiplexer using Si photonic wire waveguides", Jpn J. Appl. Phys. Vol. 43, 2004, pp. L673–L675

17. T. Fukazawa et al., "H-Tree-type optical clock signal distribution circuit using a Si photonic wire waveguide", Jpn J. Appl. Phys. Vol. 41, December 2002, pp. L1461–L1463

18. A. Sakai et al., "Low loss ultra-small branches in a silicon photonic wire waveguide", IECE Trans. Electron. Vol. E85-C (4), April 2002, pp. 1033–1038

19. A. Sakai et al., "Estimation of polarization crosstalk at a micro-bend in Si-photonic wire waveguide", IEEE J. Lightwave Technol. Vol. 22 (2), February 2004, pp. 520–525

20. K.K. Lee et al., "Fabrication of ultralow-loss Si/SiO_2 waveguides by roughness reduction", Opt. Lett., Vol. 26, December 2001, pp. 1888–1890

21. D.K. Sparacin et al., "Oxidation kinetics of waveguie roughness–minimization in silicon microphotonics", IPR '03, paper ITuC4-1

22. R. Orobtchouk, "Optical interconnects on SOI: a front end approach", Proceedings of the IEEE 2002 International Interconnect Technology Conference, 2002, San Fransisco

23. R. Orobtchouk, "Compact building blocks for optical link on SOI technology", ECIO 05, Grenoble, April 2005

24. R.U. Ahmad, F. Pizzuto, G.S. Camarda, R.L. Espinola, H. Rao, and R.M. Osgood, Jr., "Ultracompact corner-mirrors and t-branches in silicon-on-insulator" IEEE Photon. Technol. Lett., Vol. 14 (1), January 2002, pp. 65–67

25. R.L. Espinola, M.-C. Tsai, James T. Yardley, and R.M. Osgood, Jr., "Fast and low-power thermooptic switch on thin silicon-on-insulator", IEEE Photon. Technol. Lett. Vol. 15 (10), October 2003, pp. 1366–1368

26. H. Yamada, T. Chu, S. Ishida, and Y. Arakawa, "Optical directional coupler based on Si-wire waveguides", Group IV Photonics '04, Hong Kong, FA3

27. Refractive index silicon

28. P.K. Tien, "Light waves in thin films and integrated optics", Appl. Opt. Vol. 10, 1971 p. 2395

29. F.P. Payne and J.P.R. Lacey, "A theoretical analysis of scattering loss from planar optical waveguides," Opt. Quantum Electron. Vol. 26, 1994, 977
30. F. Grillot et al., "Size influence on the propagation loss induced by sidewall roughness in ultrasmall SOI waveguides", IEEE Photon. Technol. Lett. Vol. 16 (7), July 2004, pp. 1661-1663, Doerr pol diversity
31. D. Taillaert et al., "A compact two-dimensional grating coupler used as a polarization splitter", IEEE Photon. Technol. Lett. Vol. 15 (9), 2003, pp. 1249–1251
32. J. McCaulley et al., "Temperature dependence of the near-infrared refractive index of silicon, gallium arsenide and indium phosphide", Phys. Rev. B, Vol. 49 (11), 1994, pp. 7408–7417
33. Mode-coupling analysis of multipole symmetric resonant add/drop filters, "Khan, M.J.; Manolatou, C.; Shanhui Fan; Villeneuve, P.R.; Haus, H.A.; Joannopoulos, J.D.; Quantum Electronics, IEEE Journal of, Vol. 35, Issue: 10, October 1999, pp. 1451–1460
34. Manolatou et al., "High-density integrated optics", IEEE J. Lightwave Technol. Vol. 17, September 1999, pp. 1682–1692
35. W.L. Barnes et al., "Surface plasmon subwavelength optics", Nature, Vol. 424, August 2003, pp. 824–830
36. P. Dumon et al., "Temperature behaviour of silicon-on-insulator photonic wires", to be published
37. "SOITEC's UNIBOND(r) process," Microelectron. J. Vol. 27 (4/5), 1996, p. R36
38. R. Baets, W. Bogaerts, D. Taillaert, P. Dumon, P. Bienstman, D. Van Thourhout, J. Van Campenhout, V. Wiaux, J. Wouters, and S. Beckx, Low loss nanophotonic waveguides and ring resonators in silicon-on-insulator, Proceedings of the International School of Quantum Electronics , 39th course "Microresonators as building blocks for VLSI photonics", 709, Italy, 2003, pp. 308–327
39. H. Yamada, T. Chu, S. Ishida, et al., "Optical directional coupler based on Si-wire waveguides", IEEE Photon. Technol. Lett. Vol. 17 (3), March 2005, pp. 585–587
40. Fukazawa et al., "Low loss intersection of Si photonic wire waveguides", Jpn. J. Appl. Phys. Vol. 43 (2), 2004, pp. 646–647
41. D. Taillaert et al., "An out-of-plane grating coupler for efficient butt-coupling between compact planar waveguides and single-mode fibers", IEEE J. Quantum Electron. Vol. 38 (7), 2002, pp. 949–955
42. D. Taillaert et al., "Compact efficient broadband grating coupler for silicon-on-insulator waveguides", Opt. Lett. Vol. 29 (23), December 2004, pp. 2749–2751
43. B. Luyssaert, P. Vandersteegen, D. Taillaert, P. Dumon, W. Bogaerts, P. Bienstman, D. Van Thourhout, V. Wiaux, S. Beckx, and R. Baets, "A compact horizontal spot-size converter realized in silicon-on-insulator", IEEE Photon. Technol. Lett. Vol. 17 (1), 2005, pp. 73–75
44. B.E. Little, "A VLSI photonic platform", OFC '03
45. B.E. Little et al., "Ultra-compact Si-SiO2 microring resonator optical channel dropping filters", IEEE PTL, Vol. 10, 1998, pp. 549–550
46. B.E. Little et al., "Surface-roughness-induced contradirectional coupling in ring and disk resonators", Opt. Lett. Vol. 22 (1), 1997, pp. 4–6
47. C. Van Dam, PhD Thesis, TUDELFT '99
48. K. Yamada et al., "Silicon-wire-based ultrasmall lattice filters with wide free spectral ranges", Opti. Lett. Vol. 28, September 2003, pp. 1663–1664

49. R.L. Espinola, J.I. Dadap, R.M. Osgood, Jr., S.J. McNab, and Y.A. Vlasov, "Raman amplification in ultrasmall silicon-on-insulator wire waveguides", Opt. Express, Vol. 12 (16), August 2004, pp. 3713–3718
50. I. Christiaens, D. Van Thourhout, and R. Baets, "Low power thermo-optic tuning of vertically coupled microring resonators", IEE Electron. Lett., Vol. 40 (9), 2004, pp. 560–561

Photonic Crystal Microcircuit Elements

T.F. Krauss

Summary. In the context of optical interconnects and photonic circuitry, photonic crystals offer exciting opportunities. Most notably, they can be realised on the same size as electronic microcircuits and provide novel functionality on a wavelength scale. Their working principles and some of the recently introduced device concepts are described, including superprisms, slow wave structures, dispersion compensating elements and high Q cavities.

9.1 Introduction

Work on wave propagation in periodic structures dates back a long time, e.g. to 1884 (Achille Floquet), 1913 (William Lawrence and William Henry Bragg) and 1928 (Felix Bloch), to name but a few of the famous examples. It was not until 1987, however, when Eli Yablonovitch [1] formally recognised the similarity between the propagation of electromagnetic waves in periodic structures and that of the bandstructure phenomena well known from solid-state physics, thus coining the terms "photonic bandgap" and "photonic crystal". Initial work in this area was aimed at improving the properties of light emitters by controlling the process of spontaneous emission. This led to the development of rather impressive optical structures that are periodic in one, two or three dimensions. The work is now bearing fruit, with several examples of spontaneous emission control having recently been demonstrated [2,3], including the demonstration of "strong coupling", i.e. the reversible energy exchange between an optical mode and the excited state of a quantum dot embedded in a photonic crystal microcavity [4].

During the mid-1990s, another application for photonic crystals, emerged namely in photonic circuits. The idea was to use 2D or planar photonic crystals and to exploit their ability to control and manipulate guided waves on a wavelength scale, thus enabling the vision of a "photonic metropolis" [5], i.e. a network of interlinked optical pathways based on photonic crystals. Optical confinement in the vertical, i.e. the third dimension, was to be provided by a standard waveguide heterostructure utilising total internal reflection. The particular attraction of this approach is the fact that patterns can be defined using sophisticated lithographic techniques developed by the microelectronics industry, allowing structures to be generated with high fidelity. The key requirement for the realisation of this vision, i.e. a 2D photonic bandgap in waveguide-based system, was achieved in 1996 [6] with a flurry of activity following this. To date, a wide variety of photonic circuit elements have been demonstrated, ranging from low-loss single-mode waveguides [7, 8] to bends [9], splitters [10,11], couplers and Mach-Zehnder interferometers [12,13]. Even more importantly, however, is the realisation of novel functional elements, such as microspectrometers or "superprisms" and "supercollimators" [14–16], as well as devices exploiting the available dispersion control, e.g. for pulse compression [17], dispersion compensation [18] and slow wave effects [19]. Finally, the intrinsic polarisation dependence of planar photonic crystals can be used for polarisation control, such as polarisation splitters [20, 21] and rotators [22]. In summary, planar photonic crystals give rise to a new class of microphotonic circuit elements.

9.2 Stopbands and Bandgaps

Since the focus of this article is on 2D or planar photonic crystals (planar PhCs) and photonic crystal waveguides, we shall limit the discussion to 1D and 2D effects. To start off with, let us consider a 1D periodic structure as illustrated in Fig. 9.1a.

9.2.1 One-Dimensional (1D) Periodic Structures

The propagation of light through this structure is then governed by two effects (a) the Fresnel reflection experienced at each interface, given by $r = (n_2 - n_1)/(n_2 + n_1)$, with r the amplitude reflection coefficient and n_1, n_2 the respective refractive indices of the two layers and (b) the interference between multiply scattered waves that can build up to significant reflection (Bragg effect). The Bragg effect is strongest for $k = \pi/a$, i.e. if the k-vector of the incident light matches the periodicity of the system a or, in other words, if half the wavelength of the material corresponds to the period, according to $k = \pi/a \Rightarrow 2\pi/\lambda_\mathrm{m} = \pi/a \Leftrightarrow a = \lambda_\mathrm{m}/2$.

Note that the Fresnel effect is wavelength independent, whereas the Bragg effect is not. This is important for the consideration of the wavelength dependence discussed later. The combination of the two effects is most simply represented in a reflectivity vs. wavelength plot or in a bandstructure diagram (combined in Fig. 9.1b). The bandstructure diagram follows most simply from the fundamental relationship $c_\mathrm{m} = \lambda_\mathrm{m}\nu$ or $c_\mathrm{m} = 2\pi\nu \times (\lambda_\mathrm{m}/2\pi) = \omega/k$, which

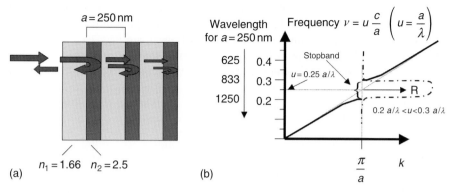

Fig. 9.1. Schematic of light propagating through a periodic structure that consists of two materials with refractive indices n_1, n_2 and periodicity a. (**a**) Indices of $n_1 = 2.5$, $n_2 = 1.66$ and a period of $a = 250\,\text{nm}$ are used for illustrative purposes, resulting in a stopband centred at $1{,}000\,\text{nm}$ wavelength. A fraction of the light is reflected at each interface. If half the wavelength of the material (λ_m) corresponds to the lattice period, or $\lambda_m/2 = a(k = \pi/a)$, the reflections add up in phase. (**b**) Resulting bandstructure diagram featuring a stopband at $k\pi/a$ for frequencies between $a/\lambda = 0.20$ and $a/\lambda = 0.30$. Superimposed is a reflectivity spectrum highlighting the fact that light can propagate outwith the stopband, but is reflected within

means that a photon propagating in a given material at speed c_m[1] will simply describe a straight line in an ω–k "dispersion" diagram. At the band edge, due to the Bragg effect, the straight line is interrupted and a stopband opens up (Fig. 9.1b). This stopband describes a frequency (or wavelength) regime for which propagation is prohibited, corresponding to a regime of high reflectivity. The width of the resulting stopband can then be described with two different models

– *Fresnel vs. Bragg.* The response of the structure is determined by the interplay between the Bragg and the Fresnel effects; for low-index contrast, the Fresnel reflection is weak and many interferences are required to build up a strong reflection. Naturally, this is strongly wavelength dependent, so the (wavelength dependent) Bragg effect dominates and the spectral bandwidth is narrow. For high-index contrast, the Fresnel reflection is strong, so the wave will only sample a few layers before being totally reflected. Therefore, the (wavelength independent) Fresnel effect dominates and the spectral bandwidth is large, i.e. a large stopband opens up.
– *Optical path.* The effective index of a given mode is determined by its field distribution in the structure. At the band edge, the field may either reside in the low- or in the high-index regions. Since the physical path of the repeating unit cell is given by the period a and fixed, the difference

---

[1] The speed of light in a material c_m is given by $c_m = c/n$, with n the refractive index of the material

between the bandedge wavelengths must be accommodated by different optical path lengths and therefore corresponds to the difference in effective refractive indices. In fact, the field redistribution can very closely map the index distribution, especially in the case of a $\lambda/4$ stack (where each layer is $\lambda/4n$ thick), to the extent that

$$\frac{\lambda_{\text{long}}}{n_{\text{high,eff}}} = \frac{\lambda_{\text{short}}}{n_{\text{low,eff}}} \Leftrightarrow \frac{\lambda_{\text{long}}}{\lambda_{\text{short}}} = \frac{n_{\text{high,eff}}}{n_{\text{low,eff}}} \approx \frac{n_{\text{high}}}{n_{\text{low}}},$$

that is the ratio of the wavelengths corresponding to the lower and upper band edges is approximately given by the ratio of the refractive indices. The higher the index contrast, the larger the bandwidth. λ_{long} and λ_{short} correspond to the long and short bandedge wavelengths, respectively; the effective indices $n_{\text{high,eff}}$ and $n_{\text{low,eff}}$ are determined by the corresponding mode distributions and are the refractive indices of the two layers. NB: The approximation does not hold for detuned structures, i.e. structures that significantly deviate from the $\lambda/4$ criterion. Such structures will also exhibit a stopband, but it may be significantly smaller than that the index ratio would suggest.

9.2.2 Two-Dimensional (2D) Periodic Structures

We can now translate these ideas into 2D in order to understand the occurrence of higher dimensionality bandgaps. The wide spectral bandwith resulting from the high-index contrast discussed earlier also translates into a wide *angular* bandwidth, i.e. for a given wavelength, a wide range of angles are reflected. This is important for the 1D/2D transition, because a 2D grating can be understood as a superposition of two or more 1D gratings, as sketched in Fig. 9.2. Depending on the direction of propagation, the effective grating period experienced will be different, so the stopband will be at a different spectral position. If the stopbands are wide enough, however, there will be an overlap, which determines the regime for which the structure reflects light irrespective of the direction of propagation. The spectral range over which this happens is referred to as the *photonic bandgap*. In the hexagonal lattice depicted in Fig. 9.2, the effective period repeats every 60°, so the largest difference between propagation directions is at 30°. The two corresponding symmetry directions, i.e. 0° and 30°, are described as Γ–M and Γ–K in analogy with solid-state physics notation (Fig. 9.2b). Combining the different directions in a single plot then yields the *bandstructure diagram* as shown in Fig. 9.2c.

9.2.3 Photonic Crystal Waveguides

The 2D bandgap material thus created can be understood as a "photonic insulator", because in the bandgap regime it does not allow any light to propagate. By adding deterministic defects to this "perfect" crystal, one can create

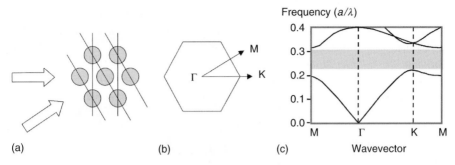

Fig. 9.2. 2D photonic bandgap. (**a**) A triangular photonic lattice can be understood as the superposition of two 1D gratings rotated at 30° with respect to one another. (**b**) Corresponding Brillouin zone representation, using the notation of symmetry points (Γ, K, M) well known from solid-state physics. (**c**) 2D photonic bandstructure of the same lattice, assuming an air-hole fill-factor of 30% and a background refractive index of 3.5. The shaded area represents the photonic bandgap, i.e. the regime where the individual stopbands (in Γ–M and Γ–K direction, respectively) overlap. NB: For simplicity, we only show the bandstructure of a single polarisation, here: TE-like polarisation, with the E-field mainly in the plane of the crystal, i.e. across the holes

channels where light can indeed propagate. The simplest channel consists of a single line of missing holes, referred to as "W1", representing a channel of one unit cell width, i.e. one line of missing holes. Similarly, one can create channels of larger width (W2, W3, etc. as well as channels of fractional width, W0.7, W1.5 and so on). These channel waveguides are the photonic equivalent of metallic waveguides used in the microwave regime, where electromagnetic waves are also guided by "perfect" reflectors.

In bandstructure terms, these waveguides create states in the bandgap. The most commonly used waveguide is the "W1" type depicted in Fig. 9.3.

These waveguides can then be joined and combined to form more complex circuits. Because of the multimode character of even the W1-type waveguide (highlighted by the existence of a first-order mode in Fig. 9.3b), it is very important to design junctions and transitions appropriately. A large body of work has been spent to develop these with the vision of using photonic crystals as a universal platform for future microphotonic circuits. Recent examples include a multielement Y-junction with ultrashort tapers, Y-junctions and bends featuring full numerical optimisation in the design process [11] as well as an entire Mach-Zehnder circuit for all optical switching, comprising multiple input ports, directional couplers and junctions [13]. It should be noted, however, that photonic crystals have yet to prove their advantage for simple guiding and routing applications over the more conventional ridge waveguides (also referred to as "wire" waveguides) of comparable size and index contrast; these tend to be much easier to design and implement as well as featuring lower propagation losses [23]. The areas where photonic crystal waveguides

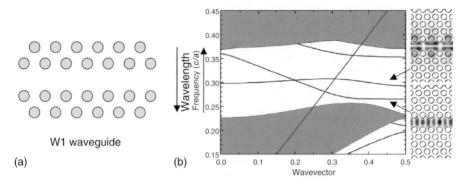

Fig. 9.3. Schematic (**a**) and bandstructure (**b**) of a W1 waveguide consisting of a single line of missing holes. The insets in (**b**) highlight the nature of the two modes of the waveguide, i.e. both a zero-order and a first-order mode can exist and the diagonal line represents the light line for air cladding. The slope of the light line is given by c/n, i.e. it is shallower and lies further to the right of the figure for a higher index cladding material

have real advantages, however, are where they offer additional functionality, such as dispersion and cavity confinement, and these areas are discussed in Sect. 9.4–9.7. To exploit these advantages, it is imperative to design suitable transitions between classical and photonic crystal waveguides, e.g. using the numerical methods discussed in [11].

9.2.4 The Light Line

When creating line and point defects in a 2D photonic crystal, one must not forget that the structure, although only 2D periodic, is still a 3D object. The concept of the "photonic insulator" works very well in the plane, but does not apply in all the three dimensions; for the third dimension, we typically rely on total internal reflection. This is acceptable for many types of waveguides, if for example we use modes with k-vectors that lie mainly in the plane of the photonic crystal. We should be aware of the limitation imposed by total internal refection, however, which is typically described by the "light line" of the system. The light line in a dispersion diagram represents the refractive index of the cladding.[2] Any mode above the light line may couple to the cladding and is therefore described as "leaky"; only modes below the light line are truly confined and intrinsically loss less.[3]

---

[2] In the case of asymmetric waveguides, the higher cladding index is relevant

[3] Intrinsically loss less means that modes do not suffer losses in a perfect crystal; if the crystal structure exhibits roughness and/or imperfections, radiation losses still occur

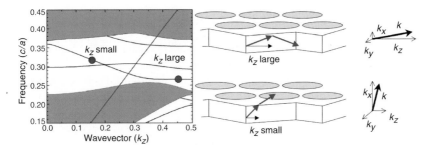

Fig. 9.4. A mode propagates in z-direction through a photonic crystal in the y–z plane. The modulus of the wavevector $\boldsymbol{k}$ of the mode is given by $|\boldsymbol{k}| = n_{\mathrm{eff}} 2\pi/\lambda$ with n_{eff} being its effective refractive index given by the field distribution, and l its free-space wavelength. k_z is the component of $\boldsymbol{k}$ in propagation direction. k_z can be determined from the bandstructure diagram. For the two operating points highlighted here, $|\boldsymbol{k}|$ is almost constant, whereas k_z varies significantly. This means that for k_z small, the structure has to accommodate the same (large) $\boldsymbol{k}$-vector with a smaller k_z. It can only do this by increasing k_x and k_y, because of the relationship $k^2 = k_x^2 + k_y^2 + k_z^2$ between the components. While k_y can grow large due to the existence of a bandgap in the y–z plane, k_x is limited by total internal reflection. Once k_x increases beyond the value corresponding to total internal reflection, the mode can leak out

This effect can be understood in terms of k-vectors, using the mode of a W1 waveguide as an example (Fig. 9.4). We should also recall that the wavevector $\boldsymbol{k}$ plotted in the bandstructure diagram corresponds to the wavevector in direction of the periodicity, i.e. k_z in the example.

The significance of the light line limitation depends on the type of structure, especially its length. For short interaction lengths (typically a few tens of micrometres), the leakage may not be noticeable at all; this is clearly demonstrated in [7], where the transmission for a 25 μm long W1 waveguide is almost identical above or below the light line; the same article, however, shows that the difference for longer structures (up to 2 mm long) is severe, with a substantial difference in loss. Whether the light line imposes a limitation or not therefore depends on the specific application and length of structure. For W3 waveguides in an InP/InGaAsP heterostructure, for example, which operate entirely above the light line, acceptable losses of 2 dB mm^{-1} have already been achieved [24].

If it is necessary to operate below the light line, the key limitation is the available bandwidth (as highlighted by the spectrally narrow regime of low loss in [7]). In order to maximise the bandwidth, one can use the lowest possible cladding index, i.e. air. This requires the use of free-standing membranes. While impressive results have been achieved, such membranes suffer from mechanical instabilities and poor thermal and electrical conductivities, which impose a new set of limitations, especially for actively tuned, switched or amplified devices.

9.3 Wave Propagation in Periodic Structures

The earlier discussion mainly refers to the occurrence of stopbands and bandgaps, i.e. regimes where the propagation is forbidden and the photonic crystal is used as an insulator. Away from the bandgap, however, modes can indeed propagate through the structure subject to the limitations imposed on it by the periodicity. They then assume the form of a *Bloch* mode [25].

9.3.1 Bloch Modes

The theorem first proposed by Achille Floquet and later extended to multi-dimensional structures by Felix Bloch states that normal modes in periodic structures can be expressed as the superposition of plane waves whose wave-vectors are related by momentum conservation,

$$k_n = k_0 \pm nG, \quad G = \frac{2\pi}{a}, \quad n = \text{integer}. \tag{9.1}$$

where G is referred to as the lattice vector and a the period of the structure. Please note that k_0 does not describe the vacuum wavevector, but instead some arbitrary initial wavevector; the suffix "0" highlights the fact that k_0 has not been Bragg scattered, i.e. it has been Bragg scattered "0" times. A set of these waves (the k_n's) together then forms a Bloch mode, which is characterised by a single group velocity. Paradoxically, this means that a Bloch mode is a mode with many phase velocities (many k_n's), yet only a single group velocity.

What does this actually mean? Imagine a plane wave impinging on a periodic structure. In order for the wave to "fit" into this structure, it needs to have higher order spatial components. It gains them by interference with other plane waves. These plane waves are the k_n's that arise from Bragg scattering of the original wave, so a Bloch mode is an interference pattern between multiply Bragg-scattered plane waves [25]. This pattern is reminiscent of the interference pattern one obtains between plane waves, as illustrated in Fig. 9.5.

Another simple example for a Bloch mode is the standing wave that forms in a multilayer stack. When the Bragg condition is met, i.e. when the period of the structure corresponds to half the wavelength, the forward and backward propagating waves interfere constructively and form a standing wave. This can be understood as the superposition of an original wave k_0 and a wave that has been Bragg scattered once, k_1, as follows:

$$k_0 = \frac{\pi}{a}, \quad k_1 = \frac{\pi}{a} - G = -\frac{\pi}{a}. \tag{9.2}$$

In this case, k_0 and k_1 interfere to form a Bloch mode with a group velocity of zero, since standing waves do not transport energy.

There have been several observations of Bloch modes in waveguides, e.g. [26, 27], which can be explained in the same way. Their intriguing aspect (an aspect that, in fact, applies to Bloch modes in general) is that they have a

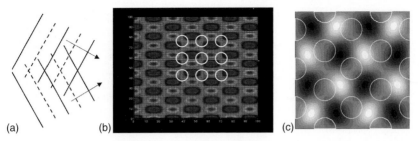

Fig. 9.5. Schematic illustration of the similarity between two-wave interference and a Bloch mode. (a) Two plane waves propagating at an angle with respect to one another generate a multilobed interference pattern. (b). This pattern can fit into a suitable periodic structure (indicated by the *white circles* added for illustration purposes). (c) Bloch mode in a real photonic crystal. This is for H-polarisation (i.e. TE-like mode) near the K-point. The mode consists of an interference pattern between different Bragg-scattered plane waves in order to fit into the structure, The similarity between (b) and (c) is obvious

periodicity which is larger than the lattice. This apparently contradicts the simple description used in Fig. 9.5, where the field distribution is periodic with the lattice. How can this be understood?

It is again an analogy with an interference pattern that yields the answer. In the examples above, a wavevector at the Brillouin zone boundary had been chosen; in this case, the mode pattern really follows the lattice periodicity because the interfering waves are in phase. If we move away from the Brillouin zone boundary, i.e. for $k < \pi/a$, the two interfering waves are no longer in phase, but develop a beating pattern instead. This is illustrated in Fig. 9.6. Here, a forward propagating plane wave (large amplitude in Fig. 9.6a) and a Bragg-scattered wave (small amplitude in Fig. 9.6a) interfere and form the beating pattern shown in Fig. 9.6b. This is remarkably similar to the equivalent Bloch mode shown in Fig. 9.6c. Again, the basic physics of the apparently very complicated Bloch mode propagation can be explained by two-beam interference alone.

Based on this understanding of the nature of Bloch modes, one can recognise several important consequences for wave propagation in periodic structures.

The group velocity of a Bloch mode is highly dependent on its propagation direction. This is a simple consequence of the need for the wave to acquire different spatial components in order to fit into the lattice. In some directions, a plane wave may propagate through the structure almost unperturbed, whereas in other directions, it may form a complicated interference pattern that, as a whole, only propagates slowly through the crystal. This dependence of the Bloch mode's group velocity on propagation direction and wavelength is exploited in the superprism and supercollimator structures described in Sect. 9.4.

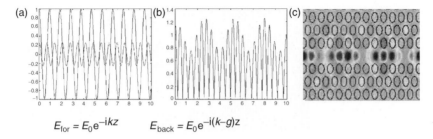

$$E_{for} = E_0 e^{-ikz} \qquad E_{back} = E_0 e^{-i(k-g)z}$$

Fig. 9.6. Comparison of the interference between a forward and a Bragg-scattered plane wave with the equivalent Bloch mode. (**a**) Field distribution of the forward and backward propagating plane wave. The amplitude of the backward propagating wave is reduced due to the lower coupling coefficient resulting from the phase mismatch between the two waves. (**b**) Intensity pattern resulting from the interference between the two waves. (**c**) Field distribution of the equivalent Bloch mode calculated via the supercell method and taking all contributing plane waves into account. The envelopes in (**b**) and (**c**) are identical and consist of six lobes

Considering the same phenomenon in a different way, one can see that the group velocity of a Bloch mode can vary between zero and the speed of light in the medium. This gives rise to opportunities for slow wave structures, e.g. optical delay lines, optical buffers and such like, discussed in Sect. 9.5.

Since k depends intrinsically on the wavelength, Bloch modes can be very dispersive, i.e. their propagation constant can change rapidly with wavelength. This can also be used in applications such as dispersion compensation and pulse compression that are highlighted in Sect. 9.6.

9.4 Superprism and Supercollimator

The wavelength dependence of the Bloch modes can be exploited in a wavelength selective device, i.e. in a type of microspectrometer. Such a device is known as a "superprism" [14–16], which is based on the strong dependence of the propagation direction on the input wavelength near a band edge. Conversely, there are other regimes where the Bloch modes change very little with either input direction or wavelength; these are known as "collimation" regimes.

9.4.1 Superprism

Figure 9.7 illustrates the operation of a superprism. A multiwavelength input beam (represented by the greyscale on the input arrow) is separated into its wavelength components (different output arrows). The wavelengths on the output side represent the values obtained experimentally [16], so an angular separation of $10°/20\,\text{nm}$ (and better) can be obtained. While a wavelength

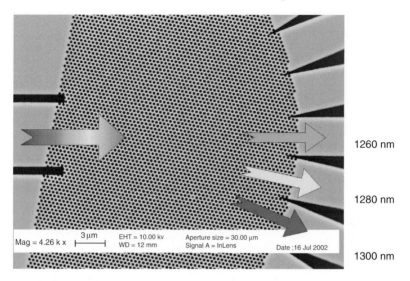

1260 nm

1280 nm

1300 nm

Fig. 9.7. SEM micrograph of a superprism. The overlaid arrows are for illustrative purposes only and correspond to the results obtained in [16]

resolution of 20 nm appears very low, e.g. compared to the separation of <1 nm achieved in commercial arrayed waveguide gratings (AWGs), the advantage here lies in the size: the entire device occupies only $10 \times 10\,\mu$m, as compared to millimetres, or even centimetres, for typical AWGs.

The superprism operates near the band edge of the photonic crystal where the effective index of the lattice changes rapidly with wavelength. As the wavelength changes, the band becomes increasingly flat, so the group velocity decreases. The propagation direction of a Bloch mode in a photonic lattice is given by its group velocity (i.e. by its Poynting vector), which points in the direction normal to the equifrequency contour, or $\nu_g = \nabla_\kappa \omega(\kappa)$ [28]. Therefore, the superprism operates in a regime where the band has strong curvature, e.g. close to a band edge, as illustrated in Fig. 9.8. The bandstructure (Fig. 9.8a) represents the crystal by showing the major symmetry axes, whereas the wavevector diagram (Fig. 9.8b) represents the different crystallographic axes as they occur in real space, so they repeat every 60°. In other words, the wavevector diagram is obtained by "slicing" the bandstructure horizontally, at a given frequency. The resulting lines on a wavevector diagram are therefore called "equifrequency contours".

The origin of the superprism phenomenon is the fact that the equifrequency contours can change dramatically with a change in frequency, as shown in Fig. 9.8b [29]. In the example shown, the band becomes very flat for frequencies higher than the operating point around $a/\lambda = 0.248$ and the equifrequency contour curves strongly ($a/\lambda = 0.254$ in (b)). For lower frequencies, the band behaves almost like an isotropic medium (e.g. $a/\lambda = 0.238$ in (b)).

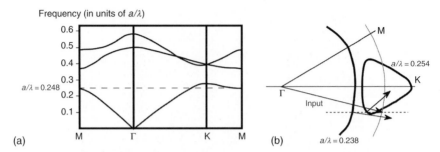

Fig. 9.8. (a) Bandstructure and (b) wavevector diagram for a triangular lattice of holes with 40% air fill-factor and $a = 320\,\text{nm}$, realised in a GaAs/AlGaAs heterostructure ($n \approx 3.2$)

This is also clear from the bandstructure: for the lower frequency, allowed states exist in both Γ–M and Γ–K direction, so propagation irrespective of direction is allowed; the magnitude of the wavevector in both Γ–M and Γ–K is almost identical, hence the resulting equifrequency contour is almost a circle. For higher frequencies, a stopband opens up in Γ–M; the allowed states therefore cluster around Γ–K, as seen by the small "island" that forms around the K-point in Fig. 9.8b).

The angular dispersion is then obtained as follows. The input wave travels in the unpatterned semiconductor waveguide, which is isotropic and therefore represented by a circular surface (dotted line in Fig. 9.8b). At the interface, the parallel wavevector is conserved, so all possible solutions have to lie on the dashed line. The intersection with the equifrequency contour then determines the direction of the output wavevector. For $a/\lambda = 0.238$, the output is almost collinear with the input wavevector and there is little change in the propagation direction. For $a/\lambda = 0.254$, the shape of the dispersion surface has changed dramatically, with a correspondingly large change in the output direction.

The above description only serves to highlight the superprism effect (also illustrated in Fig. 9.9) rather than trying to present a well-engineered device; such a device requires further research. The dependence of the operation on incident angle, for example, remains an issue, causing crosstalk [30]. A third issue is the interface reflectivity, caused by inefficient excitation of the relevant Bloch mode. All the three parameters need to be optimised, which suggests that more than one type of lattice should be involved, in the same way as one requires multiple equations to determine multiple parameters. Recent work suggests that using gradual interfaces [31] or modified interface-holes [32] helps to address the excitation problem, but more complex structures will need to evolve in order to address the problem in its entirety.

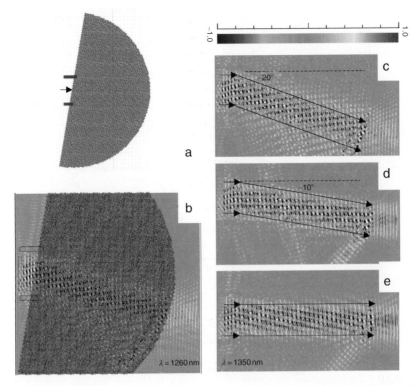

Fig. 9.9. Illustration of the superprism effect using FDTD simulation. The two key points to note are the change of propagation direction with wavelength and the clear signature of the Bloch modes in (c–e), i.e. the formation of an "envelope" described in Sect. 9.3. (a) Outline of the PhC structure for FDTD simulation. (b–e) Field distribution for H-mode (TE-like) at different wavelengths. Cases (b) and (c) are identical, except that the Phc outline is shown in (b) for illustration. A Gaussian beam with a width of 5 μm is launched 2 μm away from the input interface. The propagation direction changes from −20° (c) to −10° (d) and then to 0° (e) when the wavelength changes from 1260 to 1290 nm and 1350 nm

9.4.2 Supercollimator

As much as wavelength and/or input-angle dependent propagation may be desirable for some applications, there are other cases where beams need to be collimated in a circuit. For this purpose, we can use the fact that for some operating points, the propagation direction of the Bloch mode changes very little irrespective of input direction. The corresponding equifrequency contour is very flat, as shown in Fig. 9.10. Such a regime is referred to as the "supercollimator" [33] regime, or more recently also as "diffractionless" propagation [34].

The next, apparently obvious, step is the combination of a superprism and a supercollimator element to create a microspectrometer with good beam

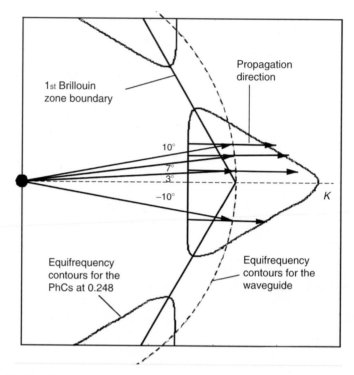

Fig. 9.10. Equifrequency contours at frequency 0.248 for the different incident angles of 3°, 7°, 10° and −10°. This graph clearly indicates that in the incident angle range of ±10°, the light in the PhC will propagate along Γ–K direction, irrespective of input angle, representing the case of collimation

quality. This approach has a fundamental flaw; however, While the collimator changes the direction of the Bloch mode, i.e. the Poynting or S-vector, it does not change the k-vector distribution. This means that an input beam, say with a ±10° angular distribution as in Fig. 9.10, may be redirected by the collimator, but the different angular components (or different k-vectors) that constitute the beam are conserved. Therefore, at the next interface, the different k-vectors will behave as if the collimator had not been there. Unlike a lens, the collimation effect only exists inside the particular photonic lattice, but is lost outwith.

In order to create a good microspectrometer, we therefore need to design a composite structure that advantageously combines the effects of both the k- and the S-vector dispersions.

9.5 Slow Waves

The theory of relativity tells us about the fundamental upper limit of the speed of light, i.e. $3 \times 10^8 \, \mathrm{m\,s^{-1}}$. Is there also a lower limit? Is there some fundamental lower boundary? Can light stand still? At first sight, the lower boundary seems to be close to the upper one, as the speed of light in any given medium is given by c/n, with c the speed of light in vacuum and n the refractive index of the respective medium. In order for the speed to go to zero, n would have to go to infinity. Considering that the highest refractive indices of typical optical materials are around 4, this seems rather excessive. Therefore, when people refer to "slow light", they typically refer to a slow *group* velocity, which is the speed at which the wave transports energy. In contrast, the above expression c/n describes the *phase* velocity of light, which is the speed at which the wave oscillates. Many types of waves may indeed have very high phase velocities, i.e. they oscillate at or near the speed of light, yet they may have very low group velocities and transport very little energy. The simplest example for such a wave is a standing wave, generated by two counterpropagating plane waves beating against one another. Such a wave has zero group velocity. Therefore, in terms of group velocity, slow speeds commonly occur, e.g. at the band edge of every Bragg mirror. The lower boundary for the phase velocity of light is therefore c/n, as mentioned above, but the lower boundary or the group velocity is indeed zero; light can be made to stand still and oscillate in place.

Making use of slow waves in a controllable and useful fashion is an active area of research; because increased light–matter interaction can be achieved in this way. The interest in this area of research is justified by the fact that strong light–matter interaction is otherwise difficult to achieve. This is because typically, lightwaves do not interact. In an optical fibre, for example, many different signals coexist without interacting a property that is widely exploited in optical telecommunications. If, in contrast, one does want the lightwaves to interact, e.g. for switching, optical logic or to extract gain from a material, this virtue becomes a major hindrance. One then wants to extend the interaction time between an optical pulse and the host material, e.g. to extract more gain from a laser, or to accumulate a larger phase shift in an interferometer. Slow light structures allow exactly that: increasing the light–matter interaction time.

In order to better understand the nature of slow waves in photonic crystal waveguides, we recently studied the mode propagation in these waveguides with a phase-sensitive near-field scanning optical microscope (PS-NSOM) [35]. The PS-NSOM gives insight into the nature of the different modes propagating in the waveguide. We observed a wide range of modes with different group velocities exhibiting both positive and negative phase velocities as well as a range of higher Bloch harmonics [27]. The most striking result, however, was the direct observation of a very slow mode, traveling at about 1/1,000th of the speed of light in vacuum (Fig. 9.11). This clearly demonstrates the opportunity for slow wave structures based on photonic crystal waveguides. By selectively exciting the slow mode (i.e. only the mode on the left of Fig. 9.11a, not the one on the right) using a suitable injector, the incoming pulse could be delayed in its entirety.

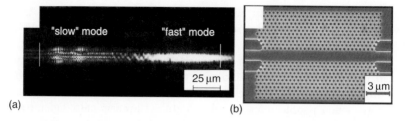

(a)   (b)

Fig. 9.11. Ultraslow light in photonic crystal waveguides. (**a**) Direct observation of pulse breakup in a W3 photonic crystal waveguide. The incoming 120 fs pulse couples to both the fundamental, "fast", mode (identified on the *right* of the picture) and a very slow higher order mode (on the (*left* of the picture). The vertical lines indicate the beginning and end of the photonic crystal section. (**b**) The type of waveguide used in the experiment. NB: The waveguide is only shown for illustrative purposes; as indicated by the scalebar, the structure used in (**a**) is eight times longer than the one shown in (**b**)

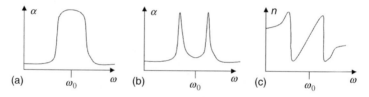

Fig. 9.12. Principle of slowdown via electromagnetically induced transparency using a simple two-level atom as an example. (**a**) The absorption characteristics of the original atom. (**b**) Absorption characteristics of bleached atom and (**c**) Resulting refractive index distribution

A related but different approach, based on coupled cavities was recently proposed [19]. Here, the signal is injected into a chain of cavities. The delay occurs because the signal takes longer time to tunnel between the respective cavities than it would to propagate through a normal waveguide. The chain of cavities is branched, so different signals can be "parked" in different branches. By detuning the cavities placed at the entrance/exit points to these branches, the light can be stored there at will – assuming that there is no loss in the cavity chains. If this scheme could be demonstrated experimentally, it would offer the possibility of real optical memory elements.

It is a useful exercise to compare these photonic crystal-based results to the ones obtained by other means. The most effective way of generating slow light, in fact, is via electromagnetically induced transparency (EIT). In simple terms, EIT works by bleaching an atomic transition (Fig. 9.12). Consider a simple two-level atom; its characteristic absorption centred at ω_0 is shown in Fig. 9.12a. By exposing the atom to an external laser beam centred at ω_0, the transition is saturated to the extent that it no longer absorbs at line centre and the double-peaked absorption characteristics shown in (b) is obtained. The corresponding phase function, or refractive index distribution, is shown in (c).

Because the refractive index varies so rapidly around ω_0, the group index, which is given by the first derivative of the phase index, can be very high indeed. As a consequence, propagation speeds down to $17\,\mathrm{m\,s^{-1}}$ $(c/18,000,000)$ has been measured in ultracold atomic gases [36]. In solid-state materials, which are ultimately more useful for practical applications, high slowdown factors have also been observed now, e.g. down to $c/31,000$ [37]. How does slow light in photonic crystals compare to this? Since the slowdown factors are so much higher, should we not be pursuing the EIT approach instead? The following table compares the two different approaches:

	EIT quantum dots [37]	photonic crystals First-order mode in W1 (Fig. 9.3b)
slowdown factor	10^4–10^5	100
bandwidth	GHz	THz
tunability	yes	maybe
loss	intrinsic absorption	roughness scattering
power requirement	medium (nonlinear optics)	low (linear optics)
oper. wavelength	determined by absorption line	solely determined by design

Several observations can be made:

– Although the EIT approach allows larger slowdown factors, its bandwidth is considerably lower. This is a critical limitation if one were to use this effect to slow down real data streams that may have a bandwidth of tens of $\mathrm{Gbit\,s^{-1}}$. Such bandwidths could not "fit" spectrally into an EIT-based delay line.

– EIT allows tuning of the slowdown factor by changing the power on the pumping laser. Although this is a useful tuning mechanism, it requires moderately high powers. The effect in the photonic crystal waveguide, in contrast, is entirely linear.

– Losses and operation wavelength in EIT are determined by the nature of the absorption line. In photonic crystals, there are no intrinsic losses, i.e. the effect is entirely dielectric,[4] and the wavelength can be determined freely.

The above discussion shows that photonic crystals offer several advantages. Once the effect is better understood, also in terms of the ability to tune it, photonic crystal based optical delay lines, buffers and possibly even memory elements may become a real possibility.

---

[4] Although roughness scattering does occur

9.6 Dispersion Compensation

When light passes through a waveguiding material, it inevitably experiences the fact that the material response varies with the wavelength of the excitation – the well-known effect of dispersion, or chromatic dispersion, to be more specific. Chromatic dispersion causes multichromatic signals to drift apart and short pulses to be separated into their spectral components. It also limits processes such as parametric wavelength conversion and second harmonic generation that require a fixed phase relationship between signals of different wavelength. Dispersion therefore tends to be an undesirable process. Furthermore, the dispersion of any given material is difficult to adjust, as it depends on its intrinsic electronic properties. As a result, a lot of research has been conducted on materials that exhibit low dispersion, at least over a given wavelength range of interest, in order to minimise the "damage" incurred by the dispersive process. Why, therefore, should we be interested in studying structures such as photonic crystals that can exhibit particularly strong dispersion?

The answer is obvious. If the material choice is imposed by some other constraint, such as transparency window, refractive index or availability of gain or nonlinearities, one needs to introduce some other mechanism to control the dispersion. Photonic crystals offer this mechanism, because their response can be adjusted almost independently of any other material parameter, as long as the required refractive index contrast can be achieved. To illustrate, photonic crystal fibres offer dispersion control over a wide range of applications such as endlessly single mode operation or supercontinuum generation [38].

What levels of dispersion can be achieved with photonic crystals? The high dispersion arises from the fact that the properties of the Bloch modes may be highly frequency dependent, as already illustrated by the superprism. In bandstructure terms, this dispersion is described by the curvature of a given band, and referred to as the group velocity dispersion (GVD). The GVD is given by the second derivative of the bandstructure diagram at a given point,

$$\text{GVD} = \frac{2\pi c}{\lambda^2} \left(\frac{\mathrm{d}^2 k}{\mathrm{d}\omega^2} \right) \tag{9.3}$$

and expressed in units of $\mathrm{ps\,nm^{-1}\,km^{-1}}$. Optical fibres operate in the range of 0–$100\ \mathrm{ps\,nm^{-1}\,km^{-1}}$ and semiconductor waveguides up to $10^3\ \mathrm{ps\,nm^{-1}\,km^{-1}}$ (e.g. Si [39]). The highest experimentally demonstrated value in photonic crystal waveguides reported thus far is $10^7\ \mathrm{ps\,nm^{-1}\,km^{-1}}$ (Fig. 9.13) [17], although values as high as $4 \times 10^8\ \mathrm{ps\,nm^{-1}\,km^{-1}}$) have already been predicted numerically [18]. Considering a typical interaction length of 1 mm in photonic crystal waveguides, which appears reasonable in view of the recently reported losses of $7\,\mathrm{dB\,cm^{-1}}$ [8], the latter value would correspond to $400\,\mathrm{ps\,nm^{-1}}$. Values of this order of magnitude compare well with competing technologies, such as planar waveguide-based ring-resonators ($\pm 1{,}000\ \mathrm{ps\,nm^{-1}}$) [40] or fibre Bragg gratings.

Fig. 9.13. Illustration of pulse compression in a photonic crystal waveguide [17]. A chirped pulse, represented by the greyscale, enters the very dispersive waveguide and, by selectively delaying the wavelength components with respect to one another, is recompressed

How much dispersion compensation is needed, however? A typical fibre span of 50 km at 20 ps nm^{-1} km^{-1} accumulates 1,000 ps nm^{-1}. At 10 Gbit s^{-1}, electronic techniques can compensate for dispersion of this magnitude due to recent advances in digital signal processing techniques [41]. Since the impact of dispersion scales with the bit rate squared, however, optical techniques offer clear advantages once systems move to 40 Gbit s^{-1} and beyond. Photonic crystals are a possible contender here as they offer true on-chip dispersion compensation. Due to their small size, the latency between channels is also minimised, which means that the dispersion compensation element causes minimal time delay between a compensated and an uncompensated channel. Overall, while a lot of research needs to be done before photonic crystal dispersion compensating elements can be considered for real systems, the promise is clearly there.

9.7 Photonic Crystal Cavities

The strong confinement offered by the photonic bandgap leads to some of the most compelling applications for photonic crystals: the realisation of high Q cavities. The fact that light can be confined in a very small space with very high spectral purity (i.e. narrow spectral linewidth or high Q) allows one to realise the original goal of photonic bandgap research: the control of spontaneous emission.

9.7.1 Purcell Effect and Strong Coupling

The equation governing the interaction between an emitter and a cavity was first proposed by Purcell [42] in 1946, and applies when the linewidth of

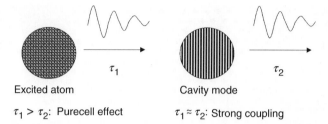

Excited atom   Cavity mode

$\tau_1 > \tau_2$: Purecell effect $\tau_1 \approx \tau_2$: Strong coupling

Fig. 9.14. Purcell effect and strong coupling in photonic crystal microcavities. The atomic emitter transfers energy to the cavity with a characteristic lifetime τ_1, which is then transferred to the outside world at the cavity photon lifetime τ_2. Typical lifetimes are nanoseconds for τ_1 and femtoseconds for τ_2 (a few optical cycles for low Q cavities), so there is no interaction. If Q/V goes up, τ_1 decreases due to the Purcell effect and τ_2 increases due to the longer photon lifetime in a high Q cavity. Once the two are approximately equal, they interact strongly ("strong coupling") and start to exchange energy ("Rabi oscillations")

the emitter is equal or narrower than that of the cavity, and the two are in resonance

$$F_\text{P} = \frac{3}{4\pi^2} \frac{Q}{V/\lambda_\text{m}^3}, \qquad (9.4)$$

where F_P stands for the Purcell factor and λ_m is the effective wavelength in the medium, so V/λ_m^3 can be understood as the normalised volume of the cavity in multiples of the wavelength. The dominant factor is therefore Q/V; the smaller the real and the spectral space over which the emitter and the cavity interact, the stronger the effect. The factor F_P then describes the change in decay rate over an emitter in "free space". Please note that F_P can be either smaller or larger than unity, so the cavity may both suppress or enhance the emission. Large Purcell factors in excess of 100 or even 1,000 can be achieved in photonic crystals [43,44], so the radiative lifetime may be shortened by several orders of magnitude. In fact, photonic crystals offer the highest Q/V factor of any known type of cavity. Once the Purcell factor is large enough for the radiative lifetime of the emitter to approach that of the photon lifetime in the cavity, the two may interact and the system enters the "strong coupling" regime. This regime can be understood as a system of two coupled oscillators that exchange energy at the characteristic Rabi frequency, which is determined by the coupling coefficient between the two. Figure 9.14 schematically describes the different regimes.

9.7.2 High Q Cavities in 2D Photonic Crystals

How is it possible to achieve these high Q-factors in what appears to be an "open" cavity? Do we not need full 3D confinement? The answer lies in balancing the two effects of Bragg reflection by the photonic crystal, and total internal reflection by the semiconductor slab (Fig. 9.15). The key is cavity

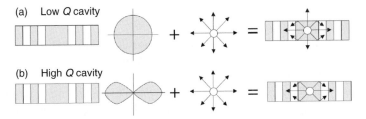

Fig. 9.15. Realisation of high Q cavities in planar photonic crystals. **(a)** If the emission pattern of the cavity (indicated by the *circle*) has out-of-plane components that lie outwith the total internal regime (i.e. above the *light line*), radiation generated by an emitter can escape and the cavity Q is low. **(b)** If the emission pattern (indicated by the *dipole shape*) lies below the light line, the radiation is contained inside the cavity by total internal reflection and subject to the full Bragg reflection of the photonic crystal mirrors

emission pattern: By creating a cavity mode with an emission pattern that has no component normal to the slab, all the emission are contained in the device by total internal reflection and thereby subject to the full strength of the Bragg reflection of the 2D lattice (Fig. 9.15). Such an emission pattern can be created by k-space or Fourier space engineering of the cavity mode [45], a method that has demonstrated good agreement with recent experimental results [43, 44]. An alternative explanation that may be even more powerful is based on impedance matching and slow wave effects [46]. Either method helps the understanding of how very high Q factors can be obtained with photonic crystal microcavities, but the difference between low and high Q cavity designs is very subtle, thereby emphasising the requirement of high fabrication accuracy.

9.8 Conclusions

The uniqueness of photonic crystals derives from two basic properties: Bloch modes and confinement. The Bloch modes lead to the dispersive phenomena, such as superprism, pulse compression and slow light, whereas the confinement allows us to create tiny waveguides and cavities with amazing Q/V values.

Of the two properties, the Bloch modes have so far been less successfully exploited. This is in part to do with the fact that their rich behaviour is more difficult to understand. With improved understanding, and the further technological advances demonstrated by the recently demonstrated low propagation losses, the realisation of devices such as microspectrometers, dispersion compensators and slow light elements leading to optical buffers and possibly even optical memories appears in reach. Furthermore, the dispersion control allows phase matching for parametric nonlinear applications. If such phase matching can be achieved with tolerable losses on the required length scale,[5] novel nonlinear devices will appear.

Another point to make about photonic crystals is that their function is determined by their shape, rather than some intrinsic material property. Therefore, a desired property can, in principle, be obtained for any given wavelength due to the scalability of electromagnetic waves. This advantage becomes clear when looking at properties such as chromatic dispersion, which is intrinsic to a given material. Photonic crystals can override this intrinsic material property and enable the creation of flat, negative or positive dispersion slopes at any desired wavelength, as already demonstrated in photonic crystal fibres [38]. The same applies to slow light: Whereas the EIT-based generation of slow light is intrinsically fixed by the absorption line of the chosen system, photonic crystals allow one to create slow light at any desired wavelength.

Although the above description has described many applications of 2D photonic crystals, ranging from the "optical insulator" phenomenon to real control of the spontaneous emission process in high Q cavities, it would be wrong to assume that there is no need for 3D photonic crystals. 3D crystals offer additional unique properties: They allow the control of spontaneous emission over a broad bandwidth, i.e. not just in high Q cavities. Furthermore, as far as waveguide applications are concerned, they offer the opportunity of exploiting the entire range of available k-vectors, i.e. not just those that lie below the light line. This would open up a much wider operating range and remove the light line limitation.

Acknowledgments

I hope to have clarified the basics of photonic crystal operation and conveyed some of the excitement of a very rich field that continues to attract researchers from a wide range of disciplines. I would like to thank my coworkers for their contributions, especially my research group at St Andrews, as well as collaborators in the Ultrafast Photonics Collaboration in the UK and collaborators in Ghent, Belgium as well as Twente, The Netherlands. I also acknowledge funding by the Engineering and Physical Sciences Research Council of the UK as well as the European Community.

References

1. E. Yablonovitch, Phys. Rev. Lett. **58**, pp. 2059–2062 (1987)
2. S.Ogawa, M. Imada, S. Yoshimoto, M. Okano and S. Noda, Science **305**, pp. 227–229 (2004)
3. P. Lodahl, A.F. van Driel, I.S. Nikolaev et al., Nature **430**, pp. 654–657 (2004)

---

[5] Typical nonlinear coefficients are weak, hence require long interaction lengths. Nonlinear devices used in practise, such as those based on $LiNbO_3$, tend to be on a cm lengthscale

4. T. Yoshie, A. Scherer, J. Hendrickson et al., Nature **432**, pp. 200–203 (2004)
5. J. Joannopoulos, P. Villeneuve and S. Fan, Nature **386**, pp. 143–149 (1997)
6. T.F. Krauss, S. Brand and R.M. De La Rue, Nature **383**, pp. 699–702 (1996)
7. S.J. McNab, N. Moll and Y. Vlasov, Opt. Exp. **11**, pp. 2927–2939 (2003).
8. Y. Sugimoto, Y. Tanaka, N. Ikeda et al., Opt. Exp. **12**, pp. 1090–1096 (2004)
9. L.H. Frandsen, A. Harpøth, P.I. Borel et al., Opt. Exp. **12**, pp. 5916–5921 (2004)
10. R. Wilson, T.J. Karle, I. Moerman and T.F. Krauss, J. Opt. A **5**, pp. S76–S80 (2003)
11. M. Ayre, T.J. Karle, L. Wu et al., IEEE J. Sel. Areas Commun. (JSAC), **23**, pp. 1390–1395 (2005)
12. E.A. Camargo, H.M.H. Chong and R.M. De La Rue, Opt. Exp. **12**, pp. 588–592 (2004)
13. H. Nakamura, Y. Sugimoto, K. Kanamoto et al., Opt. Exp. **12**, pp. 6006–6014 (2004)
14. H. Kosaka, T. Kawashima, A. Tomita et al., Phys. Rev. B **58**, pp. R10096–R10099 (1998)
15. M. Notomi, Phys. Rev. B, **62**, pp. 10696–10705 (2000)
16. L. Wu, T. Karle, M. Steer and T.F. Krauss, IEEE J. of Quantum Electron. **38**, pp. 915–918 (2002)
17. T.J. Karle, Y.J. Chai, C.N. Morgan et al., IEEE JLT **22**, pp. 514–519 (2004).
18. A. Petrov, K. Preusser-Mellert, G. Böttger et al., PECS-5, Kyoto, Japan, March 2004
19. M.F. Yanik and S. Fan, Phys. Rev. Lett. **92**, Art. No. 083901
20. S.Y. Kim, G.P. Nordin, J.B. Cai and J.H. Jiang, Opt. Lett. **28**, pp. 2384–2386 (2003)
21. L.J. Wu, M. Mazilu, J.-F. Gallet et al., Opt. Lett. **29**, pp. 1620–1622 (2004)
22. M.V. Kotlyar, L. Bolla, M. Midrio et al., Opt. Exp. **13**, pp. 5040–5045 (2005)
23. Y.A. Vlasov and S.J. McNab, Opt. Exp. **12**, pp. 1622–1631 (2004)
24. M.V. Kotlyar, T. Karle, M.D. Settle et al., Appl. Phys. Lett. **84**, pp. 3588–3590 (2004)
25. P.StJ. Russel, Phys. World, pp. 37–42, (Augst 1992)
26. M. Loncar, D. Nedeljkovic, T.P. Pearsall et al., Appl. Phys. Lett. **80**, pp. 1689–1691 (2002)
27. H. Gersen, T.J. Karle, R.J.P. Engelen et al., Phys. Rev. Lett. **94**, p. 123901 (2005).
28. P. Yeh, J. Opt. Soc. Am. **69**, p. 742 (1979).
29. L. Wu, M. Mazilu, and T. F. Krauss, IEEE J of Lightwave Technol. **21**, pp. 561–566 (2003)
30. T. Baba and T. Matsumoto, Appl. Phys. Lett. **81**, pp. 2325–2327 (2002)
31. J. Witzens, M. Hochberg, T. Baehr-Jones and A. Scherer, Phys. Rev. E **69**, Art. No. 046609 (2004)
32. T. Baba and D. Ohsaki, Jpn. J. Appl. Phys. **40**(Pt.1), pp. 5920–5924 (2001)
33. H. Kosaka, T. Kawashima, A. Tomita et al., Appl. Phys. Lett. **74**, pp. 1212–1214 (1999)
34. D.M. Pustai, S.Y. Shi, C.H. Chen, A. Sharkawy and D.W. Prather, Opt. Exp. **12**, pp. 1823–1831 (2004)
35. H. Gersen, T.J. Karle, R.J.P. Engelen et al., Phys. Rev. Lett. **94**, Art. No. 073903 (2005)
36. L.V. Hau, S.E. Harris, Z. Dutton and C.H. Behroozi, Nature **397**, pp. 594–598 (1999).

37. P.-C. Ku, F. Sedgwick, C.J. Chang-Hasnain et al., Opt. Lett. **29,** pp. 2291–2294 (2004)
38. J.C. Knight, Nature **424**, pp. 847–851 (2003)
39. H.K. Tsang, C.S. Wong, T.K. Liang et al., Appl. Phys. Lett. **80**, pp. 416–418 (2002)
40. C.K. Madsen, IEEE JLT **21**, pp. 2412–2420 (2003)
41. M.G. Tyler, IEEE PTL **16**, pp. 674–676 (2004)
42. E.M. Purcell, Phys. Rev. **69**, p. 681 (1946)
43. Y.Akhane, T. Asano, B.S. Song and S. Noda, Nature **425**, pp. 944–947 (2003)
44. B.S. Song, S. Noda, T. Asano and Y. Akhane, Nature Mater. **4**, pp. 207–210 (2005)
45. O. Painter, J. Vuckovic and A. Scherer, J. Opt. Soc. Am. B **16**, pp. 275–285 (1999)
46. C. Sauvan, P. Lalanne, J.P. Hugonin, Nature **429**, p. U1 (2004)

10

On Chip Optical Waveguide Interconnect: The Problem of the In/Out Coupling

R. Orobtchouk

Summary. This chapter is intended to give a description of different optical couplers used for the injection of light in a silicon waveguide. It starts with an introductory part, which provides a general overview of the coupling problem. The next four parts present the theory of the different couplers that use a butt coupling method like the 3D taper, the tips taper and the guide to guide coupler. The last three parts are devoted to the transverse coupling techniques which are the grating and the prism couplers.

10.1 Introduction

An optical system on chip is constituted by several basic integrated optics elements. The first element is the light source provided with a modulator which can be internal or external so as to transform an electric signal into an optical signal. This source can be either integrated into the device or external. In the case of an external source, the light will be injected into the optical system on chip through an optical coupler. The second element is the waveguide which transports the optical signal in the circuit. Splitters and bends are used to distribute the signal at N points of the circuit. The optical signal in these N points is converted to an electric signal by an integrated photodetector.

The silicon is a material of choice to develop this kind of optical system on chip because it is widely used in the industry of the microelectronics. We can benefit from the maturity of the manufacturing technology of CMOS integrated circuits for a massive production of low-cost optical circuits. The hybridization of CMOS integrated circuits (IC) with optical components is possible and can occur at various levels.

The first approach consists in making the optical components at the same level of the CMOS IC: the front end approach. The material used to develop this approach is the Silicon-on-Insulator (SOI) [1–3]. SOI substrate is made of a thin silicon top layer separated from the silicon substrate by a buried oxide layer. According to the transparency of these layers in the near infrared range

($\lambda = 1.2$–$1.7\,\mu$m) and to the large refractive index difference between silicon ($n = 3.5$) and SiO_2 ($n = 1.45$), light can be guided in the thin silicon top layer. The strong refractive index contrast between the core and the cladding material of the waveguide allows the realization of very compact components. Waveguides cross-section have sizes lower than a micrometer.

The second approach, also called the above IC approach, consists in realizing the optical circuits above the electronic circuits. A first possibility consists in realizing the optical components in dielectrics (doped silica [4,5], silicon nitride [6–8] or silicon oxynitride on oxide [8,9]), or polymers [10,11] materials which are deposited above the electronic components. The second possibility consists in separating the manufacturing of the electrical and optical circuits. The optical devices are made on a separate wafer and this optical plate is put together on the CMOS IC by a flip chip or wafer bonding process. In that case, all the materials previously mentioned are useful and also polycrystalline silicon material [12] can be used.

The main drawback of silicon is that we cannot integrate a light source favoring a solution using an external light source. It raises the problem of the external coupling of this source with the main optical element of the optical system on chip: the waveguide.

Dielectrics or polymer waveguides have a weak refractive index difference between the cladding and the core layers that leads to cross-sections of the waveguide in a range of 1–10 μm. These sizes are comparable to those of a standard optical fiber or a tapered fiber and in that case the coupling does not raise dry problem.

On the other hand in the case of silicon waveguides, the strong contrast between the refractive index of the silicon and the silicon oxide layers imposes sizes of waveguides lower than a micrometer. In that case the coupling of the light in such a waveguide is a hard problem. This chapter is dedicated to the description of the different couplers used between an optical fiber and a silicon photonics device.

10.2 Coupling Efficiency

The butt coupling method is the most attractive solution of an industrial point of view because we can benefit from its maturity and from the wide development in the field of the optical telecommunications.

For this kind of device, the coupling efficiency is given by the overlapping integral:

$$\eta = \eta_{\mathrm{F}} \frac{\left| \int E_{\mathrm{fiber}}(x,y)\, E^*_{\mathrm{waveguide}}(x,y) \mathrm{d}S \right|^2}{\int |E_{\mathrm{fiber}}(x,y)|^2\, \mathrm{d}S \int |E_{\mathrm{waveguide}}(x,y)|^2\, \mathrm{d}S}, \tag{10.1}$$

where E_{fiber} and $E_{\mathrm{waveguide}}$ are the transverse modal component of the fiber and waveguide electric fields. η_{F} is the transmission coefficient of the waveguide facet for a normal incident beam.

In the case of a normal incident beam between two mediums, the transmission coefficient is given by:

$$\eta_F = \frac{4 n_{inc} n_0}{(n_{inc} + n_0)^2},$$
(10.2)

where n_{inc} and n_0 are, respectively, the refractive index of the incident material and effective index of the waveguide mode.

In most part of the applications, the used waveguide has to be monomode. In that case, the guided mode and the optical fiber beam can be approximated by gaussian beams. With this approximation, an analytical expression of (10.1) is obtained [13]. This analytical expression gives the tolerance alignments in the case of a beam waist variation, a defocusing effect, a lateral shift, and an angular detuning between the incident optical fiber and the waveguide.

$$\begin{cases} \dfrac{\eta_1}{\eta_F} = \left(\dfrac{2 w_{inc} w_0}{w_{inc}^2 + w_0^2} \right)^2 & \text{(beam waist variation)} \\[2ex] \dfrac{\eta_2}{\eta_F} = \dfrac{1}{1 + \left(\dfrac{\Delta z}{k_0 n w_0^2} \right)^2} & \text{(defocusing)} \\[2ex] \dfrac{\eta_3}{\eta_F} = \exp\left(-\dfrac{\Delta x^2}{w_0^2} \right) & \text{(lateral shift)} \\[2ex] \dfrac{\eta_4}{\eta_F} = \exp\left(-\dfrac{1}{4} k_0^2 n^2 w_0^2 \Delta\theta^2 \right) & \text{(angular detuning)}, \end{cases}$$
(10.3)

where w_{inc} and w_0 are, respectively, the waists of the fiber incident beam and of the waveguide. Δz, Δx, and $\Delta\theta$ are, respectively, the defocusing distance, the lateral shift, and the angular detuning between the optical fiber and the waveguide.

Figure 10.1 shows the different effects involving a decrease in the coupling efficiency. By using (10.3) in the case of a beam waist variation, we see that when we place end to end a single mode fiber of 5.6 μm beam waist radius and a silicon waveguide of 0.16 μm mode radius, the efficiency of coupling is at most 0.3%, if we neglect the Fresnel reflection loss at the waveguide facet.

A solution is to use a tapered fiber or a focus lens. In that case, the diffraction limit imposes that the minimum radius of the incident beam is about 1 μm. The efficiency of coupling is at most 10%.

A solution of the problem of coupling is to integrate spot-size-converter at the input and the output facets of the waveguide to have the same beam waist radius. The tolerance alignments for a loss at 1 dB are reported in Table 10.1.

Current fiber attach technique where developed [14–19]. These techniques use selective etching of the silicon with V or U grooves which allow a good positioning of the fiber with the input waveguide. The achievable accuracy of the passive alignment technique is typically of ± 0.5 μm and do not involve a loss greater than 1 dB according with Table 10.1.

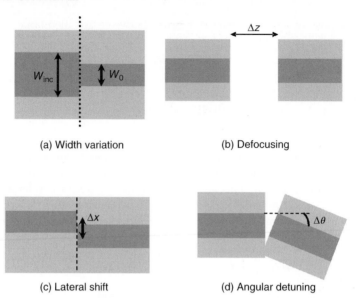

(a) Width variation   (b) Defocusing

(c) Lateral shift   (d) Angular detuning

Fig. 10.1. Different configurations involving a decrease in the coupling efficiency (**a**) different widths, (**b**) defocusing effect, (**c**) lateral shift, and (**d**) angular detuning

Table 10.1. Tolerance alignments for a loss at 1 dB, respectively, for beam waist radius of 5.6 and 1 μm

beam waist radius	defocusing	lateral shift	angular detuning (deg)
5.6 μm	± 33 μm	± 2.7 μm	± 2.4
1.0 μm	± 1.1 μm	± 0.48 μm	± 13.6

The simplest spot size converter that can be used to expand the waveguide width of the silicon waveguide is a lateral taper. In this case, the problem of the overlapping of the modes in the vertical direction still remains. To obtain a further improvement in coupling, it is necessary to increase the height to the silicon waveguide. These kinds of devices are called 3D tapers.

10.3 3D Taper

These devices were developed for applications in the optical telecommunications on III–V substrate to solve the problem of the coupling of light between a DFB laser source and an optical fiber. A state of the art of the main realizations on III–V substrate was made in the paper [20].

The simulation tools used for the design the 3D tapers are the mode solvers and propagative methods as the beam propagation method (BPM), the method of lines (MOL), the finite difference time domain method (FDTD), and the mode matching method.

Mode solvers are essentially used to find the propagation constant of guided modes in longitudinally invariant waveguides. The knowledge of the propagation constant and the optical field distributions of the different modes that are supported by the waveguide is essential to display the coupling efficiency between two waveguides.

A review of the different methods used to develop a mode solver is made by Vassallo [21]. The different methods can be classified in two categories: the global methods and the transverse resonant methods. Global methods include mainly finite element methods, finite difference methods. The most used transverse resonant methods are: the effective index method, the spectral index method, the method of lines, and the modal transverse resonance method.

We can notice that the effective index method is the only method that is not a full vectorial method. This method is very useful in case we want to simplify a 3D problem and to come down to a 2D problem when the propagative methods are much more consuming in time and memories. The propagative methods which are mentioned have been described in Scarmozzino's review [22]. A comparison between different propagative schemes for the simulation of 2D taper is realized by Haes et al. [23]. 3D simulations using the spectral index [24], the beam propagation [25,26], and the FDTD [27] methods are reported in the literature.

The most powerful method to simulate the propagation of an optical field through an arbitrary waveguide is the mode matching method. In this method, the waveguide is approximated by a stack of longitudinally invariant waveguide section as depict in Fig. 10.2.

Any field in a cross-section of an optical waveguide is written as a sum of the modes of the invariant longitudinal waveguide. This sum includes the guided modes and the radiation modes. The propagation of each mode through the section is described by a phase factor that gives an elementary

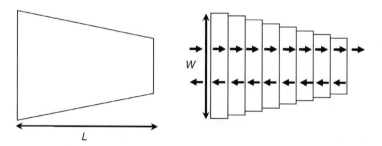

Fig. 10.2. Concatenation of the taper in the longitudinal direction. Any field in a cross-section is written as a sum of the modes of the invariant longitudinal waveguide. These modes are called the normal local modes

matrix describing the propagation of the backward and forward modes in the waveguide. At the end of a section, these modes all couple to the modes of the next section. By using the properties of the continuity of the transverse components of electric fields and its first derivation, we obtain a matrix describing the passage between two neighboring waveguides. We can so connect the field at the beginning of the taper to the field at the end of the taper by a product of all these matrices. Also the power reflected back into the modes of the preceding section is taken into account, making the mode matching technique a bidirectional algorithm. The concept of the local normal modes was first introduced by Louisell and described in detail by Burns [28].

It allows to define an adiabaticity criteria for the length and profile optimizations of the taper [29, 30]. This method is available in the commercial software FIMMPROP [31].

The commonly used adiabatic condition is derived from the physical argument that the local taper length-scale $(W/(\mathrm{d}W/\mathrm{d}z))$ must be much larger than the beat length (L_b) between the fundamental mode (with propagation constant β_1) and the dominant coupling mode (with propagation constant β_2) for power loss to be small [29].

The beat length between the modes is given by

$$L_\mathrm{b} = \frac{2\pi}{\beta_1 - \beta_2}. \tag{10.4}$$

The adiabatic condition is given by

$$W/\frac{\mathrm{d}W}{\mathrm{d}z} \gg L_\mathrm{b}. \tag{10.5}$$

It is more useful to define the local angle taper $(\Omega = (\mathrm{d}W/W))$. Then, from (12.4), we find that the local angle taper must satisfy:

$$\Omega \langle\langle \frac{W}{L_\mathrm{b}}. \tag{10.6}$$

We can notice that the beat length will vary with the waveguide width. We see from (10.6) that the limit on the taper angle will change along the longitudinal distance, resulting in a taper with nonlinear profile. The design of a linear taper imposes that the constant angle taper must be less than the lowest local angle taper.

A schematic view of the 3D tapers that achieves low loss coupling to single mode fiber is given in Fig. 10.3. The two mains solutions developed for the silicon photonic are the pseudovertical taper (Fig. 10.3a) and the gradual horizontal and vertical taper (Fig. 10.3b). The first solution can be realized with a standard lithography process contrary to the second solution where a gray scale lithography step is needed.

The principal realizations of this kind of taper are described in papers [18, 32, 33]. Losses of 0.5 dB for a pseudovertical taper from a 12 μm by 12 μm

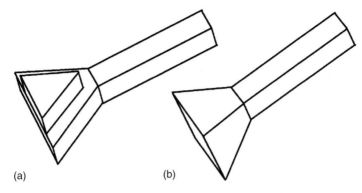

Fig. 10.3. Schematic view of the 3D taper made on SOI waveguide with low loss coupling to single mode fiber, (**a**) pseudovertical taper and (**b**) gradual horizontal and vertical taper

input to $4\,\mu m$ by $5\,\mu m$ waveguides have been demonstrated [32]. The problem of this kind of taper is that the taper length is about 1 or 2 mm, resulting in a noncompact device. It can be used only for the realization of a discrete circuit.

10.4 Tips Taper

Other solution to realize an efficient coupling is to use an inverse taper called tips taper. A schematic view of the tips taper is given in Fig. 10.4a. We have reported the evolution of the mode profile for a quadratic tips taper.

At the input of the device, the cross-section of the waveguide is very weak and it is gradually increased up to the size of the waveguide used for the optical signal transport. Because of the small waveguide cross-section, the mode is less confined in the core layer and the mode effective index tends to the value of the cladding material refractive. In these conditions, the coupling efficiency is increased because the good overlapping between the incident beam of the optical fiber and the waveguide mode, and the reflectivity of the facet is decreased. The cross-section of the waveguide is then gradually widened, so that the mode is more and more confined in the core layer.

The design of such a coupler involves first the knowledge of the input cross-section in such way to obtain the same mode profile as the optical fiber beam.

Figure 10.5 shows the mode profile of a strip silicon waveguide of, respectively, 0.14 and $0.3\,\mu m^2$. Calculation is performed at $\lambda = 1.55\,\mu m$ with a full vectorial finite difference method [34]. Symmetric and transparent [35] boundaries conditions are added to reduce the calculation time.

The mode radius of the lower strip silicon waveguide is closed to that of a standard optical fiber and then a high coupling efficiency can be achieved.

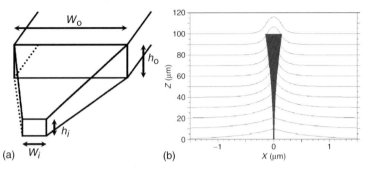

(a) W_i (b) $X\,(\mu m)$

Fig. 10.4. (a) Schematic view of a tips taper, (b) Evolution of the mode profile in the tips taper. At the input of the device, the cross-section of the waveguide is very weak and the mode is near the cut of condition. The effective index value of the mode is closed to the value of the refractive index of the cladding material and the mode profile is enlarged

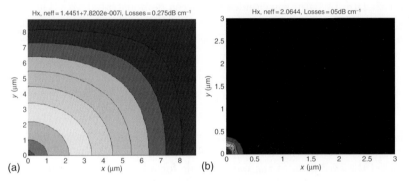

Fig. 10.5. Major part of the transverse magnetic field of a strip silicon waveguide modes. The cross-section is about $0.14\,\mu m^2$ (a) and $0.3\,\mu m^2$ (b)

The $0.3\,\mu m^2$ cross-section of the waveguide at the end of the taper is necessary to preserve a monomode condition of propagation. Because of the weak difference of the two cross-sections, the taper length is very short ($\approx 50\,\mu m$).

A 3D FDTD modeling of this taper have been made by Vivien [36], resulting on a coupling losses lower than $0.2\,dB$ (coupling efficiency of 95%) and an alignment tolerance larger than $4\,\mu m$ at $\pm 1\,dB$.

Almeida et al. [37] have measured a coupling efficiency of 12% for a $40\,\mu m$ long Si nanotaper in a SiO_2 cladding. The cross-section at the beginning and the end of the tips are, respectively, of $0.12\,\mu m$ by $0.22\,\mu m$ and $0.45\,\mu m$ by $0.22\,\mu m$. The coupling efficiency is increased up to 40% by Shoji et al. [38] by using a nanotaper with a polymer cladding waveguide. Because of the thickness of the buried silicon oxide of the SOI material of $2\,\mu m$ and the presence of the silicon substrate, the guided mode becomes asymmetric and the overlapping with incident beam is reduced. The role of this cladding waveguide is to reduce the asymmetry of the guided mode in the presence of the silicon substrate.

10.5 Guide to Guide Coupler

Another way to transfer light in a waveguide is to use a guide to guide coupling effect. This effect occurs when two waveguides are closed to each other. On Fig. 10.6, we have reported the evolution of the mode profile along the propagation direction for a structure composed by two twin waveguides. We can observe a mode beating. The light passes gradually from a waveguide to the other one and conversely. The well-known coupled mode theory gives a complete explanation of the guide to guide coupling effect [30, 39, 40]. This effect is illustrated in Fig. 10.7.

At the beginning of the coupling region, the incoming mode excites equally the two normal modes of the twin waveguides structure. The two normal

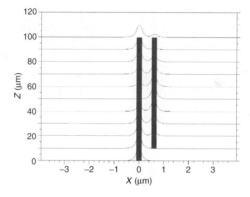

Fig. 10.6. Evolution of the mode profile when two identical waveguides are closed to each other

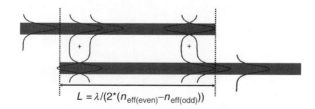

$$L = \lambda/(2^{*}(n_{\text{eff(even)}} - n_{\text{eff(odd)}}))$$

Fig. 10.7. Basic concept of the guide to guide effect. The incoming mode excites equally the even and odd modes of the twin waveguides. These two modes propagate with different phase velocities until the end of the coupling region. A total transfer of the light occurs only when the phase shift between the two modes is equal to π. In this case all the light is in the silicon waveguide

modes are the even and odd modes composed by a superposition of the amplitudes of the two fundamental modes of the unperturbed waveguides. All of the light power is on the low refractive index waveguide. They propagate with different phase velocities until the end of the coupling region. Their superposition at the output determines the power of light that is transferred to the outgoing mode in the silicon waveguide. A total transfer of the light occurs only when the phase shift between the two modes is equal to π. In this case, the electromagnetic field in the structure is given by the subtraction of the even and odd modes. All of the light power is in the silicon waveguide. The length required for a total transfer of light is given by:

$$L = \frac{\lambda}{2(n_{\mathrm{eff}}\mathrm{even} - n_{\mathrm{eff}}\mathrm{odd})}. \tag{10.7}$$

As the two waveguides have not the same geometries and refractive index, a strong coupling effect occurs only if the two waveguides have the same effective index.

In the case of two different waveguides, the coupling effect decrease rapidly and the coupling efficiency is not total. For a nonidentical waveguide a complete transfer of the light occurs only when the phase matching condition is verified. This means that the two unperturbed waveguides have the same propagation constants.

A way of obtaining this phase matched condition is to draw the diagram representing the evolution of the propagation constants of the modes in both waveguides versus the width of one of the guides. The anticrossing point gives the correct value of the waveguide width that is needed to obtain a complete transfer of light between the two waveguides. This effect can be used to obtain a strong coupling efficiency between an optical fiber and a submicronic silicon waveguide.

The basic concept of the guide to guide coupler is illustrated in Fig. 10.8. The light from the fiber is first injected in a polymer or dielectric waveguide

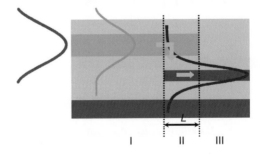

I   II   III

Fig. 10.8. Schematic view of a guide to guide coupler. The light is first injected in a dielectric waveguide with a low refractive index contrast between the core and the cladding. The size of the guided modes is closed to the size of the tapered fiber mode, that is given an high coupling efficiency. The light is transferred in the silicon waveguide using a guide to guide effect on top of an SOI wafer. As the refractive index difference between the core and the cladding layer of the waveguide is smaller than the SOI waveguide, the mode is wider and then the overlapping with the beam of the fiber is increased. Light is transferred to the submicronic silicon waveguide by a guide to guide coupling effect. The design of the coupler involves the knowledge of the cross-section of the two waveguide.

The first condition to satisfy is that the low refractive index waveguide must have a good overlapping with the beam of the fiber. A waveguide of $2\,\mu m^2$ cross-section $(n_{core} = 1.6)$ is compatible with a commercial tapered fiber. In order to have the correct geometry of the silicon waveguide, we plot the diagram giving the effective index of the twin waveguides modes versus the size of the silicon waveguide. This diagram is shown in Fig. 10.9.

A strong coupling effect occurs at the anticrossing point of this diagram. This gives the correct value of the silicon waveguide cross section $(0.23\,\mu m^2)$. According to (10.9), the difference between the two effective indexes at the anticrossing point gives the value of the coupling length needed for a total transfer of light in the submicronic silicon waveguide.

A straight line on a logarithmic scale is obtained for the plot of the coupling length versus the spacing length between the two waveguides. A linear regression gives a simple relation:

$$L = 10^{Cx+D}, \tag{10.8}$$

where x stands for the spacing between the two waveguides and C and D are, respectively, equal to 1.0418 and 1.0067. A spacing length of $0.5\,\mu m$ gives a coupling length of $21\,\mu m$. This result leads to a compact spot size converter.

When it is not possible to have the same propagation constant for the modes in the two waveguides, a total transfer of light can occur only by adding a grating between the two waveguides. This grating coupler is described in Sect. 10.6.

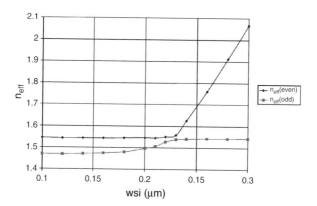

Fig. 10.9. Diagram of the twin waveguides effective index versus the size of the silicon waveguide

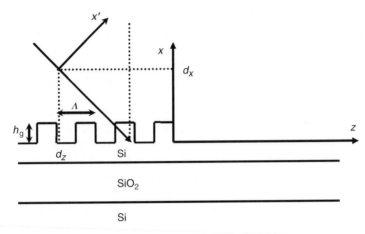

Fig. 10.10. Schematic view of the grating coupler

10.6 Grating Coupler

The grating coupler is illustrated in Fig. 10.10. It is constituted by elementary patterns distributed periodically on the surface of the waveguide in whom we want to inject the light of an incident beam.

The basic principle of the grating coupler can be simply understood by considering the fact that it consists of elementary patterns of size lower than the wavelength of the incident beam.

An elementary pattern diffracts the incident light in all the directions of space. For a grating coupler, the elementary patterns are distributed regularly on the surface of a material, which involves that all these patterns will diffract the light. As the optical paths of the diffracted beams are not identical, owing to the fact that the patterns are not localized at the same place, these beams will give rise to interferences which, in most cases, are destructive. The diffracted beams exist only in the case of constructive interferences and their directions of propagation are given by the Bragg relation:

$$k_0 n \sin\theta_j = k_0 n_1 \sin\theta_i \pm j\frac{2\pi}{\Lambda}, \tag{10.9}$$

where k_0 represents the wave vector in vacuum, n and n_1, respectively, represents the refractive indexes of the incident and diffracted mediums, θ_i and θ_j the angles of the incident and diffracted beams in the order j and Λ the grating period.

For a particular value of the incidence angle, the diffracted beam ($n \sin\theta_j$) corresponds to the effective index of the guided mode (n_{eff}). Under these conditions a part of the power of the incident beam is transferred to the guided mode.

The first stage of optimization of the grating couplers consists in choosing the grating period such that it exits only the reflected and transmitted beams in the specular direction in the air (diffraction order 0) and the diffracted beam in the +1 or −1 diffraction order corresponding to the guided mode coupling, respectively, in a forward or backward direction in the waveguide. The fact of reducing the number of diffracted beams in the air makes it possible to increase the amount of power which will be transferred in the waveguide.

An approximate value of the grating period can be simply obtained by calculating the effective index of the guided mode of the nonperturbed structure and by applying the Bragg relation. If the Bragg relation makes it possible to explain qualitatively the coupling of light in a waveguide assisted by a grating, it does not give the coupling efficiency. This one will depend on other parameters which are: the modulation depth of the grating h_g and the thicknesses of the dielectric multilayer stack which constitute the waveguide. These parameters should also be determined to obtain an optimum value of the coupling efficiency.

To calculate these parameters, rigorous methods are available in literature: integral method [41], differential method [42,43], modified coordinate [44,45], rigorous coupled wave analysis (RCWA) method [46,47].

The RCWA method seems to be the more useful method for the case of a rectangular dielectric grating. Some improvements of the method are to overcome numerical problems occurring for calculations with the TM polarisation [48–51] and to the hill conditioning of the matrix caused by the increases of an exponential term produced by the propagation of the high order diffractive waves in a layer [52,53]. These methods give the diffraction efficiency in reflection and transmission of the planar stack layer structure with an infinite diffraction grating.

An example of calculation is given in Fig. 10.11. It corresponds to the plot of the diffraction efficiency in reflection and transmission in the specular direction (0 diffraction order) and the power of the guided modes versus the incident angle.

One can observe on this figure a sharp variation of the reflectivity and transmission. These variations are known as Wood anomalies [54]. The origins of these anomalies are explained by the presence of the guided mode [55,56] and then the reflectivity and transmission in amplitude can be approximated by (10.10):

$$f(k_z) = f_0 \frac{(k_z - k_{z,z})}{(k_z - k_{z,p})}, \tag{10.10}$$

where $k_{z,p}$, $k_{z,z}$, and f_0 are, respectively, the complex pole equal to the propagation constant of the guided mode, a complex zero and f_0 is the reflection or the transmission coefficient in amplitude off resonance.

We can notice that the curve giving the evolution of the guided mode power versus the incident angle has a lorentzian shape. The position of maximum of lorentzian gives the coupling angle of the guided mode. It is connected to the real part of the complex pole as:

$$Re(k_{z,p}) = k_0 n_1 \sin\theta_i. \tag{10.11}$$

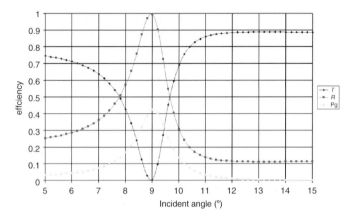

Fig. 10.11. Evolution of the diffraction efficiency in transmission T_0 and in reflection R_0, as well as the module of the electromagnetic field of the guided mode corresponding to the diffraction order -1 (Pg) according to the incidence angle. The resonance angle is equal to $9.1°$ and the half-width of the lorentzian ($\Delta\theta$) representing the variation of the module of the electromagnetic field of the guided mode is equal to $1.6°$ $L_\mathrm{d} = 7.3\,\mathrm{\mu m}$

The half-width at the maximum of the resonance ($\Delta\theta$) is connected to the imaginary part of the complex pole by the formula:

$$\mathrm{Im}(k_{z,\mathrm{p}}) = k_0 n_1 \cos\theta_\mathrm{i} \Delta\theta. \tag{10.12}$$

The complex pole can be obtained if we considered an output grating coupler as shown in Fig. 10.12. We have reported in Fig. 10.12, the spatial evolution of the guided power through a waveguide with an input and output grating coupler. Light injected by the input grating is propagated on the waveguide with some losses and is arrived on the output grating. As the guided mode become leaky, the guided power decrease exponentially. The decreasing length at $1/eL_\mathrm{d}$ is the parameter of the outcoupling effect.

The imaginary part of the guided mode propagation constant is linked to L_d as:

$$L_\mathrm{d} = \frac{1}{2\mathrm{Im}(k_{z,\mathrm{p}})}. \tag{10.13}$$

L_d has also an influence on the coupling. When the phase matching condition is satisfied, light is injected into the waveguide. The guided power is growing as the amplitude of the incident gaussian beam increase. In addition, because of the presence of the grating, the guided mode is leaky and a part of the light which propagates in the guide has the possibility to run away in the air. The result is a delay of the maximum value of the guided power compared to the maximum value of the incident beam. After this value, the guided power decreases because the amplitude of the incident beam decreases.

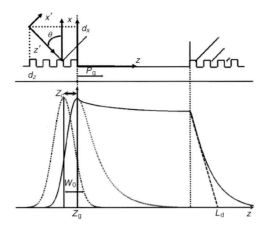

Fig. 10.12. Coupling mechanism of an input and output prism coupler and spatial evolution of the guided power

The theoretical method used to study the coupling phenomenon of an incident beam of a gaussian shape in a waveguide assisted by a grating is based on the plane wave decomposition as reported by several authors [56–58].

In this model, we consider that the grating is infinite and the determination of the coupling efficiency of coupling requires the calculation of the guided mode profile in the multilayer stack at the coupling angle besides the parameters of the grating coupler and of the beam that are, respectively, the decoupling length and the waist radius.

For an incident gaussian beam, it can be shown [58, 59] that an optimum coupling efficiency of 80% can be achieved when the two conditions are satisfied:

$$\begin{cases} w_0 = 1.368 \, L_\mathrm{d} \cos\theta_\mathrm{i} \\ L_\mathrm{d} z_\mathrm{c} = 1 \end{cases} \tag{10.14}$$

where w_0, θ_i, L_d and z_c are, respectively, the waist radius of incident gaussian beam, the incident angle, the decoupling length of the grating, and the position of the incident beam compared to beginning of the unperturbed waveguide.

The limitation of the maximum coupling efficiency value is explained by the overlapping of the exponential shape of the output beam compared to the gaussian beam shape of the incident beam.

In the case of a gaussian beam resulting from a single mode optical fiber ($w_0 = 5.6\,\mu\mathrm{m}$), (10.14) gives the value of the decoupling length ($L_\mathrm{d} = 7.8\,\mu\mathrm{m}$) to have a maximum coupling efficiency.

In Fig. 10.13, we have reported the evolution of the decoupling length versus the depth of the grating. The shape of this curve has been explained by Tamir and Peng [59]. According to Fig. 10.13, this value of the decoupling length is obtained for a grating depth of $0.1\,\mu\mathrm{m}$.

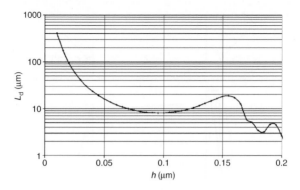

Fig. 10.13. Evolution of the decoupling length versus the depth of the grating

Equation 10.10 gives the response of an infinite extend grating to an incident plane wave. The response to an incident gaussian beam is obtained with the help of a plane wave decomposition [57, 58]. According to this decomposition, the incident beam for a TE polarization can be expressed as a Fourier decomposition:

$$E_y(x', z') = \int_{-k_0 n_1}^{+k_0 n_1} p(k_{x'}) \exp[i(k_{x'}x' + k_{z'}z')]\mathrm{d}k_{x'}, \qquad (10.15)$$

where $p(k_{x'}) = \int_{-\infty}^{+\infty} \exp[-x'^2/w_0^2]\exp[-ik_{x'}x']\mathrm{d}x'$, w_0 is the waist of the gaussian beam and $k_{z'}$ is related to $k_{x'}$ as $k_{z'} = \sqrt{n_1^2 k_0^2 - k_{x'}^2}$.

A change of coordinates allows to obtain the value of the field in the plane of the grating.

$$\begin{cases} x' = \sin\theta\,(x - \mathrm{d}x) + \cos\theta\,(z - \mathrm{d}z) \\ z' = -\cos\theta\,(x - \mathrm{d}x) + \sin\theta\,(z - \mathrm{d}z) \end{cases} \text{and} \begin{cases} k_{x'} = -k_x\sin\theta + k_z\cos\theta \\ k_{z'} = k_x\cos\theta + k_z\sin\theta. \end{cases}$$
$$(10.16)$$

This change of coordinates gives the reflective and transmitted field in the plane of the grating as:

$$f(z) = \int_{-n_1 k_0}^{+n_1 k_0} f(k_z)\, p\,(-k_x\sin\theta + k_z\cos\theta)\left(\frac{\sin\theta}{k_x} + \cos\theta\right)$$
$$\times \exp\left[i\,(k_z z - k_x\mathrm{d}x - k_z\mathrm{d}z)\right]\mathrm{d}k_z. \qquad (10.17)$$

For a strongly divergent beam, one proceeds by fast Fourier transform (FFT) algorithm [60] or a fractional fourier transform algorithm [61] to obtain a numerical solution of (10.15) and (10.17).

The reflection and transmission coefficient in intensity for a TE polarization is obtained by the use of the pointing vector:

$$F(z) = \frac{Re\left\{f(z)\frac{\mathrm{d}}{\mathrm{d}z}f^*(z)\right\}}{\int\limits_{-\infty}^{+\infty} Re\left\{E_y(z)\frac{\mathrm{d}}{\mathrm{d}z}E_y^*(z)\right\}\mathrm{d}z}. \tag{10.18}$$

We can notice that the coupling efficiency η of the grating coupler is obtained by considering the energy balance criterion:

$$\eta = 1 - \int\limits_{-\infty}^{0} R_{\mathrm{g}}(z)\,\mathrm{d}z - \int\limits_{-\infty}^{0} T_{\mathrm{g}}(z)\,\mathrm{d}z - \int\limits_{0}^{+\infty} R_{\mathrm{w}}(z)\,\mathrm{d}z - \int\limits_{0}^{+\infty} T_{\mathrm{w}}(z)\,\mathrm{d}z, \tag{10.19}$$

where η is the coupling efficiency, R and T are the reflectivity and the transmission in the grating and waveguide areas denoted, respectively, by the subscript g or w. This calculation is illustrated on the Fig. 10.14, where we plot the evolution of the reflectivity and the transmission versus to the incident angle of the beam.

The position of the incident beam with respect to the beginning of the unperturbed waveguide at $z = 0$ in Fig. 10.10 can be modified by changing the value of d_z in (10.17).

In Fig. 10.14 a, the lateral shift d_z of the incident beam is chosen in such way as all the light is totally on the grating area, what corresponds to the case of an infinite grating. In that case, the energy balance is verified and there is no light injected in the waveguide.

According to (10.14), the lateral shift of the incident beam corresponded to the position of a maximum coupling efficiency in Fig. 10.14b. In this case, the lack in the energy balance gives the coupling efficiency versus the incident angle. A maximum value of 56% is obtained for an incident angle of 9.2° and

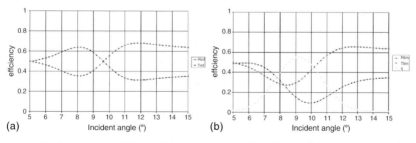

Fig. 10.14. Evolution of the reflectivity and transmission of the grating coupler in the case of an infinite (**a**) and finite (**b**) grating coupler for a gaussian beam resulting from a monomode optical fiber. In the case of an infinite grating, the energy balance criterion is satisfied. For the grating of finite extend, the lack in the energy balance criterion gives the coupling efficiency

the acceptance of the resonance ($\Delta\theta = 3.8°$) is increased compared to the case of an incident plane wave ($\Delta\theta = 1.6°$).

This expression enables us to study the variations of the coupling efficiency according to the waist size of the incident beam, to its position compared to the device and to the angular detuning between the incident angle and the resonance angle of the grating coupler.

For example, in the case of a gaussian beam resulting from a single mode optical fiber ($w_0 = 5.6\,\mu m$), an optimal coupling efficiency is obtained for decoupling length of $7.8\,\mu m$.

The alignment tolerances at ± 1 dB for a detuning angle, a defocusing and lateral displacement are, respectively, equal to $\pm 2.0°$, $\pm 240\,\mu m$, and $\pm 2.6\,\mu m$.

The same procedure can be used in the case of a TM polarization. Also, it is possible to obtain the tolerance alignments for detuning a lateral or transversal shift of the incident beam, respectively, by varying the parameters θ_i, d_z, and d_x.

It should be noted that the maximum value which can be obtained is 80%. In general, the coupling efficiency of a coupler with conventional grating is lower. This is due to the fact that a part of energy is lost in the reflected and transmitted beam and does not take part in the coupling phenomenon. This optimal value of 80% is obtained when there is only one reflected or transmitted beam.

To increase the coupling efficiency up to 80%, two solutions has been proposed and tested for SOI waveguide. The first solution is to eliminate the reflected or the transmitted beam by varying the thickness of the silicon layer and the depth of the grating [58, 59]. The second solution consists of the deposition of metallic mirror on top of the grating [64].

In order to exceed the limit of 80%, the shape of the decoupling beam must be closed to the shape of the gaussian beam. The different solutions proposed earlier use a blazed grating [65, 66], a focused grating [67–71], or a double relief grating [72]. The focused grating is used for the realization of an integrated-optic disk pickup device [68].

We can notice that these results are obtained in the case of planar structures. To have a strong coupling effect between an optical fiber and a submicronic silicon waveguide, a taper is used.

Solutions involving this kind of coupler are reported in literature [73–77]. In order to reduce the length of the taper, a solution with a segmented spot size converter has been proposed and tested [78]. Also a solution with a 2D grating are used to have a polarization independent photonic integrated circuit [79].

10.7 Guide to Guide Assisted by Grating Coupler

The concept of this coupler is illustrated in Fig. 10.15. Light of a tapered fiber is first injected by a butt coupling method in the upper low refractive index waveguide. When the effective index of the modes that propagate in the two

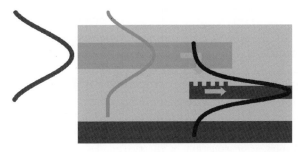

Fig. 10.15. Schematic view and basic principal of guide to guide assisted by grating coupler

waveguides are very different, a coupling effect can be occurred with the help of a grating. This effect is called a guide to guide process assisted by the grating. It was first studied using the coupled mode theory [80–84] and more recently with the Floquet–Bloch theory [85–88].

The design of such coupler consists first in the determination of the cross-section of the input waveguide to have a good overlapping with the beam of a standard or tapered optical fiber. The effective index method [87,88] is used to simplify the 3D problem in a planar multilayer stack for the determination of the grating period.

The propagation constants and the complete field distribution of the two modes are calculated with a grating solver based on the RCWA method [46]. According to the Floquet–Bloch theory, the modes in the coupler can be seen as an infinite sum of the diffraction order of the grating. The complete field pattern is obtained by the summation of all space harmonics. Dispersion curves of the two modes are found as a function of the grating period. Figure 10.16 shows an example of the propagation constants of the two modes as a function of the grating period where Fig. 10.16a, b correspond to the real and the imaginary parts of the propagation constants of the two modes, respectively. There are two space harmonics which have significant amplitudes ($n = 0$ and 1 for the mode in the low refractive index contrast waveguide and $n = 0$ and -1 for mode in the silicon waveguide). Out of the resonance, the field distribution of these two harmonics is similar to the modes of the two unperturbed waveguide. At resonance, when the difference between the real parts of the longitudinal propagation constants is minimized, the magnitude of the two modes are nearly identical, in which case the coupling between the two waveguide is maximum. As shown in Fig. 10.16a, the resonant condition corresponds to a grating period of $0.738\,\mu\text{m}$. With such grating period, the space harmonic ($n = 2$) of the grating occur in the air that induced radiation losses as shown in Fig. 10.10b. In this case, the optimum coupling length is given by the relation [86]:

$$\tan\left(\frac{\delta}{2}L_{\text{c}}\right) = \frac{\delta}{2\alpha}, \tag{10.20}$$

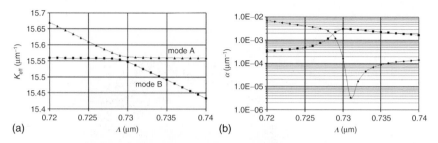

Fig. 10.16. Propagation constant of the two modes in a guide to guide assisted by grating coupler versus the grating period Λ. The anticrossing point gives the optimum value of the grating period ($\Lambda = 0.729\,\mu m$) for a maximum coupling efficiency

where α is the modal attenuation leakage of the second diffractive order of the grating in the air. For this planar coupler, a coupling length of $L_c = 200\,\mu m$ produces an optimum coupling efficiency of $\eta = 64\%$.

We can notice that for these kinds of couplers can be eliminated by the use of a contra-propagative coupling effect. According to the Bragg relation, the grating period is of $0.29\,\mu m$. Because of this short period, there is no radiation effect in the air and consequently no losses. The advantages of this coupler are compactness and high compatibility with the CMOS technology.

This kind of couplers have been used for the light transfer between a glass waveguide and a semiconductor waveguide [85] and also when a fiber is directly posed on the top of the grating [90]. Because of this long coupling length of a few millimeters, it is not a compact solution.

In order to have a more compact coupling solution, new geometries with a dual grating [91] or an ARROW coupler [92] have been proposed. ARROW means antiresonant reflection optical waveguide [93]. The ARROW coupler is made with the standard planar CMOS processes because the ARROW waveguide uses the dielectric cap layer of the microelectronic circuits and the grating is etched in the silicon.

The drawback is that the ARROW waveguide has low propagation losses only for the TE polarization making the coupler selective in wavelength and polarization.

10.8 Prism Coupler

In the case of the prism coupler, one searches to inject light in a waveguide when the incident beam is above the waveguide. If one only considers the waveguide, the Fresnel relations show us that whatever is the incident angle, the refracted angle in the guiding layer cannot correspond to the case of the guided mode. Then, there is no possibility for light to be injected into the waveguide.

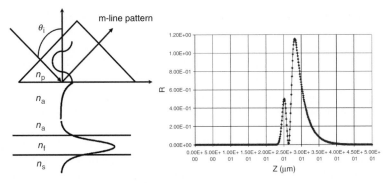

Fig. 10.17. (a) Basic principles of the prism coupler, (b) M line pattern obtained when the incident light is injected in the waveguide

The coupling by a prism is illustrated in Fig. 10.17. The coupling by prism consists in approaching the waveguide, a material of strong refractive index n_p, comparable with that of the core layer n_f of the waveguide. The incident beam is in total reflection condition at the base of the prism and the evanescent part of the incident beam will interact with the evanescent part of the guided mode.

The operating mode of the prism coupler can be explained by the fact that the total reflection of the incident beam at the base of the prism will be frustrated. A part of the energy of the incident beam will be transferred in the guide through the air gap by an optical tunnel effect. A coupling of the light occurs when the phase condition is satisfied:

$$k_0 n_\mathrm{p} \sin \theta_\mathrm{i} = k_0 n_\mathrm{eff}. \tag{10.21}$$

Tien and Ulrich described for the first time the total internal reflection prism coupling phenomenon [94–96].

In a more rigorous way, one can use the plane wave analysis to describe the prism coupling effect. The propagation of an incident plane wave in a multilayer stack is calculated by a matrix representation [97,98]. For the same reasons as for the grating coupler, the reflection coefficient at the basis of the prism can be approximated by (10.10). The plane wave decomposition gives the solution of the shape of the reflected beam in the case of an incident gaussian beam.

For an incident gaussian beam, it can be shown [99,100] that a maximum coupling efficiency of 80% can be achieved when the two conditions of (10.14) are satisfied.

The limitation of the maximum coupling efficiency value is explained by the overlapping of the exponential shape of the output beam compared to gaussian beam shape of the incident beam. The maximum coupling efficiency can be increased to 100% in the case of cylindrical prisms [101]. Because of the varying gap of these prisms, the output beam has a shape that is closed to the gaussian shape of the incident beam.

We can notice that the competitive effect between the coupling and the leakage of the guided power on the input prism results on a m-line in the reflected beam pattern [102–104].

The m-line can be calculated with the same formalism as those developed for the grating coupler (10.17). This kind of calculation is illustrated in Fig. 10.17b.

Measurement of the position of this dark line on the reflected beam gives the real part of the propagation constant. This experimental method has been used for the material characterization and the loss properties of thin films [105, 106].

The main problem of the prism coupler is that it cannot be used for a monolithic integration of optical devices. It is for this reason that the coupling technique is generally used for the characterization of planar waveguides in laboratory.

Acknowledgments

The author gratefully acknowledges T. Benyattou for the many useful discussions, proof reading, and results presented in this chapter. The author also acknowledges the organizers of the 3rd Optoelectronic and Photonic Winter School of Trento in Silicon photonics Prof. G. Guillot and L. Pavesi for providing the opportunity to write this chapter.

References

1. K.K. Lee, D.R. Lim, H.C. Luan, A. Agarwal, J. Foresi, L. Kimerling, "Effect of size and roughness on light transmission in a Si/SiO2 waveguide: Experiment and model", Appl. Phys. Lett. Vol. 77, p. 1617, 2000
2. W. Bogaertsa, P. Dumona, D. Taillaerta, V. Wiauxb, S. Beckxb, B. Luyssaerta, J. Van Campenhouta, D. Van Thourhouta, R. Baets, "SOI nanophotonic waveguide structures fabricated with deep UV lithography", Photon. Nanostruct. – Fundam. Appl. Vol. 2, pp. 81–86, 2004
3. P. Dumon, W. Bogaerts, V. Wiaux, J. Wouters, S. Beckx, J. Van Campenhout, D. Taillaert, B. Luyssaert, P. Bienstman, D. Van Thourhout, R. Baets, "Low-loss SOI photonic wires and ring resonators fabricated with deep UV lithography", IEEE Photon. Technol. Lett. Vol. 16, pp. 1328–1330, 2004
4. M. Kawachi, "Recent progress in silica-based planar lightwave circuits on silicon", IEE Proc. Optoelectron. Vol. 143, pp. 257–262, 1996
5. Y.P. Li, C.H. Henry, "Silica-based optical integrated circuits", IEE Proc. Optoelectron. Vol. 143, pp. 263–280, 1996
6. N. Daldosso, M. Melchiorri, F. Riboli, M. Girardini, G. Pucker, M. Crivellari, P. Bellutti, A. Lui, L. Pavesi, "Design, fabrication, structural and optical characterization of thin Si3N4 waveguides", IEEE J. Lightwave Technol. Vol. 22, p. 1734, 2004

7. M.J. Kobrinsky, B.A. Block, J.F. Zheng, B.C. Barnett, E. Mohammed, M. Reshotko, F. Robertson, S. List, I. Young, K. Cadien, "On-chip optical interconnects", Intel Technol. J. Vol. 8, pp. 129–139, 2004

8. F. Ay, A. Aydinli, "Comparative investigation of hydrogen bonding in silicon based PECVD grown dielectrics for optical waveguides", Opt. Mater. Vol. 26, pp. 33–46, 2004

9. B.H. Larsen, L.P. Nielsen, K. Zenth, L. Leick, C. Laurent-Lund, L.U. Aaen Andersen, K.E. Mattsson, "A low-loss, silicon-oxynitride process for compact optical devices", Proceeding of ECOC conference, 2003

10. C.-Y. Chao, and L.J. Guo, "Reduction of surface scattering loss in polymer microrings using thermal-reflow technique", IEEE Photon. Technol. Lett. Vol. 16, pp. 1498–1500, 2004

11. A.V. Mule, R. Villalaz, T.K. Gaylord, J.D. Meindl, "Photopolymer-based diffractive and MMI waveguide couplers", IEEE Photon. Technol. Lett. Vol. 16, pp. 2490–2492, 2004

12. A.M. Agarwal, L. Liao, J.S. Foresi, M.R. Black, X. Duan, L.C. Kimerling, "Low-loss polycrystalline silicon waveguides for silicon photonics", J. Appl. Phys. Vol. 80, pp. 6120–6123, 1996

13. W. Joyce, B. Deloach, "Alignment of gaussian beams", Appl. Opt., Vol. 23, pp. 4187–4196, 1984

14. K. Tatsuno, et al., "Fiber-pigtail-detachable plastic miniDIL transmitter module with a tool-free Optical Connector", J. Lightwave Technol. Vol. 21, pp. 1066–1070, 2003

15. R. Hauffe et al., "Methods for passive fiber chip coupling of integrated optical devices", IEEE Trans. Adv. Packaging, Vol. 24, pp. 450–455, 2001

16. M.A. Rosa et al., "Self-alignment of optical fibers with optical quality end-polished silicon rib waveguides using wet chemical micromachining techniques", IEEE J. Sel. Top. Quantum Electron. Vol. 5, No. 5, pp. 1249–1254, 1999

17. M. Uekawa, H. Sasaki, D. Shimura, K. Kotani, Y. Maeno, T. Takamori, "Surface mountable silicon microlens for low cost lasers modules", IEEE Photon. Technol. Lett. Vol. 15, pp. 945–947, 2003

18. J.T. Kim, K.B. Yoon, C.G. Choi, "Passive alignment method of polymer PLC devices by using a hot embossing technique", IEEE Photon. Technol. Lett. Vol. 16, pp. 1664–1665, 2004

19. M. Salib, L. Liao, R. Jones, M. Morse, A. Liu, D. Samara-Rubio, D. Alduino, M. Paniccia, "Silicon Photonics", Intel Technol. J. Vol. 8, pp. 141–160, 2004

20. I. Moerman, P.P. Van Daele, P.M. Demeester, "A review on fabrication technologies for the monolithic integration of tapers with III–V Semiconductor devices", IEEE J. Sel. Top. Quantum Electron. Vol. 3, pp. 1308–1318, 1997

21. C. Vassallo, "1993–1995 Optical mode solvers", Opt. Quantum Electron." Vol. 29, pp. 95–114, 1997

22. R. Scarmozzino, A. Gopinath, R. Pregla, S. Helfert, "Numerical techniques for modeling guided wave photonic devices", IEEE J. Sel. Top. Quantum Electron. Vol. 6, pp. 150–162, 2000

23. J. Haes, R. Baets, C.M. Weinert, M. Gravert, H.P. Nolting, M.A. Andrade, A. Leite, H.K. Bissessu, J.B. Davies, R.D. Ettinger, J. Ctyroky, E. Ducloux, F. Ratovelomanana, N. Vodjdani, S. Helfert, R. Pregla, F.H.G.M. Wijnands, H.J.W.M. Hoekstra, G.J.M. Krijnen, "A comparison between different propagative schemes for the simulation of tapered step index slab waveguides", J. Lightwave Technol. Vol. 14, pp. 1557–1569, 1996

24. P. Sewell, T.M. Benson, P.C. Kendall, "Rib waveguide spot-size transformers: modal properties", J. Lightwave Technol. Vol. 17, pp. 848–856, 1999
25. B.M.A. Rahman, W. Boonthittanont, S.S.A. Obayya, T. Wongcharoen, E.O. Ladele, K.T.V. Grattan, "Rigorous beam propagation analysis of tapered spot-size converters in deep-etched semiconductor waveguides", J. Lightwave Technol. Vol. 21, pp. 3392–3398, 2003
26. O. Mitomi, N. Yoshimoto, K. Magari, T. Ito, Y. Kawaguchi, Y. Suzuki, Y. Tohmori, K. Kasaya, "Analyzing the polarization dependence in optical spot-size converter by using a semivectorial finite-element beam propagation method", J. Lightwave Technol. Vol. 17, pp. 1255–1262, 1999
27. L. Vivien, S. Laval, E. Cassan, X. Le Roux, D. Pascal, "2-D taper for low-loss coupling between polarization-insensitive microwaveguides and single-mode optical fibers", J. Lightwave Technol. Vol. 21, pp. 2429–2433, 2003
28. W.K. Burns, A.F. Milton, Chap. 3, "Waveguide transitions and junctions", in "Guided Wave Optoelectronics", Ed. T. Tamir, Springer Series in Electronics and Photonics Vol. 26, pp. 89–144, 1990
29. T. Bakke, C.T. Sullivan, S.D. Mukherjee, "Polymeric optical spot-size transformer with vertical and lateral tapers", J. Lightwave Technol. Vol. 20, pp. 1188–1197, 2002
30. T.P. Felici, D.F.G. Gallagher, "Improved waveguide structures derived from new rapid optimization techniques", Proceeding of Photonics West Conference, Sanjose, Paper 4986-48, 2003
31. www.photond.com
32. I. Day et al., "Tapered silicon waveguides for low insertion loss highly efficient high speed electronic variable attenuators", IEEE OFC, March 24–27, 2003
33. M. Fritze, J. Knecht, C. Bozler, C. Keast, J. Fijol, S. Jacobson, P. Keating, J. Leblanc, E. Fike, B. Kessler, M. Frish, C. Manolatou, "3D mode converters for SOI integrated optics", Proceeding of the IEEE International SOI Conference, pp. 165–166, 2002
34. P. Lusse, P. Stuwe, J. Schule, H.G. Unger, "Analysis of vectorial mode fields in optical waveguides by a new finite difference method", J. Lightwave Technol. Vol. 12, pp. 487–494, 1994
35. G.R. Hadley, R.E. Smith, "Full vector waveguide modeling using an iterative finite difference method with transparent conditions", J. Lightwave Technol. Vol. 13, pp. 465–469, 1995
36. L. Vivien, S. Laval, E. Cassan, X. Le Roux, D. Pascal, "2-D taper for low-loss coupling between polarization-insensitive microwaveguides and single-mode optical fibers", J. Lightwave Technol., Vol. 21, pp. 2429–2433, 2003
37. V.R. Almeida, R.R. Panepucci, M. Lipson, "Nanotaper for compact mode conversion", Opt. Lett. Vol. 28, pp. 1302–1304, 2003
38. T. Shoji, T. Tsuchizawa, T. Watanabe, K. Yamada, H. Morita, "Low loss mode size converter from 0.3/spl mu/m square Si wire waveguides to singlemode fibres", Electron. Lett. Vol. 38, pp. 1669–1670, 2002
39. D. Marcuse, "Directional couplers made of nonidentical asymmetric slabs. Part I: Synchronous couplers", J. Lightwave Technol. Vol. 5, pp. 113–118, 1987
40. D. Marcuse, "Bandwidth of forward and backward coupling directional couplers", J. Lightwave Technol. Vol. 5, pp. 1773–1777, 1987
41. D. Maystre, "Rigorous vector theories of diffraction gratings", in *Progress in Optics*, Ed. E. Wolf, North Holland, Amsterdam, Vol. XXI, 1984

42. P. Vincent, "Differential methods", in Electromagnetic Theory of Gratings, Ed. R. Petit, Vol. 22 of Topics in Current Physics, Springer, Berlin Hiedelberg New York, Chap. 3, pp. 101–122, 1980

43. K. Ogawa, W.S.C. Chang, B.L. Sapori, F.J. Rosembaum, "A theoritical analysis of etched grating couplers for integrated optics", IEEE J. Quantum Electron. QE-9, pp. 29–42, 1973

44. J. Chandezon, M.T. Dupuis, G. Cornet, D. Maystre, "Multicoated gratings: a differential formalism applicable in the entire optical region", J. Opt. Soc. Am. A, Vol. 72, pp. 839–846, 1982

45. L. Li, J. Chandezon, G. Granet, J.P. Plumey, "Rigorous and efficient grating analysis method made easy for optical engineers", Appl. Opt. Vol. 38, pp. 304–313, 1999

46. M.G. Moharan, T.K. Gaylord, "Rigorous coupled wave analysis of planar grating diffraction", J. Opt. Soc. Am. A, Vol. 71, pp. 811–818, 1981

47. M.G. Moharan, T.K. Gaylord, "Diffraction analysis of dielectric surface relief gratings", J. Opt. Soc. Am. A, Vol. 72, pp. 1385–1392, 1982

48. P. Lalanne, "Improved formulation of the coupled wave method for two dimensional gratings", J. Opt. Soc. Am. A, Vol. 14, pp. 1592–1598, 1982

49. L. Li, "Use of Fourier series in the analysis of discontinuous periodic structures", J. Opt. Soc. Am. A, Vol. 13, pp. 1870–1876, 1996

50. L. Li, "New formulation of the fourier modal method for crossed surface relief gratings", J. Opt. Soc. Am. A, Vol. 14, pp. 2758–2767, 1997

51. P. Lalanne, J.P. Hugonin, "Numerical performance of finite difference modal methods for the electromagnetic analysis of one dimensional lamellar gratings", J. Opt. Soc. Am. A, Vol. 17, pp. 1033–1042, 2000

52. E. Popov, M. Neviere, "Grating theory: new equations in fourier space leading to fast converging results far TM polarization", J. Opt. Soc. Am. A, Vol. 17, pp. 1773–1784, 2000

53. M.G. Moharam, D.A. Pommet, E.B. Grann, T.K. Gaylord, "Stable implementation of the rigorous coupled wave analysis for surface relief gratings: enhanced transmittance matrix approach", J. Opt. Soc. Am. A, Vol. 12, pp. 1077–1086, 1995

54. M.G. Moharam, D.A. Pommet, E.B. Grann, T.K. Gaylord, "Formulation of stable and efficient implementation of the rigorous coupled wave analysis of binary gratings," J. Opt. Soc. Am. A, Vol. 12, pp. 1068–1076, 1995

55. R. Petit, "Electromagnetic Theory of Grating", Springer, Berlin Hiedelberg New York, 1980

56. M. Nevière, "The homogeneous problem", in Electromagnetic Theory of Gratings, Ed. R. Petit, Vol. 22 of Topics in Current Physics, Springer, Berlin Hiedelberg New York, Chap. 5, pp. 123–157, 1980

57. L. Li, M.C. Gupta, "Effect of beam focusing on the efficiency of planar waveguide grating couplers", Appl. Opt. Vol. 29, pp. 5320–5325, 1990

58. D. Pascal, R. Orobtchouk, A. Layadi, A. Koster, S. Laval, "Optimized coupling of a gaussian beam into an optical waveguide using a grating coupler: comparison of experimental and theoretical results", Appl. Opt. Vol. 36, pp. 2443–2447, 1997

59. T. Tamir, S.T. Peng, "Analysis and design of grating couplers", Appl. Phys. Vol. 14, pp. 235–254, 1977

60. D.H. Bailey, P.N. Swarztrauber, "A fast method for the numerical evaluation of continuous fourier and laplace transforms", SIAM J. Sci. Comput. Vol. 15, pp. 1105–1110, 1994

61. D.H. Bailey, P.N. Swarztrauber, "The fractional fourier transform and applications", SIAM Rev. Vol. 33, pp. 389–404, 1991

62. T.W. Ang, G.T. Reed, A. Vonsovici, A.G.R. Evans, P.R. Routley, M.R. Josey, "Effects of grating heights on highly efficient unibond SOI waveguide grating couplers", IEEE Photon. Technol. Lett. Vol. 12, pp. 59–61, 2000

63. D. Pascal, R. Orobtchouk, "Efficient light coupling into sub-micrometer rib and strip waveguides", Integrated Photonics Research, Vancouver, 17–19 Juillet 2002, Proc. TOPS 78, IBSN 1-55752-722-9

64. R. Orobtchouk, A. Layadi, H. Gualous, D. Pascal, A. Koster, S. Laval, "High efficiency light coupling in a submicron SOI waveguide", Appl. Opt. Vol. 39, No. 31, pp. 5773–5777, 2000

65. T.W. Ang, G.T. Reed, A. Vonsovici, A.G.R. Evans, P.R. Routley, M.R. Josey, "Highly efficient unibond silicon-on-insulator blazed grating couplers", Appl. Phys. Lett. Vol. 77, pp. 4215–4216, 2000

66. T. Laao, S. Sheard, L. Ming Z. Hanguo P. Prewett, P., "High-efficiency focusing waveguide grating coupler with parallelogramic groove profiles", J. Lightwave Technol. Vol. 15, pp. 1142–1148, 1997

67. T. Suhara, K. Okada, T. Saso, H. Nishihara, H. "Focusing grating coupler in AlGaAs optical waveguide", IEEE Photon. Technol. Lett. Vol. 4, pp. 903–905, 1992

68. S. Ura, T. Suhara, H. Nishihara, J. Koyama, "An integrated-optic disk pickup device", J. Lightwave Technol. Vol. 4, pp. 913–918, 1986

69. P. Kipfer, M. Collischon, H. Haidner, H. Schäfer, J. Schwider, "A novel design for beam forming grating couplers", Optik – Int. J. Light Electron Opt. Vol. 112, pp. 76–80, 2001

70. Y. Sohn, Y. Park, D. Suh, H. Ryu, M.C. Paek, "Focusing grating coupler for blue laser light", IEEE Photon. Technol. Lett. Vol. 16, pp. 162–164, 2004

71. K. Kintaka, J. Nishii, Y. Imaoka, J. Ohmori, S. Ura, R. Satoh, H. Nishihara, "A guided-mode-selective focusing grating coupler", IEEE Photon. Technol. Lett. Vol. 16, pp. 512–514, 2004

72. L. Li, "Analysis of planar waveguide grating couplers with double surface corrugations of identical period", Opt. Commun. Vol. 114, pp. 406–412, 1995

73. R. Orobtchouk, N. Schnell, T. Benyattou, S. Lardenois, D. Pascal, A. Cordat, S. Laval, E. Cassan, A. Koster, D. Bouchier, N. Bouzaida, B. Dal'zotto, B. Florin, M. Heitzmann, Z. Lamouchi, D. Louis, L. Mollard, D. Renaud, J. Gautier, "Optical interconnects on SOI: a front end approach", Interconnect Technology Conference, Proceedings of the IEEE 2002 International, pp. 83–85, 2002

74. D. Taillaert, W. Bogaerts, P. Bienstman, Member, T.F. Krauss, P. Van Daele, I. Moerman, S. Verstuyft, K. De Mesel, R. Baets, "An out-of-plane grating coupler for efficient butt coupling between compact planar waveguides and single-mode fibers", IEEE J. Quantum. Electron. Vol. 38, pp. 949–954, 2002

75. E. Cassan, S. Lardenois, D. Pascal, L. Vivien, M. Heitzmann, N. Bouzaida, L. Mollard, R. Orobtchouk, S. Laval, "Intra-chip optical interconnects with compact and low-loss light distribution in silicon-on-insulator rib waveguides", Interconnect Technology Conference, Proceedings of the IEEE 2003 International, pp. 39–41, 2003

76. W. Bogaerts, P. Dumon, D. Taillaert, V. Wiaux, S. Beckx, B. Luyssaert, J. Van Campenhout, D. Van Thourhout, R. Baets, "SOI nanophotonic waveguide structures fabricated with deep UV lithography", Photon. Nanostruct. – Fundam. Appl. Vol. 2, pp. 81–86, 2004

77. L. Vivien, F. Grillot, E. Cassan, D. Pascal, S. Lardenois, A. Lupu, S. Laval, M. Heitzmann, J.-M. Fedeli, "Comparison between strip and rib SOI microwaveguides for intra-chip light distribution", Opt. Mater. Vol. 27, pp. 756–762, 2005

78. W. Bogaerts, D. Taillaert, B. Luyssaert, P. Dumon, J. Van Campenhout, P. Bienstman, D. Van Thourhout, R. Baets, V. Wiaux, S. Beckx, "Basic structures for photonic integrated circuits in silicon-on-insulator", Opt. Express, 2004

79. D. Taillaert, H. Chong, P.I. Borel, L.H. Frandsen, R.M. De La Rue, R. Baets, "A compact two-dimensional grating coupler used as a polarization splitter", IEEE Photon. Technol. Lett. Vol. 15, pp. 1249–1251, 2003

80. D. Marcuse, "Directional couplers made of nonidentical asymmetric slabs. Part II: Grating-assisted couplers", J. Lightwave Technol. Vol. 5, pp. 268–273, 1987

81. W.P. Huang, H.A. Haus, "Power exchange in grating-assisted couplers", J. Lightwave Technol. Vol. 7, pp. 920–924, 1989

82. D. Marcuse, "Radiation loss of grating-assisted directional coupler", IEEE J. Quantum Electron. Vol. 26, pp. 675–684, 1990

83. W.P. Huang, J. Hong, Z.M. Mao, "Improved coupled-mode formulation based on composite modes for parallel grating-assisted co-directional couplers", IEEE J. Quantum Electron. Vol. 29, pp. 2805–2812, 1993

84. B.E. Little, "A variational coupled-mode theory including radiation loss for grating-assisted couplers", J. Lightwave Technol. Vol. 14, pp. 188–195, 1996

85. S. Nai-Hsiang, J.K. Butler, G.A. Evans, L. Pang, P. Congdon, "Analysis of grating-assisted directional couplers using the Floquet-Bloch theory", J. Lightwave Technol. Vol. 15, pp. 2301–2315, 1997

86. J.K. Butler, S. Nai-Hsiang, G.A. Evans, L. Pang, P. Congdon, "Grating-assisted coupling of light between semiconductor and glass waveguides", J. Lightwave Technol. Vol. 16, pp. 1038–1048, 1998

87. V.M.N. Passaro, "Optimal design of grating-assisted directional couplers", J. Lightwave Technol. Vol. 18, pp. 973–984, 2000

88. R.M. Knox, P.P. Toulios, "Integrated circuits for the millimiter through optical frequency range", presented at the Symposium on Submillimiter Waves, Polytechnic Institut of Brooklyn, pp. 497–516, 31 March – 2 April, 1970

89. R.K. Varshney, A. Kumar, "A simple and accurate modal analysis of strip loaded optical waveguides with various index profiles", J. Lightwave Technol. Vol. 6, pp. 601–606, 1988

90. C. Bulmer, M. Wilson, "Single mode grating coupling between thin-film and fiber optical waveguides", IEEE J. Quantum Electron. Vol. 14, pp. 741–749, 1978

91. G.Z. Masanovic, V.M.N. Passaro, G.T. Reed, "Dual grating-assisted directional coupling between fibers and thin semiconductor waveguides", IEEE Photon. Technol. Lett. Vol. 15, pp. 1395–1397, 2003

92. R. Orobtchouk, N. Schnell, T. Benyattou, J. Gregoire, S. Lardenois, M. Heitzmann, J.M. Fedeli, "New ARROW optical coupler for optical interconnect", Proceedings of the IEEE 2003 International Interconnect Technology Conference, 2–4 June 2003, pp. 233–235, 2003

93. J.M. Kubica, "Numerical analysis of InP/InGaAsP ARROW waveguide using transfert matrix approach", J. Lightwave Technol. Vol. 10, pp. 767–771, 1992

94. P.K. Tien, R. Ulrich, R.J. Martin, "Modes of propagating light waves in thin deposited semiconductor films", Appl. Phys. Lett. Vol. 14, pp. 291–294, 1969

95. P.K. Tien, R. Ulrich, "Theory of prism-film coupler and thin film light guides", J. Opt. Soc. Am. Vol. 60, pp. 1325–1336, 1970

96. R. Ulrich, "Theory of the prism-film coupler by plane wave analysis", J. Opt. Soc. Am. Vol. 60, pp. 1337–1350, 1970

97. F. Abeles, Ann. Phys. Vol. 5, p. 596, 1950

98. M. Born, E. Wolf, *Principal of Optics*, Pergamon, New York, p. 50, 1959

99. R. Ulrich, "Optimum excitation of optical surface waves", J. Opt. Soc. Am. Vol. 61, pp. 1467–1477, 1971

100. W.A. Pasmooij, P.A. Mandersloot, M.K. Smit, "Prism coupling of light into narrow planar optical waveguides", J. Lightwave Technol. Vol. 7, pp. 175–180, 1989

101. A.V. Chelnokov, J.-M. Lourtioz, "Optimised coupling into planar waveguides with cylindrical prisms", Electron. Lett. Vol. 31, pp. 269–271, 1995

102. R. Reinisch, J. Fick, P. Coupier, J.L. Coutaz, G. Vitrant, "Grating and prism couplers: radiation pattern and m-line", Opt. Commun. Vol. 120, pp. 121–128, 1995

103. S. Monneret, P.H. Chantome, F. Flory, "M-line technique: prism coupling measurement and discussion of accuracy for homogeneous waveguides", J. Opt. A: Pure Appl. Opt Vol. 2, pp. 188–195, 2000

104. J. Naden, G. Reed, B. Weiss, "Analysis of prism-waveguide coupling in anisotropic media", J. Lightwave Technol. Vol. 4, pp. 156–159, 1986

105. Y.T. Kim, D.S. Kim, D.H. Yoon, "PECVD SiO_2 and SiON films dependant on the rf bias power for low-loss silica waveguide", Thin Solid Films, Vol. 475, pp. 271–274, 2005

106. Au Vonsovici, A. Koster, R. Orobtchouk, D. Pascal, "Optical characterisation of SIMOX material at $g\lambda = 1.3\,\mu m$ using attenuated total reflection, Fall Meeting of Material Research Society Conference Proceeding, pp. 721–724, 1994

Si Microphotonics for Optical Interconnection

K. Wada, J.F. Liu, S. Jongthammanurak, D.D. Cannon, D.T. Danielson,
D.H. Ahn, S. Akiyama, M. Popovic, D.R. Lim, K.K. Lee, H.-C. Luan, Y.
Ishikawa, X. Duan, J. Michel, H.A. Haus, and L.C. Kimerling

Summary. Silicon microphotonics is to integrate photonic functionalities on Si
chips to extend Si chip performances and eventually realize electronic and photonic
integrated circuits (EPICs). The current trend for photonic integration on silicon is
making all photonic devices compatible to complementary metal oxide semiconductor (CMOS) technology. The first part of this Chapter describes a fractal optical
clocking architecture as a new breakthrough to enhance clocking speed far beyond
the Semiconductor Technology Roadmap. The possibility of faster clocking by orders
of magnitude is demonstrated. The second part deals with CMOS compatibilities of
photodetectors for on-chip monolithic integration. Defect- and strain-engineered Ge
photodetectors on Si are broadband and have C- and L-bands detection capability.
Finally a potential market for optical interconnection on ∼1 m distance scales is
discussed.

11.1 Introduction

Si large-scale integrated circuits (LSIs) together with optical fiber communications have been the building blocks of the internet society. Now, there
appears a strong demand to merge optical communication with a chip into
an electronic and photonic integrated circuit (EPIC) for ubiquitous society.
To converge electronics and photonics on a chip, the III–V platform used
in InP-based integrated photonics is ready to adopt the long haul communication technologies for on-chip integration. This is because the photonic
component technologies are well developed on the III–V platform. However,
EPICs should be on Si chips because there is no electronics comparable with
Si-LSIs that are of high figure of merit (FOM, defined in Sect. 11.2). The
Si platform requires that all photonic devices be compatible to complementary metal oxide semiconductor (CMOS). Recently, we have prototyped an
EPIC that would potentially breakthrough the performance limit of Si-LSIs
in terms of on-chip optical interconnection. CMOS technologies are one of the
significant resources of mankind, and CMOS compatibility makes EPICs not
only have high FOM, but also have short turn-around-time from laboratory

to fabrication. The present article reviews Si microphotonics [1], and on-chip optical interconnection [2] and CMOS-compatible Ge photodetectors [3–6].

11.2 Fundamental Limits of Si-LSIs

FOM has been employed for technology evaluation, and usually is defined as

$$\text{FOM} = \frac{\text{Performance}}{\text{Cost} \times \text{Power}}. \tag{11.1}$$

Performance in the case of Si-LSIs is speed. It is the sum of switching rate of a transistor and signal transmission rate between transistors. The switching rate already reached 1 THz in 2000 by means of the size scaling with the minimum feature size. However, the scaling is accompanied with a negative impact in capacitance and resistance increase of metal interconnects. The number of transistors will soon exceed one billion on a chip. This explosion in density of transistors/chip has made the transmission rate a dominant speed-limiter. Therefore, the scaling does not improve the signal transmission rate and the system performance of LSIs. This is the so-called interconnection bottleneck, indeed stagnating the clocking rate around 3 GHz in 2005.

Cost in FOM is dependent on the production yield which is strongly affected by multilevels of metal interconnects. A couple of years ago, LSIs were manufactured with the 0.13 μm technology node where six levels of the metal layers were involved. In the current 90 nm node, the interconnect layer has increased to as many as eight levels. The requirement for precise alignment should significantly lower the production yield.

Lastly, *power* in FOM exponentially increases with the performance increase. The higher the power, the higher the performance. This has incurred a so-called heat penalty due to high current density. Because of resistance increase of long metal interconnects, Joule heating is unavoidable. Together with subthreshold current of CMOS transistors the chips have now been heated up like "hot plates." Heat penalty is thus the most stringent issue in the current LSIs.

Photonics has a unique capability of overcoming these challenging issues because there is no resistance and capacitance for light. If we could hybridize photons for signal messengers and electrons for signal processors on a chip, the interconnection bottleneck should be solved. In addition, we should also be free of the long-standing design rule in electronics, i.e., the 50 Ω impedance matching. This might be the most significant contribution of photonics to Si-LSIs.

11.3 Material and Device Diversity in Photonics

As can be seen in the Table 11.1, electronics consists of only transistor and interconnects, and that has promoted on-chip integration. However, in photonics there are many devices currently used in long haul optical communication.

Table 11.1. Electronics and photonics – diversity of devices and materials

	devices	materials	
electronics	transistors	Si, Al, SiO_2, Si_3N_4	
	interconnects	Al, Cu	
photonics	light emitter	InGaAsP	(III–V)
	interconnect	Si, Si_3N_4,SiO_xN_y	(Si)
	MUX (filters)	SiO_2	(Si)
	detector	InGaAs	(III–V)
	modulator	$LiNbO_3$	(other)
	amplifier	InGaAsP	(III–V)
	isolator	YIG:Bi	(Other)

To implement photonic functionality monolithically on a chip, all these photonic devices must be integrable on Si. In other words, the diversity of devices in photonics is very large, and that is a clear difference from electronics supporting our digital world. This naturally leads into the other difference: the diversity of materials employed in photonics. In other words, photonic devices have been developed on the optimum materials. Some devices such as light emitter, amplifier, and detectors are on III–Vs, while others are on $LiNbO_3$ and yttrium iron garnet (YIG). In contrast, the basic materials set in electronics have been CMOS-compatible; Si, SiO_2, Si_3N_4, and Al. Recently, Cu and low k insulators have been added. The photonics materials are not CMOS-compatible at all. Thus, it is impossible to monolithically integrate photonics on the Si platform using Si CMOS technology. To overcome this problem with monolithic integration of photonic functionalities, there should be two approaches:

1. Making all photonic devices using Si CMOS-compatible materials, i.e., reduction of material diversity.
2. Making a few (hopefully one) new CMOS-compatible device(s) meeting all photonic functionality, i.e., reduction of device diversity.

Currently, the world trend is the former to reduce materials diversity. This approach of Si microphotonics can also be called as "Si CMOS-compatible photonics." In other words, Si microphotonics should be categorized into Si CMOS Photonics and Si CMOS-compatible photonics. Our Ge work described later is a typical example of Si CMOS-compatible photonics, and the modulator reported by Intel [7], and Si Raman lasers reported by UCLA [8] and by Intel [9] are others.

We have chosen an optical clocking chip as the first example of Si CMOS-compatible photonics. The optical clocking chip is one of the simplest, since the chip consists of only waveguide and photodetector, and is one of the most urgent because of stagnation of clocking rate at $\sim$3 GHz. Therefore, a high-speed clocking chip is the most appropriate target for first prototyping of Si microphotonics.

11.4 Optical Interconnection

11.4.1 Interconnection Bottleneck

Figure 11.1 shows the interconnection bottleneck in semiconductor technology issued by the Semiconductor Industry Association (SIA) in the late 1990s. This figure clearly warned of the performance limit of central processor units (CPUs). It should be noted that the delay may be categorized into two components: switching (gate) delay and transmission delay of signals. The switching delay does not depend on interconnect, as can be readily understood. However, the transmission delay does strongly depend on the materials because the delay is nothing but charging and discharging time of the condensers associated with interconnects. Accordingly, the transmission delay scales with the resistance and capacitance product (RC) time constant. Based on the traditional metal interconnect system of Al and SiO_2, the minimum transmission delay occurs around the 250 nm node, and the system gets slower when the minimum feature size is smaller, as in Fig. 11.1. A new metal interconnect system, i.e., Cu and low dielectric constant (k) materials, has been employed to avoid this problem. Cu has lower resistivity than Al ($\sim$0.5 of Al), and the low k materials have dielectric constants about a half of SiO_2. Thus, the Cu and low k system should have a gain from four times less RC time constant unless the interconnect size changes. This shifts the minimum delay at the 180 nm node. However, the transmission delay increases below the node. Further efforts have been invested to solve this problem using thicker metals and other approaches, but it is strongly suggested that we are very close to the end of the metal interconnection.

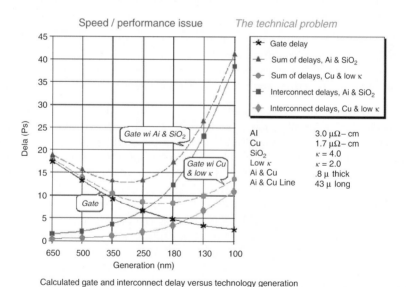

Calculated gate and interconnect delay versus technology generation

Fig. 11.1. Delays of LSIs versus technology nodes

Light does not have resistance and capacitance and should be free from charging and discharging times. This indicates optical interconnection would be capable of higher speed data transmission. The speed in Si is about $10^8 \, \mathrm{m \, s^{-1}}$, and thus propagates $\sim$0.1 mm in 1 ps. Let light do the job, as described in Sect. 11.4.2. Light has also colors which enable high transmission rates, as will be discussed in Sect. 4.5.

11.4.2 Optical H-tree

The clock speed of CPUs has shown to be stagnated at $\sim$3 GHz. Faster clocking can be done by giving more power on the chips, however, Joule heating drives the chip temperature to unacceptably high levels and causes malfunction and deterioration of the chip performance. In other words, performance improvement must be accompanied by heat dissipation due to power increase, called the heat penalty. Si-LSIs are rapidly moving to the Si on insulator (SOI) platform to achieve a faster switching rate. The substrate, however, is not friendly to heat dissipation. It is a poor thermal conductor due to the presence of the SiO_2 layer, and hence the heat penalty becomes more and more stringent.

To achieve a higher clock rate, we have proposed to implement the fractal optical H-tree architecture shown in Fig. 11.2. This is a hybrid system with optical and electrical H-trees. The optical H-tree is on the top, while the electrical H-tree is at the bottom. The optical clocking signal is to be injected from the lower part of the chip, and then propagates in the waveguide with an H-tree shape distribution. Finally the signals are detected by the photodetectors at the end of the waveguides. When the distance from in-port waveguide to the photodetectors is designed all to be the same by virtue of the H-tree architecture, the timing of signal arrival is guaranteed to be chronically the same. Once an optical signal arrives at the photodetectors, an electric pulse

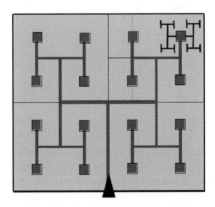

Fig. 11.2. Optical clocking H-tree and electronic H-tree. *Thick line* is waveguide, *square* is photodetector. Electrical H-trees are under the optical H-trees

is generated to transmit it to the electrical H-tree underneath. When the optical H-tree is double, as in Fig. 11.2, the longest interconnect in the electrical H-tree is 1/4 in length in comparison to the case without an optical H-tree. Since both resistance R and capacitance C are proportional to the length, the RC product should be 1/16. This simple calculation gives us 16 times faster clocking available if everything else is the same. This is a clear advantage of optical H-tree.

11.4.3 Waveguide

In order to implement an optical clock architecture on a chip, we need optical interconnects, i.e., a waveguiding platform. Planer lightwave circuits (PLCs) [10] are the first commercially available platform to integrate photonic components. They are also referred to as "Silica bench." Although it employed hybrid integration, there is a significant progress compared to on-board assembly of optical components. PLCs are based on silica fiber technology where the index contrast between the core and cladding is extremely small, as designated in Fig. 11.3, where the minimum waveguide-bending radius is plotted against index contrast. It is advantageous that PLCs are free of the challenge of efficient coupling to a single mode fiber. However, PLCs require large bending radii–up to several cm- to reduce bending losses to 0.1 dB per turn or less. The large bending radii make it difficult to confine optical interconnects on a chip of $1 \, \text{cm}^2$. On-chip optical interconnection therefore needs high-index contrast (HIC) platforms where the optical field intensity is strongly confined in the core, allowing sharp bends without significant bending losses. It should be remarked that several Si CMOS materials have been employed for HIC integrated optics. As shown in Table 11.1, Si, Si_3N_4, and SiO_2 are CMOS

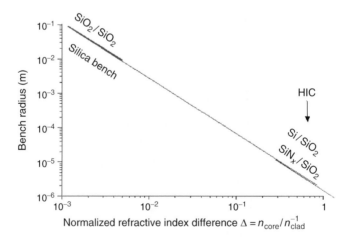

Fig. 11.3. Waveguide bending radius versus index contrast for $0.1 \, \text{dB}/90°$ transmission loss

materials and their refractive indices are 3.5, 2.0, and 1.5, respectively. Accordingly, we could keep the bending losses lower even though the bending radii are as small as 10 μm. In addition, the CMOS materials are transparent at the optical communication wavelength around 1.2–1.7 μm.

As stated above, Si microphotonics is advantageous in monolithic integration of photonics on Si CMOS platform using HIC waveguides capable of sharp bends. The next requirement for optical H-tree architectures is a low loss and even splitting ratio at Y-splitters. The loss arises mainly from scattering due to waveguide sidewall roughness and is nearly proportional to the third power of the index difference, as shown in Fig. 11.4 [11]. This trend implies a design principle for waveguide materials. The bending radius (expressed as minimum device size in Fig. 11.4) gets smaller with the index difference, while the scattering loss becomes larger with the index difference. Using silicon oxynitride materials, the index difference is controllable from 0.01 to 0.5 according to its alloy composition. If a bending radius of 1 mm is acceptable, the index difference can be 0.02 or larger (Fig. 11.3), with transmission loss per right-angle bend remaining below 10^{-2} dB cm^{-1}.

Another alternative approach that enables small bending radii may employ an "Air-trench bend" [12]. The latter is a unique low-loss waveguide bend design applicable to low-index contrast systems such as PLCs. A typical structure is shown in Fig. 11.5. The idea is to replace the cladding material on the side of waveguide core with air, only in the waveguide bending regions, by creating trench structures. This introduces HIC at the bends in the lateral dimension. A sharp bend without appreciable radiation losses is then permitted, since the index difference between the core and air can be expected to be higher than about 0.5. The taper parts of the air-trench bends are necessary to minimize loss due to mode mismatch at the input and output of the bend.

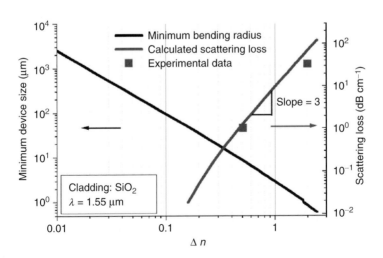

Fig. 11.4. Design diagrams of waveguide materials

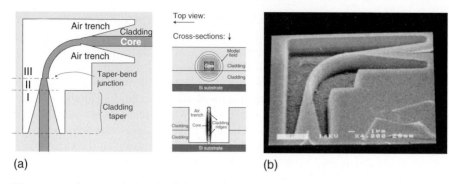

Fig. 11.5. Air trench bends: (**a**) schematic diagram of the plan view and cross-sectional views, (**b**) scanning electron microscopic image (SEM)

The tapers provide a gradual transformation of the guided mode shape connecting a low-index contrast waveguide to an air-trench bend where the mode is more strongly confined in the lateral direction. The mode field profiles are superimposed on the cross-sections in Fig. 11.5a, b. This approach enables bending radii on the order to 10 μm assuming that the SiO_xN_y core and SiO_2 cladding have an index difference of 0.1. Experimental characterization of fabricated structures showed a loss of 0.1 dB for 9 μm-radius right-angle bends.

11.4.4 Optical Clocking Chip

We have successfully prototyped the optical clocking chip for high-speed processors. Away from 1550 nm, we have used 850 nm as a clocking wavelength, since vertical coupling surface emitting laser diodes (VCSEL) have been around for many years and are capable now of 10 GHz operation with no external modulators. Si can be used as a photodetector material for 850 nm. The waveguide transparency should be obtained with Si_3N_4 or silicon oxynitride.

Figure 11.6 shows the optical clocking chip prototyped [2]. Here, the photodetectors are made of Si because of 850 nm operation. This chip is eventually the world's first EPICs. In other words, all devices used in electronic and photonic integration are built on an Si chip using Si CMOS-compatible materials. The waveguides employed an silicon oxynitride core where the index difference is 0.05 from the cladding. This chip includes a 64 fanout of the optical H-tree, which should potentially enable 64 times faster operation than the currently available CPUs.

11.4.5 On-chip Wavelength Division Multiplexing

Table 11.2 summarizes the goals of the SIA roadmap in the late 1990s.

The goal of performance was 6 GHz in 2006. Only ∼3 GHz clock speed is available in 2005. As described above, optical clocking should be the ultimate clue for fast clocking. The next issue is too many layers of metal interconnect

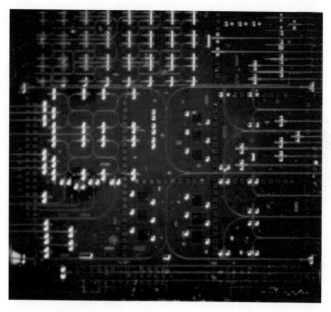

Fig. 11.6. Optical clocking chip. Used by permission from Intel Corporation as published in [13]

Table 11.2. Semiconductor Technology Roadmap

	1999	2006
interconnect length	1.5 km/chip	10 km/chip
clock	1 GHz	6 GHz
level of metals	6 layers	8 layers
power/chip	90 W	170 W

layers. It is currently eight and makes it hard to increase the production yield as discussed earlier in this review. In electronics each transistor should be connected with a metal interconnect which should not be connected to other transistors unless otherwise required. This results in a large amount of metal interconnects on a chip such as 10 km/chip as in Table 11.2. However, photonics can transmit many signals in one waveguide using wavelength division multiplexing (WDM). Although WDM requires additional device besides optical clocking, i.e., demultiplexing and multiplexing (DEMUX/MUX), one bus waveguide for carrying multiple wavelengths can practically serve the function of multiple levels of metal interconnects as follows.

Figure 11.7 shows microring resonator wavelength filters and their transmission spectra [14]. Light coming in from the "In" port can couple into the ring resonators 1–4 when their circumferences correspond to an optical path length that is an integer-multiple of the wavelength of the light. Thus, the bus waveguide has four-ring resonators of different radii as shown here should function as DEMUX. The transmission power spectra indicate four wavelengths dropped in the optical communication wavelength range via DEMUX of

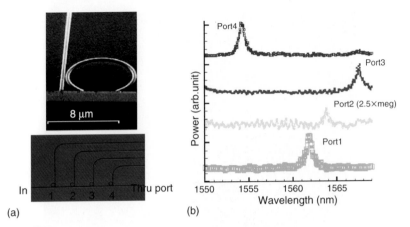

Fig. 11.7. Scanning electron micrograph (SEM) and filtering spectra of micro-ring resonators with a DEMUX function

microring resonators. Since free spectral range (FSR) of these ring resonators is large owing to the small ring diameter, each ring resonator has functioned as one channel dropping filter in the optical communication wavelength range, shown in Fig. 11.7. Here, the FSR is about 20 nm in wavelength, which is impossible to achieve with ring resonators in low-index contrast.

The resonance of a single ring resonator must be wide enough to drop a single modulated wavelength signal without spectral distortion. Then the resonance roll off may be too gradual to sufficiently reject the next adjacent wavelength channel in a dense WDM spectrum. More wavelength-selective, flat-top spectral responses may be obtained by coupling multiple ring resonators to create higher-order resonances may be obtained by coupling multiple ring resonators to create a higher-order filter. Figure 11.8 shows examples of higher-order ring resonator filters with up to four cascade rings [14]. It is clearly shown that passband flatness and out-of-band rolloff of the filter response increase with the number of rings. The transmission spectrum becomes rectangle-shaped at four rings because of resonant coupling. This flat top characteristics of the DEMUX is of importance when the chips are used in uncool conditions like Si-LSI chips. The resonant wavelength shifts with the chip temperature due to the temperature dependence of refractive index of the ring resonators. The flat top characteristics can allow a given wavelength to drop even though the center of filtering wavelength shifts around with a temperature fluctuation.

11.4.6 Heat Penalty

To get higher performance of digital chips, the simplest solution is high current operation. Hence, Joule heating is increasingly problematic, since it may incur malfunctioning digital chips. Liquid cooling of CPUs may be a short-term

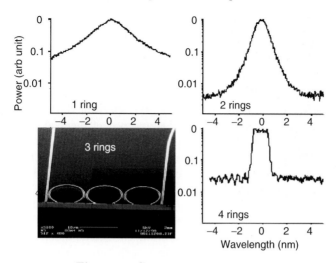

Fig. 11.8. Cascade-ring resonators

solution. Light does not experience resistance, or cause heat generation when used as a signal messenger.

Optical interconnection should be employed, not because it enhances the system performances and it eliminates too many levels of interconnection, but because it is the fundamental approach to get rid of heat penalty of current Si-LSIs.

11.4.7 Mode Converter of Off-Chip Light Source

Clocking wavelength is 850 nm of VCSELs that are not CMOS-compatible. Thus, we need mode converters that couple light from off-chip light sources to chips. This brings about two good results: First, the light emitter consumes a high power, heating the chips up. The off-chip light source concept thus allows heat isolation of chips from hot light sources. Second, state of the art technology of a light source is always available when it is off-chip. However, fiber couplers are needed. Currently, effective index coupling is quite promising and 1 dB/point has been shown as a typical coupling loss [15].

11.5 Ge Photodetector on Si

11.5.1 Ge Instead of Si

We have used 850 nm for optical clocking due to the following three reasons:

– VCSEL have been around for many years and are capable now of 10 GHz operation with no external modulators. GaAs-based VCSEL emit 850 nm.

– Si pin diodes can be photodetectors on a chip. Photodetectors currently used in long haul optical communications are based on pin InGaAs, which is not CMOS-compatible.
– Silicon nitride or silicon oxinitride serve as waveguiding core materials at 850 nm.

However, there is a challenging issue; the speed of the Si detectors. Si is an indirect semiconductor, and the absorption coefficient at 850 nm is mid-hundreds/cm, as in Fig. 11.9. This small absorption needs a long optical path to achieve high responsivity, resulting in slow operation due to a large RC time constant. Generally speaking the speed of photodetectors can be faster at a cost of responsivity or vise-versa.

Ge is an excellent candidate to replace Si photodetectors. Ge is an indirect semiconductor as well, but has the direct transition at 0.805 eV, corresponding to 1550 nm. At 850 nm the absorption coefficient of Ge is mid-tens of thousands/cm as in Fig. 11.9. Thus, Ge has the absorption coefficient more than two orders of magnitude larger than Si. Further more, Table 11.3 shows comparison of electronics properties of Ge and Si [16]. Mobility of Ge is much higher than that of Si, especially hole mobility. Ge is CMOS-compatible as well.

There are, however, two issues in Ge:

1. Defects in Ge hetero-epilayers on Si
2. The direct bandgap of Ge is wider than that of InGaAs. InGaAs can detect the whole C- and L-bands of optical communication, but Ge cannot.

These issues have been solved engineering thermal mismatch between Ge and Si, as described in Sect. 11.5.2.

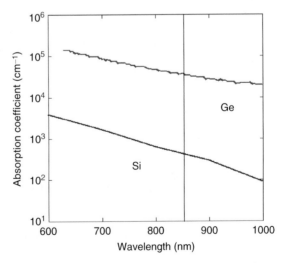

Fig. 11.9. Absorption coefficients of Si and Ge

Table 11.3. Electronic properties of Ge and Si

	mobility $(cm^2V^{-1}s^{-1})$	
	electron	hole
Ge	3,900	1,900
Si	1,500	600

11.5.2 Defect Engineering of Ge Heteroepilayers on Si

The lattice mismatch between Ge and Si is 4%, and should lead a high density of dislocations in Ge epi layers. Thus, the most of previous reports employed compositionally graded SiGe layers to buffer the mismatch. The buffer layers were, however, as thick as several μm and incurred challenging issues in monolithic integration of photodetectors with interconnects. We have successfully obtained a high-quality Ge directly grown on Si without an SiGe buffer layer, and dislocation-free selective Ge mesa on Si [3]. The issues to overcome were islanding growth and dislocations.

Because of the lattice mismatch, Ge epilayers never uniformly grow on Si to leave Ge islands. Here, ultra-high-vacuum chemical vapor deposition (UHV-CVD) has been used. Low temperature growth is necessary to avoid such islanding. Surface diffusion of Ge atoms should be limited at low temperature, allowing uniform epitaxy. Once flat epilayers were formed, Ge uniformly grow on the thin uniform Ge buffer layer even at high temperature and eventually gets thicker uniform epilayers. This low–high two-step growth has been invented in the course of GaAs on Si research a couple of decades ago, and now successfully adopted to Ge on Si for the first time.

Dislocations are generated in the flat Ge epilayers in terms of mismatch of lattice constants between Ge and Si. Typical dislocation density in $1\,\mu m$ thick as-grown Ge on Si is $\sim 1 \times 10^9\,cm^{-2}$. A transmission electron microscopic (TEM) image of cross-sectional Ge on Si is shown in Fig. 11.10. Threading dislocations and strained interlayer between Ge and Si are clearly revealed. Dislocations should be removed from the Ge epilayers since they frequently provide recombination centers in pin diodes, resulting in a large leakage current and a low responsivity.

High temperature annealing is found to be effective in dislocation reduction in the Ge epilayers. The mechanism has been modeled in terms of annihilation of dislocations and/or dislocation reactions due to thermally induced dislocation motions [3]. The dislocation motion is due to difference in linear expansion coefficients between Ge and Si. High- and low-temperature cyclic annealing reduces dislocation density to $\sim 1 \times 10^7\,cm^{-2}$. A cross-sectional TEM image confirms such dramatic reduction of threading dislocations, as in Fig. 11.11 [17]. A typical annealing condition was 900°C and 700°C for 10 minutes each.

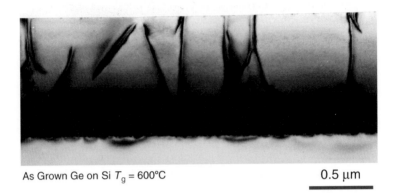

As Grown Ge on Si $T_g = 600°C$ 0.5 μm

Fig. 11.10. Cross-sectional TEM image of as-grown Ge on Si

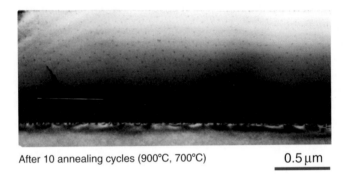

After 10 annealing cycles (900°C, 700°C)   0.5 μm

Fig. 11.11. Cross-sectional TEM microscope image of postgrowth annealed Ge on Si

Dislocation density is further reduced to $\sim 1 \times 10^6\,\mathrm{cm}^{-2}$ when Ge is grown selectively into mesas shown in Fig. 11.12. Even dislocation-free Ge has been grown on Si after postgrowth annealing when the mesas are $10 \times 10\,\mu\mathrm{m}^2$ in area as in Fig. 11.12b.

11.5.3 Strain engineering of Ge on Si

After growth of a high quality of Ge epilayers on Si, the epilayers were subjected to the high–low cyclic annealing. At high temperature the layer gets fully relaxed, and tensile strain is accumulated in the layer during cooling. This is because the lattice expansion coefficient of Ge is slightly larger than that of Si [3]. Hence, tensile strain is generated in the epilayers due to thermal mismatch of these lattice constants and shrinks the energy gap of Ge.

Strain-induced shrinkage of bandgaps between the conduction and light-hole valence bands (C–LH) and the conduction and heavy-hole valence bands (C–HH) are expressed as [18]

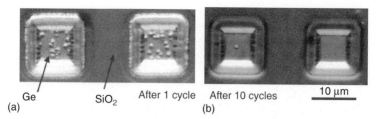

Fig. 11.12. Selective growth of Ge on Si. Selective epitaxial growth and postgrowth annealing have eliminated dislocations entirely on some mesas

$$E_g^{C-HH} = E_g^0 - \delta E_{hy} + (1/2)\,\delta E_{sh},\tag{11.2}$$

and

$$E_g^{C-LH} = E_g^0 - \delta E_{hy} - (1/4)\,\delta E_{sh} + (1/2)\,\Delta - (1/2)\sqrt{\Delta^2 + \Delta \cdot \delta E_{sh} + (9/4)\,(\delta E_{sh})^2},\tag{11.3}$$

where E_g^0 is the direct bandgap of unstrained semiconductor, δE_{hy} and δE_{sh} are the hydrostatic and shear deformation energies, and Δ is the spin–orbital splitting energy. For the biaxially strained (001) layer, δE_{hy} and δE_{sh} are written using the in-plane strain $e_{//}$ as

$$\delta E_{hy} = -2a \cdot (1 - c_{12}/c_{11})\,e_{//},\tag{11.4}$$

and

$$\delta E_{sh} = -2b \cdot (1 + 2c_{12}/c_{11})\,e_{//},\tag{11.5}$$

where a and b are the hydrostatic and shear deformation potentials, and c_{12} and c_{11} are the elastic constants of Ge. Recently, we have obtained accurate a and b in the course of strain-engineering of Ge epi on Si [19]. The revised values are: $a = -8.97 \pm 0.16\,\text{eV}$; $b = -1.88 \pm 0.12\,\text{eV}$.

Figure 11.13 shows the calculated strain dependencies of band edges. Under no strain the Ge bandgap is 0.805 eV for the direct bandgap and 0.664 eV for the indirect bandgap. With tensile strain, the conduction band bottom $E_c(\Gamma)$ for the direct transition and $E_c(L)$ for indirect transition moves downward in an energy scale. The valence band top, on the other hand, becomes nondegenerated under strain, and splits into two E_{HH} and E_{LH}. E_{LH} moves upward with tensile strain, resulting in shrinkage of bandgap. Although E_{HH} moves downward, tensile strain also shrinks the bandgap. Thus, both C–HH and C–LH bandgaps shrink with tensile strain.

Cooling down to room temperature from high temperature after growth and/or postgrowth annealing should accumulate tensile strain. The amount of strain is up to a few tenths percentage in Ge according to the temperature dependence of linear expansion coefficients of Ge and Si. Here, it is assumed that Si substrate obeys its temperature dependence. The amount of tensile strain has been measured by x-ray diffractometer. The strain was indeed $\sim$0.2%, suggesting the bandgap shrinkage by 20 meV according to Fig. 11.13.

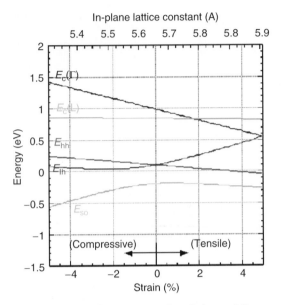

Fig. 11.13. Strain versus bandedges of Ge

Bulk Ge shows cut-off at 1550 nm in absorption coefficient as shown in Fig. 11.14. The gradual reduction of the absorption coefficient at longer wavelength is due to its indirect bandgap edge of Ge. In contrast, the present Ge epilayers on Si show extended absorption edge near 1,600 nm [6]. This shift is indeed 20 meV predicted in the Fig. 11.13 and extremely beneficial for signal detection of not only C-band but also L-band by Ge, as in the figure.

Figure 11.15 shows responsivity spectra of the present Ge photodetector on Si and of the theoretical calculation based on Ge absorption in literature. It is clearly shown that the responsivity edge shows "red-shift" toward 1,600 nm as expected.

It is also noted that the responsivity obtained by experiment is higher than the theoretical calculation. It is still an open question what explains the difference in responsivity. One of the possible explanations is that there is a large scattering in absorption coefficients of Ge in literature, as in Fig. 11.16 [20–23]. Especially the scattering is larger when the wavelength is below 1550 nm, where the experiment shows larger responsivity than calculation does. All these absorption coefficient data were reported in 1950s. It is strongly recommended that the Ge absorption coefficient should be measured more accurately.

The temporal responses of Ge photodetectors have measured to verify that 3 dB cut-off frequency is about 10 GHz at 980 nm with ~2-μm thick intrinsic Ge layer in pin structures [24]. The speed of the present Ge photodetector is limited by transit-time not by RC time, and therefore simple theory predicts that thinner Ge photodetector with ~0.5-μm thick intrinsic Ge should show

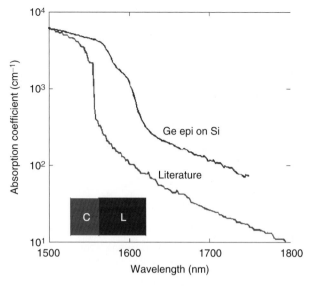

Fig. 11.14. Absorption coefficients of Ge. Bulk Ge reported shows rapid reduction of absorption coefficient at 1,550 nm, while Ge epi on Si shows extended absorption edge up to 1,600 nm

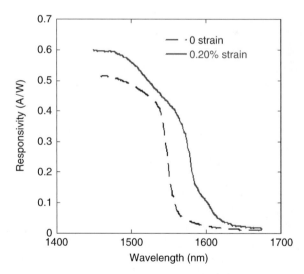

Fig. 11.15. Responsivity spectra of Ge photodetector on Si and its calculation without strain

the cut-off frequency beyond 40 GHz. Ge should be capable of fast on-chip photodetector.

The Ge on Si technology will also enhance fast optical clocking implementation on a chip.

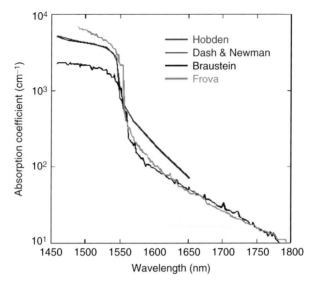

Fig. 11.16. Absorption coefficients of Ge reported so far

11.6 Summary

We have first prototyped optical interconnection on a chip; optical fractal H-tree circuits. Optical clocking is the simplest example of EPIC, requiring off-chip light source, SiO_xN_y waveguides and Si photodetectors. The waveguides and photodetectors have been implemented using CMOS processes. Defect- and strain-engineered Ge on Si technology allows broadband and high responsivity Ge photodetectors with an extended spectral response to L-band of optical communication, and better characteristics than InGaAs blow 1550 nm, i.e., especially at 850 nm range. The optical H-tree architecture together with Ge photodetectors will enable ultrafast CPUs by at least one order of magnitude.

Currently a technology gap exists at a reach of 1 m, i.e., between local area networks and chips, where a market exists for interchip, inter-server, etc. as in Fig. 11.17. This market will be quite important for Si microphotonics to play a dominant role, since it requires both low-cost and mass-productivity, where the key is monolithic integration.

Finally, it should be remarked that there have been not many attempts to reduce diversity of devices in contrast to those to reduce that of materials as described in Sect. 11.3. In other words, expansions of device functionality should be a big challenge for Si microphotonics. If failed, the impact of EPICs might be less significant than that of digital chips where transistors and interconnects are only on-chip components.

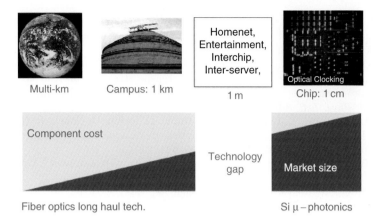

Fig. 11.17. The range of optical interconnection from long haul to on-chip

Acknowledgments

The authors appreciate Dr. Jun Fei Zheng at Intel Corp. for his excellent contribution on optical clocking research at Massachusetts Institute of Technology. The authors thank to Prof. Eugene Haller at The University of California, Berkeley for discussions on Ge optical properties. Discussions with Drs. Y. Yasha, and S. Saini, and Mr. D. Sparacin are appreciated. One of the authors (KW) express his thanks to Dr. A. Ishida for his enjoyable discussions. They also thank to Pirelli Photonics for their supports on integrated photonics research.

References

1. L.C. Kimerling, L. Dal Negro, S. Saini, Y. Yi, D.H. Ahn, S. Akiyama, D.D. Cannon, J.F. Liu, J.G. Sandland, D. Sparacin, J. Michel, K. Wada, M.R. Watts, "Monolithic Silicon Microphotonics", in *Silicon Photonics* edited by L. Pavesi, D.J. Lockwood (Springer Berlin Heidelberg New York, 2004), pp. 89–119
2. J.F. Zheng, F. Robertson, E. Mohammed, I. Young, D.H. Ahn, K. Wada, J. Michel, L.C. Kimerling, "On-chip optical clock distribution", in *Optics in Comput.* OSA, 20, 2003; M.J. Kobrinsky, B.A. Block, J.F. Zheng, B.C. Barnett, E. Muhammed, M. Reshotko, F. Robertson, S. List, I. Young, K. Cadien, "On-chip Optical Interconnections", *Intel Technol. J.* **8**, 129, 2004
3. H.-C. Luan, K. Wada, L.C. Kimerling, G. Masini, L. Colace, G. Assanto, "High responsivity near infrared Ge photodetectors integrated on Si" *Electron. Lett.* **17** 1467 (1999); L. Colace, G. Masini, G. Assanto, H.C. Luan, K. Wada, L.C. Kimerling, "Efficient high-speed near-infrared Ge photodetectors integrated on Si substrates", *Appl. Phys. Lett.* **76**, 1231, 2000
4. Y. Ishikawa, K. Wada, D.D. Cannon, H.-C. Luan, L.C. Kimerling, "Strain-induced bandgap shrinkage Ge grown on Si substrates", *Appl. Phys. Lett.* **82**, 2044, 2003

5. J.F. Liu, Douglas D. Cannon, K. Wada, Y. Ishikawa, S. Jongthammanurak, D.T. Danielson, J. Michel, L.C. Kimerling, "Silicidation-induced band gap shrinkage in Ge epitaxial films on Si", *Appl. Phys. Lett.* **84**(5) 660, 2004

6. D.D. Cannon, J. Liu, Y. Ishikawa, K. Wada, D.T. Danielson, S. Jongthammanurak, J. Michel, L.C. Kimerling, "Tensile strained epitaxial Ge films on Si(100) substrates with potential application in L-band telecommunications", *Appl. Phys. Lett.* **84**(6) 906, 2004

7. A. Lie, R. Jones, L. Liao, D. Samara-Rubio, D. Rubin, O. Cohen, R. Nicolaescu, M. Paniccia, "A high-speed silicon optical modulator based on a metal–oxide–semiconductor capacitor", *Nature* 427, 615, 2004

8. O. Boyraz and B. Jalali, "Demonstration of a silicon Raman laser", *Opt. Express* **12**, 5269, 2004

9. H. Rong, R. Jones, A. Liu, O. Cohen, D. Hak, A. Fang, M. Paniccia, "A continuous-wave Raman silicon laser", *Nature* **3346**, 1, 2005

10. T. Miya, "Silica-based planar light wave circuits: passive and thermally active devices", *IEEE J. Sel. Top. Quantum. Electron.* **6**, 38, 2000

11. K.K. Lee, "Transmission and routing of optical signals in on-chip waveguides for silicon microphtonics", The Doctoral Thesis at Massachusetts Institute of Technology, 2000

12. M. Popovic, K. Wada, S. Akiyama, H.A. Haus, J. Michel, "Air Trenches for Sharp Silica Waveguide Bends", *J. Lightwave Technol.* **20**, 1762, 2002; S. Akiyama, M. Popovic, K. Wada, J. Michel, L.C. Kimerling, H.A. Haus, "Air trench waveguide bends for high density optical integration", *Proceedings of SPIE, Photonics West*, San Jose, CA, Jan. 2004

13. M. Kobrinsky, B. Block, J.-F. Zheng, B. Barnett, E. Mohammed, M. Reshotko, F. Robertson, S. List, I. Young, K. Cadien, "On-Chip Optical Interconnects" *Intel Technol. Jl.* http://developer.intel.com/technology/itj/2004/volume08issue02/ (May 2004)

14. D.R. Lim, "Device integration for silicon microphotonics platforms", The Doctoral Thesis at Massachusetts Institute of Technology, 2000

15. K.K. Lee, D.R. Lim, D. Pang, C. Hoepfner, W.-Y. Oh, K. Wada, L.C. Kimerling, K.P. Yap, M.T. Doan, "Mode transformer for miniaturized optical circuits", *Opt. Lett.* **30**, 498, 2005

16. S.M. Sze, *Physics of Semiconductor Devices* (Wiley, New York, 1969), p. 20

17. H.C. Luan, D.R. Lim, K.K. Lee, K.M. Chen, J.G. Sandland, K. Wada, L.C. Kimerling, "High quality Ge epilayers on Si with low threading-dislocation densities", *Appl. Phys. Lett.* **75**, 2909, 1999

18. Chuang, *Physics of Optoelectronic Devices* (Wiley, New York, 1995), p. 124

19. J.F. Liu, D.D. Cannon, K. Wada, Y. Ishikawa, "Deformation potential constants of biaxially tensile stressed Ge epitaxial films on Si(100)", *Phys. Rev. B* **70**, 155309, 2004

20. M.V. Hobalen, "Direct optical transitions from the split-off valence band to the conduction band in germanium." *J. Phys. Chem. Solids* **23**, 821, 1962

21. W.C. Dash and R. Newman, "Intrinsic optical absorption of single-crystal Ge and silicon at 77 K and 300 K", *Phys. Rev.* **99**, 1151, 1955

22. R. Braustein, A.R. Moore, F. Herman, "Intrinsic optical absorption in germanium–silicon Alloys", *Phys. Rev.* **109**, 695, 1958

23. A. Frova and P. Handler, "Franz–Keldysh effect in space-charge region of a germanium p-n junction", *Phys. Rev.* **137**, A1857, 1965

24. J.F. Liu et al., unpublished

12

Silicon-Integrated Optics[*]

M. Paniccia, L. Liao, A. Liu, H. Rong, and S. Koehl

Summary. In this chapter, we introduce Intel's research efforts and recent break-throughs in the field of silicon photonics. We present an industrial perspective on the benefits of silicon photonics, a plan for the development of integrated devices, and the challenges involved with manufacturing these devices in high volumes. We then present an overview of Intel's silicon modulator work, including design, charac-terization, and scaling of the modulator. Finally, we provide an overview on Intel's research on devices based on stimulated Raman scattering (SRS), including mea-surements of continuous-wave (CW) gain and CW lasing in silicon.

12.1 Introduction

A little more than 40 years before the printing of this book, a young innovator named Gordon Moore made a simple observation about the exponential scaling of transistors in a microprocessor. This observation became a challenge that the semiconductor industry rose to meet with such consistency that it is now attributed the status of law – Moore's law. This success came at a price – namely the countless manhours, decades of research, and investment of billions of dollars into silicon processing technology.

Moore's law in the simplest sense describes the scaling of transistors, but in a wider context it describes much more. For transistors are more than just semiconductor devices. They are bits. They are intelligence. They are power, the power to direct and apply information to solve complex problems, improve standard of living, and generally make everything much more productive. Silicon transistors are the Age of Information, and what Gordon Moore really described was an exponential increase in the ability of humanity to harness and apply the resources of our world.

In fact, the rate at which we can process information has begun to outpace our ability to communicate it. We are stumbling over our words. Computer systems will soon be left waiting, starved for data. They will have all the capability to act, but not the means. This is because the copper-based inter-connect among dense information processors cannot keep up the pace. They have already failed for long-haul and metro networks. They are failing in the data center. One day they might fail even on PC motherboards.

Silicon microprocessors solved the problem of high-density information processing. Fiber optics solved the problem of high-density information trans-portation – but not for everyone. In terms of total interconnects in the world, fiber only occupies the tiniest niche. For every microprocessor in the world, there are hundreds, if not thousands, of copper interconnects connecting them to memory, chipsets, peripheral devices, and to the outside world. One day billions of optical interconnects will be needed, and today's technology is not up to the challenge.

The problem lies not in the fiber, but in the devices at either end. These devices are many things – discrete, bulky, expensive, difficult to manufacture, and difficult to assemble. They are many things, but the one thing they are not, to date, is made from silicon.

With microprocessors, silicon has proven to have none of these liabilities. It is abundant, well understood, and inexpensive to manufacture, even in very high volumes.

However, the one thing that silicon is not, to date, is a good optical ma-terial. Therein lies the irony – a semiconductor that has proven itself reliable and effective for electronics appears nearly useless for photonic applications.

Recent efforts by Intel and other researchers around the world have shown that the once-desolate world of silicon photonics now shows signs of life. The benefits of decades of investments in Moore's law for transistors have inspired some to work to expand the definition of this law to encompass not only transistors, but photonic devices.

Silicon has not been considered an optical material due to a lack of com-monly used physical effects. The two major ones are the electro-optic effect, used to make high-speed modulators in materials such as lithium niobate; and stimulated emission, used to make lasers in materials such as indium phos-phide. These effects are either nonexistent or extremely weak in silicon. In the case of stimulated emission much work has been done to coax a few more photons per electron out of the silicon lattice.

The key to unlocking silicon's optical potential may be to focus less on what the material is bad at and more on what it is good at doing. Although radiative band-to-band recombination emission is poor in silicon, stimulated Raman scattering (SRS) is uncommonly strong in silicon and can produce similar results – namely amplification and lasing. Free-carrier effects are also strong and, more importantly, take advantage of learning from well-understood complementary metal oxide semiconductor (CMOS) devices such as diodes and transistors. In fact, Intel's latest research breakthroughs rely on controlling free carriers with semiconductor devices. In the case of the silicon modulator, these charges are injected by a metal oxide semiconductor (MOS) capacitor to cause a phase shift in the optical signal. In the case of the silicon laser, a PIN diode is used to remove light-absorbing free carriers and sustain net amplification.

This chapter discusses Intel's perspective and research and its overarching goal to "siliconize" photonics. Section 12.2 discusses Intel's vision for bringing silicon photonics from the lab to the fab over the coming decade. This includes Intel's current research into enabling optical capabilities in silicon, plans for integrating devices, and issues of manufacturing photonics that are "CMOS fabrication compatible." Section 12.3 describes the first of Intel's two recent breakthroughs, a silicon modulator that has been shown to scale to 10 Gbps speeds by modulating light with free carriers. Section 12.4 details Intel's most recent breakthrough, the first silicon Raman laser to transmit CW light by removing free carriers generated by two-photon absorption (TPA).

12.2 Silicon Photonics Industrial Perspective

Silicon presents a compelling platform for integrated optics for both pragmatic and technical reasons. From an economic perspective, the benefit is clearly the infrastructure, toolsets, knowledge, and sheer capacity that have been developed for silicon microelectronics. A further benefit of this development is that it is expected to advance into the foreseeable future. Not only does this mean that nanoscale lithography will soon become available for photonics, it also means that the majority of optical devices, whose critical dimensions are rarely less than $1\,\mu m$, can be manufactured in older factories. If silicon photonics maintain a degree of compatibility with the microelectronics already in process at these facilities, photonics will find a friendlier environment to run along existing products.

In addition to drafting on microelectronic advances, silicon photonics will also benefit from the techniques under development in micromachining, such as those used to create silicon V-grooves for fiber arrays and for micro-electromechanical-systems (MEMS) devices. These techniques can be used to create silicon optical benches for hybrid assemblies or to simplify assembly through the direct attachment of fibers and external light sources. Rather than performing complex active alignments with the laser on, fibers could be dropped into a V-groove or U-groove and lasers could be attached "passively."

The greatest benefit of silicon photonics is the promise of integration of different types of devices into the silicon platform. Materials such as silica support passive devices to route light but lack high-speed active devices to manipulate light, while active materials such as indium phosphide require yield-killing regrowths to integrate different types of devices. Silicon, on the other hand, has both active and passive capabilities and the potential to integrate them with much higher yield. If CMOS compatibility is maintained, photonic devices could even be monolithically integrated with electronics.

Intel's research in silicon photonics is in the first of three phases. The first phase is to prove silicon's viability as an optical material. Silicon does have active and passive optical capabilities, but the active ones have been limited in performance. These capabilities must be scientifically extended and expanded before enough functions exist to build useful integrated photonic modules. To this end, Intel has focused most of its research on active capabilities such as tuning, detecting, switching, modulating, and amplifying light. This research has produced several scientific breakthroughs, including the high-speed modulator and the CW-Raman laser, which is detailed later.

Once the silicon photonics research community has produced a compelling set of active optical devices, Intel hopes to enter a second phase of hybrid integration. Starting with a silicon optical bench, in which electronics and photonics are mounted on top of a micromachined silicon assembly platform, researches will begin to integrate certain functions directly into the silicon. For instance, a silicon MUX and detectors could be integrated to form a multi channel receiver. Or, silicon modulators, lasers, and passive components could be integrated to create a multichannel transmitter. This will happen only when the integrated solution brings a benefit to the final module. This may be an increased performance, a smaller form factor, or a lower total cost.

The final phase of research will be to approach monolithic silicon integration – putting all the components together to form modules that process light in ways that could never be done before and at costs low enough to allow them to become ubiquitous. This may include the light source, if there is a breakthrough in silicon–emitters, and electronics if this brings a benefit and if photonic-electronic integration challenges can be addressed.

Integration of optical devices will bring several benefits. The obvious benefits are cost and size reduction. In addition, there will be less optical loss due to the elimination of fiber interfaces. Assembly and test will also be simpler and less costly since there are less independent devices. Likewise, reducing the number of devices to align means less opportunity for flawed packaging, increasing total yield. Integration of electronics with the optics could eliminate the parasitic due to packaging, further reduce the size, cost, and packaging complexity.

Many challenges will need to be addressed to make integration a reality, some less obvious than others. Some of these include film topology, fab contamination, thermal budgets, yield metrology, heat dissipation, process complexity, and fiber coupling. Since the issues have already been detailed in another work [1], this chapter only provides a survey of some of these issues.

Film topology is an issue due to the difference in feature heights for silicon waveguides and silicon transistors. State-of-the art 90 nm lithography tools have a depth of focus of only about $0.2\,\mu m$. This suffices for a transistor, which extends only $0.1\,\mu m$ above the wafer but becomes a problem when etching a waveguide that might be several microns in height. The tool simply cannot resolve feature variations of that magnitude. In order to use these advanced tools, new planarization techniques must be developed to build up a flat surface with an oxide or similar material, on which the next mask layer can be reliably developed. This will also require high aspect ratio trenches to be etched into this layer if the silicon surface needs to be accessed. Film topology is an issue for any silicon optics that are to be done in these facilities.

For optoelectronic integration thermal budgets pose a serious challenge. For a microelectronic process, the flow of processing steps is dictated in large part by the temperature. The magnitude and duration of a high-temperature procedure such as an anneal determines the changes or damage that occur in the materials that have already been deposited. High temperatures can displace dopants, change crystal structures, and melt metals. In order to produce low loss polysilicon waveguides, long anneals of several hours at up to 1100°C are necessary, forcing this to be one of the first processing steps. On the other hand, if one seeks to make use of optical polymers with the silicon devices, these must occur at the back end – even after metal depositions, since they will be damaged by temperatures over 200°C.

There are several other processing issues. The tools that exist to measure important optical parameters, such as index of refraction, do not exist in microelectronic fabs. Furthermore, wafer level testing is essential for high yields but is extremely challenging for planar photonic devices – creative methods for surface coupling must be developed. For optoelectronic devices, the optical devices will have to tolerate heat generated by the circuits themselves, which could cause temperatures to vary by tens of degrees Celsius across the chip. Finally, both optical integration and optoelectronic integration will result in increased complexity and compromises in device design. All these issues must be studied and solved to achieve high-yield, high-performance silicon photonics devices.

Finally, there remain the issues of fiber coupling and optical assembly. If standard SMF fibers are to be used it will be necessary to couple light into an 8–10 μm fiber core. This is an issue for rib waveguides, which tend to be a few microns in height and will be even more challenging for strip and photonic crystal waveguides, which tend to be only a fraction of a micron in size. Tapers and other coupling structures must be developed. Also, the fibers and any other external optics (such as an InP light source) must be attached passively to achieve low costs. Today's state-of-the-art active alignments are manual and complex and contribute significantly to the cost of optical products. If the device costs are to scale significantly, the packaging costs must scale with them.

These challenges can be overcome, but only with increased research and innovation in both academia and industry. If a compelling set of active optical functions can be developed, CMOS processing and compatibility issues can be resolved, and creative methods for optical testing and assembly are developed, then silicon photonics will be a reality. Sections 12.3 and 12.4 describe some of Intel's contributions to bring this vision closer to reality.

12.3 Silicon Modulator

An optical modulator is a key component of any optical communications link. Optical modulators with speeds greater than 1 GHz are typically fabricated from either the electro-optic crystal $LiNbO_3$ [2, 3] or III–V semiconductor compounds and multiple quantum wells [4–7]. These devices have shown modulation frequencies in excess of 40 GHz [3, 7]. The demand for low-cost solutions has prompted studies of silicon-based modulators. These are attractive from a cost standpoint because mature silicon processing technology and manufacturing infrastructure already exist and can be used to build cost-effective devices in volume. In addition, silicon photonics technology provides the possibility of monolithically integrating optical elements and advanced electronics on silicon using bipolar or CMOS technology [8].

Numerous silicon waveguide-based optical modulators have been proposed and demonstrated [9, 10]. The majority of research has focused on using the free-carrier plasma dispersion effect in a forward-biased p–i–n diode geometry. This is because unstrained pure crystalline silicon does not exhibit any appreciable electro-optic effect [11]. Since the modulation speed due to the free-carrier plasma dispersion effect is determined by the rate at which carriers can be injected and removed, the long recombination carrier lifetime in the intrinsic silicon region generally limits the modulation frequency of these silicon-based devices. Although there is literature showing theoretically that ~1 GHz modulation frequency might be achievable by reducing the waveguide dimensions to sub-micrometer range [12], the previous fastest demonstrated speed of an optical modulator based on current injection in silicon-on-insulator (SOI) was approximately 20 MHz [13, 14]. These devices are of little practical interest because today's communication networks demand GHz performance.

Recently, a novel MOS capacitor-based silicon modulator has been proposed, and it has experimentally demonstrated a modulation bandwidth of 2.5 GHz at optical wavelengths of around 1.55 μm [15]. It uses a free-carrier-based phase shifter in which a MOS capacitor is embedded in a silicon-based rib waveguide. An applied voltage will induce an accumulation of charge near the oxide of the capacitor, which will in turn modify the waveguide refractive index profile and ultimately the optical phase of light passing through it. When placed in one or both arms of a Mach-Zehnder interferometer (MZI), the phase shifter(s) can change the relative phase difference between waves passing through the two arms of the MZI. The phase modulation will thus lead to an optical intensity modulation. Because charge transport in the MOS capacitor is governed by majority carriers, device bandwidth is not limited by the relatively slow carrier recombination processes of p–i–n diode devices. As a result, this type of capacitor-based design has demonstrated high bandwidths that are unprecedented in silicon-based modulators. In the following, we review the design, fabrication, and characterization of such MOS capacitor-based silicon modulators.

12.3.1 Device Design and Fabrication

In the MOS capacitor phase shifter, the refractive index modulation occurs as a result of charge density changes in the vicinity of the gate oxide. Therefore, one would expect that the phase modulation efficiency to be strongly dependent on the phase shifter dimensions and gate position. To experimentally verify this, numerous phase shifters with different waveguide geometries, gate oxide positions, and gate oxide thicknesses have been designed and fabricated [16]. Figure 12.1 shows schematic cross-sectional views of two representative devices. Each comprises an n-type doped crystalline silicon layer (the silicon layer of the SOI wafer) and a p-type doped poly-silicon (poly-Si) layer with a thin horizontal gate oxide sandwiched between them. The poly-Si is formed by first depositing amorphous silicon using low-pressure chemical vapor deposition (LPCVD) at 550°C; this amorphous film is then crystallized into poly-Si using a high-temperature anneal. The annealing process is designed to minimize poly-Si grain boundaries, which can cause optical loss and limit electrical activation of dopants [17, 18]. The n-type silicon has an active doping concentration of $\sim 1.7 \times 10^{16}$ cm^{-3}, and the poly-Si has an active doping concentration of $\sim 5 \times 10^{16}$ cm^{-3}. These doping concentrations are chosen to target high speed phase modulation and acceptable optical transmission.

Table 12.1 summarizes and compares the relevant physical parameters of the different phase shifter designs considered. Designs B–E have 2.5-μm wide ribs and varying silicon slab thicknesses. Designs G and G1 have narrower ribs

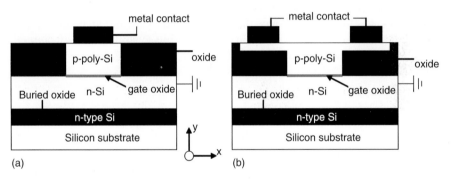

Fig. 12.1. Cross-sectional views of two representative MOS capacitor waveguide phase shifters in SOI: (**a**) design with metal contact directly above waveguide rib, and (**b**) design with metal contacts on poly-Si wing

Table 12.1. Physical parameters of the six phase shifter designs

device design	rib width (μm)	rib height (μm)	poly-Si Thickness (μm)	silicon thickness (μm)	gate oxide thickness (nm)	poly-Si Wing?
B	2.5	0.8	0.8	2.0	12	No
C	2.5	0.9	0.9	1.6	12	Yes
D	2.5	0.8	0.8	1.4	12	No
E	2.5	0.9	0.9	1.4	12	Yes
G	1.65	0.9	0.9	0.9	12	Yes
G1	1.65	0.9	0.9	0.9	6	Yes

Refer to Fig. 12.1 for schematics of device cross-sections

of 1.65 µm and thinner silicon slabs of 0.9 µm. They are different from each other in that they have different gate oxide thicknesses. The dimensions given in the table are obtained from cross-sectional scanning electron microscope (SEM) images of fabricated devices. Because the sidewalls of the waveguide ribs are not perfectly vertical, rib widths are measured at the midpoint of the ribs.

As illustrated in Table 12.1, thicknesses of the silicon and gate oxide layers are varied. By design, the position of the horizontal gate oxide in the waveguide is changed by changing the silicon layer thickness. To contact the p-type poly-Si, aluminum pads are deposited directly on top of the rib for some device designs (B and D), as shown in Fig. 12.1a. To make ohmic contacts to the metal, both the crystalline silicon and poly-Si have a high surface doping concentration of 1×10^{19} cm^{-3}. To reduce optical loss due to the metal [19] contacting the poly-Si rib and the associated high doping, other device designs contain a 0.2 µm thick top poly-Si layer, which is designated the poly-Si wing, above the silicon dioxide layers on both sides of the rib. Here aluminum contacts, along with the high concentration doping, are placed on top of this poly-Si wing, away from the rib, as shown in Fig. 12.1b. The oxide regions on either side of the rib maintain horizontal optical confinement and prevent the optical field from penetrating into the metal contact areas. Vertical optical confinement is provided by the ~0.375-µm buried oxide and an oxide cap (not shown). Modeling and testing confirm that all phase shifters are single-mode devices at 1.55 µm, the wavelength of interest.

12.3.2 Theoretical Background

In accumulation, the n-type silicon in the MOS capacitor phase shifter is grounded and a positive drive voltage, V_D, is applied to the p-type poly-Si, causing a thin charge layer to accumulate on both sides of the gate oxide. The voltage-induced charge density change ΔN_e (for electrons) and ΔN_h (for holes) are related to the drive voltage by [20]:

$$\Delta N_e = \Delta N_h = \frac{\varepsilon_0 \varepsilon_r}{e t_{ox} t} [V_D - V_{FB}], \qquad (12.1)$$

where ε_0 and ε_r are the vacuum permittivity and low-frequency relative permittivity of the oxide, e is the electron charge, t_{ox} is the gate oxide thickness, t is the effective charge layer thickness, and V_{FB} is the flat-band voltage of the MOS capacitor.

Due to the free-carrier plasma dispersion effect, the accumulated charges modify the optical properties of the silicon [11, 21]. For a wavelength of 1.55 µm, the refractive index and absorption changes due to accumulated electrons and holes were obtained from experimental absorption spectra through Kramers–Kronig analysis by Soref and Bennett as [22]:

$$\Delta n = \Delta n_e + \Delta n_h = -8.8 \times 10^{-22} \Delta N_e - 8.5 \times 10^{-18} (\Delta N_h)^{0.8}, \qquad (12.2)$$

$$\Delta\alpha = \Delta\alpha_e + \Delta\alpha_h = 8.5 \times 10^{-18}\Delta N_e + 6.0 \times 10^{-18}\Delta N_h, \qquad (12.3)$$

where ΔN_e and ΔN_h are in units of cm^{-3} and $\Delta\alpha$ is in units of cm^{-1}. The change in refractive index results in a phase shift $\Delta\phi$ of the optical mode given by:

$$\Delta\phi = \frac{2\pi}{\lambda}\Delta n_{eff}L, \qquad (12.4)$$

where L is the length of the phase shifter and Δn_{eff} is the effective index change in the waveguide before and after charge accumulation.

To model the phase shift, as well as transmission loss, a fully vectorial Maxwell waveguide solver based on the film mode matching method [23] is used to solve the waveguide mode. A complex effective index of the waveguide mode is obtained for each phase shifter design. The real part of the complex effective index is used to calculate the phase shift and the imaginary part is used to simulate the optical loss of the phase shifter.

12.3.3 DC Characterization of Phase Shifters

Two figures of merit are used to compare the device performance of the different designs: phase modulation efficiency and transmission loss. For phase modulation efficiency, (12.1) and (12.4) clearly show that drive voltage and device length are two of the primary determinants. The $V_\pi L_\pi$ product is therefore chosen, where V_π is the drive voltage swing (or $V_D - V_{FB}$) and L_π is the device length required for a π-radian phase shift. The goal is to minimize $V_\pi L_\pi$, lower the required drive, and shorten the device length. For all device designs, the same electric field strength $V_D/t_{ox} = 0.5\,V/nm$ is maintained to determine L_π. For instance, the drive voltage V_D is 6 V for a 12-nm gate device and 3 V for a 6 nm gate device. The electric field strength is monitored and kept constant to ensure that all devices operate with the same desired reliability [24]. For transmission loss, it is not enough to simply minimize the standard parameter of loss per length. The goal is to minimize loss per device length required for a π-radian phase shift: dB/L_π.

Due to the geometric asymmetry of the waveguides and the presence of the horizontal gate oxide, the phase shifters are expected to exhibit strong polarization dependence. Because the TE mode is more tightly confined inside the waveguide as compared to the TM mode, it interacts more strongly with the accumulated charges. As a result, all data given are for TE polarization.

Phase Modulation Efficiency

The phase modulation efficiency of the devices is measured using a free-space interferometric test setup in which one arm contains the device under test, and the other contains commercial liquid crystal phase modulators. Polarization of the input beam is controlled with a half-wave plate. This setup allows accurate measurement of phase shift and has good sensitivity - phase shifts as small as 0.005π can be measured.

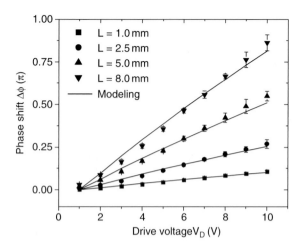

Fig. 12.2. Phase shift $\Delta\phi$ vs. drive voltage V_D of phase shifter design E at a wavelength of $\lambda = 1.55\,\mu m$ for different device lengths. The symbols represent the measured phase shifts and the solid lines are the simulated phase shifts

A representative sample of phase shift data is given in Fig. 12.2. It shows the measured and modeled phase shift of design E as a function of the drive voltage for different phase shifter lengths, $L = 1$, 2.5, 5, and 8 mm. The symbols represent experimental data and the solid curves are the modeled results. As the figure illustrates, there is good agreement between simulation and measurement for all device lengths. The phase shift is almost linearly dependent upon the drive voltage. The data also confirm the linear dependence of the phase shift on the phase shifter length for a given drive voltage, as expected from (12.4). Using data in Fig. 12.2, and taking into account the flat-band voltage of $V_{FB} = 1.25\,V$ [15], a $V_\pi L_\pi$ product of 7.8 V cm is obtained for design E. Phase shifters based on this design are used for speed measurements [15].

Figure 12.3 summarizes the measured and modeled $V_\pi L_\pi$ products of the five phase shifter designs with 12-nm gate oxide (designs B–G). Clearly, the phase efficiency of the devices is strongly affected by the waveguide dimensions and the position of the horizontal gate oxide. For designs B–E, as the total waveguide height is reduced and the gate oxide is moved closer and closer to the center of the optical mode, the interaction between the accumulated charges and the optical mode is consistently enhanced. The result is a consistent drop in the $V_\pi L_\pi$ product, from 35.6 to 7.8 V cm. For design G, which has even smaller waveguide dimensions and more centrally positioned gate oxide, its $V_\pi L_\pi$ product is 3.3 V cm, which represents a greater than two times improvement over that of design E.

In addition to strengthening the interaction between the optical mode and the charge carriers, another way to improve $V_\pi L_\pi$ is to increase the number of accumulated carriers per volt of electrical drive. One simple approach is

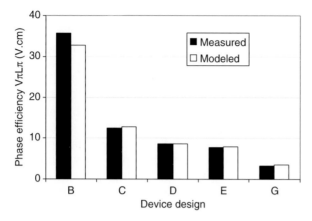

Fig. 12.3. $V_\pi L_\pi$ products of the five phase shifter designs with 12-nm gate oxide thickness. *Black bars* represent measured data and *white bars* represent modeled values

to reduce the gate oxide thickness t_{ox}. As (12.1) illustrates, to maintain the same accumulated charge density change ΔN_e and ΔN_h, the voltage swing $V_D - V_{FB}$ can be reduced when t_{ox} is thinned. The ultimate result is of course a reduction in $V_\pi L_\pi$. One has to be careful here because thinning the gate has the adverse effect of increasing capacitance, which will ultimately suppress device speed by increasing the RC time constant. To compensate, one can reduce capacitance by shrinking device dimensions. This not only shrinks the width of the gate capacitor, but because device phase efficiency improves with reduced waveguide dimensions (as shown earlier), but reduces the capacitor length as well.

Phase shifter designs G and G1 can be compared to illustrate the oxide thickness – phase efficiency dependence. By reducing the oxide thickness by 50% from 12 to 6 nm, $V_\pi L_\pi$ drops nearly 60% from 3.3 to 1.4 V cm. $V_\pi L_\pi$ is not halved, as predicted by (12.1), because as the gate oxide is thinned, it perturbs the optical mode less. Linecuts of modeled mode profiles of the two designs show that G1, with the thinner 6-nm gate, has a stronger optical field in the vicinity of the gate oxide. The result is of course an enhancement in the interaction between the optical field and charge carriers. Of all the phase shifter designs included in Table 12.1, design G1 has the best phase efficiency of 1.4 V cm. It has the smallest waveguide dimensions, the most centered gate oxide placement, and the thinnest gate oxide.

Transmission Loss

Transmission loss of the phase shifters is determined experimentally using the cutback method [25]. For each design, multiple identical devices at four different lengths are measured. Light is coupled into and out of the waveguides using lensed single-mode fibers with ∼3.3-µm spot size. An in-line polarization controller is adjusted to maintain the input beam at TE polarization.

Phase shifter designs B and D have high transmission loss because their metal contacts and high concentration doping are directly on top of the waveguide rib. Design D, for instance, has a loss of $36\,\mathrm{dB}/L_\pi$. To lower loss to $12\,\mathrm{dB}/L_\pi$, the total area that the metal and doping come in contact with the phase shifter was reduced by 98%. This is done by substituting the long, continuous metal contacts that cover the entire length of the phase shifters with multiple, short metal contacts that are spaced far apart. This approach, however, is not practical because it will significantly suppress device speed. The solution is to move the contacts and high-concentration doping away from the optical mode but close enough to maintain high-frequency response. As a result, the poly-Si wing design of Fig. 12.1b was implemented for the other phase shifters.

Figure 12.4 compares measured and modeled loss data for design D (with long and short metal contacts) and two designs with poly-Si wing. The data show that design E, which has very similar waveguide dimensions and gate oxide position as design D, has a slightly lower loss of $10\,\mathrm{dB}/L_\pi$. This result confirms that putting the metal and high concentration doping on the poly-Si wings is a very effective way of minimizing contact loss. The question now is how to reduce the remaining loss of $10\,\mathrm{dB}/L_\pi$. Modeling shows that 9% of the optical mode is in the poly-Si regions of the rib waveguide. Although significant effort was spent to improve dopant activation in poly-Si and reduce transmission loss, the poly-Si material with 5×10^{16} cm^{-3} active doping concentration still has a loss of ∼$50\,\mathrm{dB/cm}$. This poly-Si loss alone translates to $7.4\,\mathrm{dB}/L_\pi$ ($9\% \times 50\,\mathrm{dB/cm} \times 1.65\,\mathrm{cm}/L_\pi$). It is therefore clear that to improve phase shifter loss, poly-Si material loss must be reduced. Also shown in Fig. 12.4 is the loss data for design G1. It suggests that transmission loss per

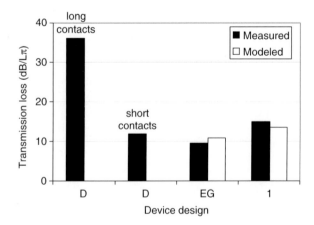

Fig. 12.4. Transmission losses of phase shifters. *Black bars* represent measured data and *white bars* represent modeled values

L_π increases with reduced waveguide dimensions and more centered gate oxide position. The reason is that although the waveguide design changes resulted in a $\sim 2\times$ increase in mode overlap with the accumulated charge carriers to improve phase efficiency, they resulted in even higher mode overlap, $\sim 2.5\times$, with the lossy poly-Si. In order to simultaneously improve phase efficiency and loss per L_π for these device designs, poly-Si loss again needs to be reduced. If the poly-Si region is replaced by single-crystalline silicon, designs E–G1 have modeled transmission losses less than $4\,dB/L_\pi$.

12.3.4 AC Characterization of Modulation Speed

To characterize the frequency response of the MOS-capacitor phase shifters, optical intensity modulation is needed. As a result, MZI structures have been fabricated with design E phase shifters imbedded in one or both waveguide arms. Design E has been chosen because it has relatively high phase efficiency and sufficiently low transmission loss. Compared to design G, it also allows better optical coupling due to its larger waveguide cross-section.

To determine the intrinsic bandwidth of the modulator, a small-signal measurement is performed using an MZI with a single 2.5-mm phase shifter. The overall length of the MZI modulator is 16 mm, which includes input and output waveguides, 3 dB splitters, and the two MZI arms. To facilitate laboratory testing, the arms have a path length difference of $\sim 16.7\,\mu m$, which allows for tuning of the interferometer bias point via small adjustments to the input laser wavelength. To ensure that the MZI operates in a high-sensitivity region, an input wavelength of $1.558\,\mu m$ was chosen and the MOS capacitor was biased into accumulation with $3\,V_{DC}$. A constant amplitude AC source was delivered to the 17 pF MOS capacitor through 50-Ω coax cable, a short section of 50-Ω PCB trace, and a simple network of resistors near the phase shifter to form an approximate broadband 50-Ω termination. Because of the nonideal source termination, the voltage arriving at the phase shifter bond pad was monitored with a 6-GHz high-impedance oscilloscope probe. The optical output of the modulator was collected in a 15-GHz high-frequency photoreceiver and measured on an electrical spectrum analyzer. Figure 12.5a shows these measured values. The normalized response of the device (photoreceiver output/on-chip voltage) is presented in Fig. 12.5b and shows that the phase shifter has an intrinsic bandwidth of greater than 2.5 GHz (as determined from the $-3\,dB$ point).

To investigate the data transmission performance of the modulator, a different MZI design is used. It has an overall length of 15 mm with two arms of nominally equal length (13 mm). Each arm comprises a 3.45-mm long high-speed RF (design E) phase shifter and two $\sim$4.75-mm long low-speed phase shifters that can be driven with DC voltages to electrically bias the MZI at quadrature. High-speed operation requires the use of a low-impedance drive circuit. A custom IC is developed and used [26]. It is manufactured using a 70 GHz-F_T SiGe HBT process and employs a push–pull emitter coupled

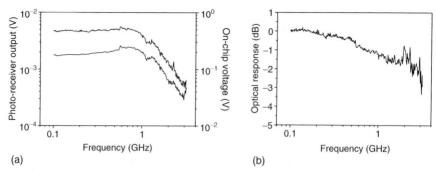

Fig. 12.5. (a) On-chip modulation voltage (V_{RMS}) and photoreceiver output of an MZI containing a single 2.5-mm phase shifter, and (b) phase shifter normalized response (photoreceiver output/on-chip voltage) showing an intrinsic bandwidth of approximately 2.5 GHz. The device was biased into accumulation with a 3 V DC bias

logic (ECL) output stage, which is wire bonded directly to the RF phase shifter; as a result, the modulator is operated in push–pull configuration [27]. First, a DC voltage of −5.2 V is applied to the n-type silicon slab. Then DC voltages applied to the 4.75-mm phase shifters are adjusted to bias the MZI at quadrature. Finally, an AC voltage swing in excess of 2 V is applied to the 3.45-mm high-speed phase shifters; the associated DC bias is chosen such that the entire AC swing is above the flat-band voltage [15].

Figure 12.6 shows the measured eye diagram of the MZI modulator at $\lambda = 1.55\,\mu m$. The data are generated using a 2.5-Gbps pseudorandom bit sequence (PRBS) source and collected using an optical plug-in module of a digital communications analyzer (DCA). It shows 2.8-dB extinction ratio (ER), 160-ps rise time, and 151-ps fall time. Testing of the driver into a test capacitive load that simulates the phase modulator shows a peak-to-peak voltage swing of 2.1 V. Calculations suggest that this output voltage should yield enough phase shift for an ER of 2.6-dB from the MZI, which is in good agreement with measurement.

12.3.5 Modulator Bandwidth Scaling

Our modulator with MOS capacitor phase shifters has demonstrated bandwidth performance previously unseen in silicon modulators, but the work is just beginning. In fact, all performance parameters discussed – efficiency, transmission loss, and frequency response – can be further improved. Given the history of silicon modulator development, improving frequency response is particularly pressing. Because the speed of the MOS capacitor-based phase shifter is governed by device resistance and capacitance, the path to higher speed involves reducing both these parameters. What needs to be done is therefore fairly obvious: reduce poly-Si and silicon resistances with higher

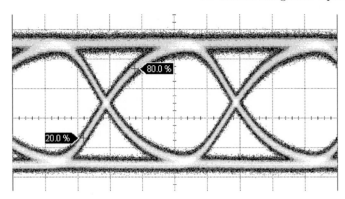

Fig. 12.6. Measured 2.5 Gbps eye diagram of MZI modulator with two 3.45-mm high-speed MOS capacitor-based phase shifters

doping and reduce capacitance by shrinking device dimensions. Both these approaches, however, translate to higher transmission loss because loss increases with (1) increased doping and (2) smaller size in that the overlap between the optical mode and the lossy poly-Si is increased.

To achieve higher modulation bandwidth, the focus should therefore be on improving poly-Si loss and dopant activation or, better yet, replacing poly-Si with a low loss and high dopant activation alternative such as single-crystalline silicon. The challenge, however, is to grow single-crystalline material on amorphous films such as the gate oxide. A process called epitaxial lateral overgrowth (ELO) is employed [28]. Using the ELO technique, crystalline silicon has been successfully grown on top of gate dielectric. Phase shifters with ELO-Si rib were designed, fabricated, and characterized [29]. They have 1.6-μm rib width, 0.65-μm rib height, 0.9-μm waveguide slab, and 11-nm gate thickness. The doping concentration is $\sim 2 \times 10^{17}\,\mathrm{cm}^{-3}$ for the Si slab and $1 \times 10^{18}\,\mathrm{cm}^{-3}$ for ELO-Si rib. These doping levels are not optimized because device resistance is likely limited by the Si slab resistance such that the ELO-Si rib doping level can be lowered without any significant impact on speed. Compared to some of the best poly-Si based MOS phase shifters discussed earlier, these ELO-Si devices have comparable phase efficiency and transmission loss: 3.4 V cm and $9\,\mathrm{dB}/L_\pi$. Nonetheless, it should be noted that these ELO-Si devices have much higher active doping concentrations to target high-bandwidth performance. If poly-Si is still used for the waveguide rib, loss is expected to be $15\,\mathrm{dB}/L_\pi$.

To understand the intrinsic bandwidth of the ELO-Si modulator, the impedance of 315-μm long phase shifter sections is measured. The data are given in Fig. 12.7. Note that test structures, with ELO-Si rib and high dopant concentrations, have capacitance of 2.4 pF and resistance of 6.5 Ω. The RC cutoff frequency, $(2\pi RC)^{-1}$, is therefore in the range of 10 GHz.

To investigate the data transmission performance of the ELO-Si modulator, the same MZI design described earlier is used: a 15-mm long device with

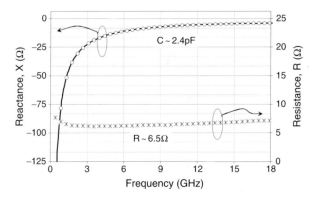

Fig. 12.7. Impedance of a 315-μm long phase-shift segment from the ELO-Si device. The reactance matches an ideal capacitor of 2.4 pF and the resistance is approximately 6.5 Ω, giving an RC cutoff of 10.2 GHz

two arms of nominally equal length of 13 mm. Optical characterization of the MZI modulator shows that it has an on-chip loss of ~10 dB, of which 3.5 dB is due to the high-speed RF phase shifters, 2.5 dB is due to the low-speed phase sections to bias the MZI, and the remaining 4 dB is likely due to combination of the voltage-induced free-carrier absorption (FCA) of (12.3) and unoptimized design of the splitters and bends. To characterize data transmission performance, the MZI is again copackaged with a custom driver IC that is similar to the one used to demonstrate 2.5 Gbps transmission. The driver operates from a single-ended power supply in the range of 3.3–3.9 V and delivers up to 1.6 V_{pp} (3.2 V_{pp} differential) into 27 pF phase shifter when operating at 8 Gbps [26]. Testing shows that a slightly lower AC voltage swing of 1.4 V is applied to the ELO-Si rib of the RF phase shifters; the associated DC bias is chosen such that the entire AC swing is above the flat-band voltage [15]. Figure 12.8a shows the measured eye diagram of the device at $\lambda = 1.55$ μm. The data are generated using a 6-Gbps $2^{32}-1$ PRBS source and collected using an optical plug-in of a DCA. The eye diagram shows 4.5 dB ER and 57 ps rise time (T_r), which gives a nominal bit rate, defined as $BR^{nom} = (3 \times T_r)^{-1}$, of 6 Gbps. Even though these ER and data rate represent a significant improvement over the previous best of 1.3 dB at 4 Gbps [26], 6 Gbps is significantly lower than the predicted 10-GHz intrinsic bandwidth of the device, as determined from the RC cutoff. Further analysis reveals that it is first the wirebond inductance (estimated to be 0.7 nH) with an LC cutoff frequency of ~3.9 GHz and then the driver itself that are limiting bandwidth performance – testing of a modulator with 20% lower RC gave the same rise and fall times. The modulators were also tested with 10-Gbps PRBS input, a representative eye diagram is given in Fig. 12.8b. Note that the eye is open and near rail-to-rail transmission with 3.8 dB ER is obtained. Note that the edge rates and jitter change gradually as data rate is increased. The use of electronic equalization in the receiver could improve eye margin at 10 Gbps.

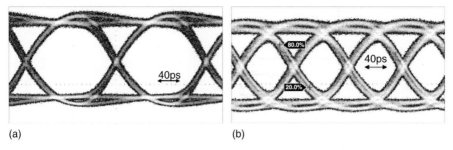

Fig. 12.8. Optical eye diagrams of modulator copackaged with CMOS driver; both eye diagrams have the same vertical and horizontal scales. (a) 6 Gbps: ER, T_r, and T_f are 4.5 dB, 57 ps, and 54 ps, respectively. (b) 10 Gbps: ER, T_r, and T_f are 3.8 dB, 55 ps, and 56 ps, respectively. The modulator speed is limited by the driver performance

Even with the much improved performance of the ELO-Si modulator, the MOS-capacitor design can be further optimized. The following discussion shows, by semiconductor device simulations, that it is possible to achieve 10-GHz modulation bandwidth and beyond in a silicon modulator with an on-chip optical loss less than 2 dB [30]. As the phase modulation frequency is determined by the rate of the carrier density modulation in the device, one can model the transient response of the phase shifter to predict the device speed. To do so, a two-dimensional (2D) device simulation package DESSIS is used. This device simulation tool employs the "box discretization" method [31] to solve the coupled Poisson equation and electron and hole continuity equations that govern charge transport in semiconductor devices. This is used to predict the electrical characteristics associated with specified physical structures and conditions. In addition, the simulation tool incorporates advanced semiconductor physics models to account for carrier statistics, doping-and field-dependent carrier mobility, and doping-dependent carrier recombination lifetimes. In the transient device modeling, a step-like drive voltage is applied to the phase shifter and the time dependence of the induced charge density at the oxide/silicon interfaces is calculated. The 3-dB modulation bandwidth Δf is then obtained from [32]:

$$\Delta f = \frac{0.35}{T_r}, \tag{12.5}$$

where T_r is the modeled rise time of the phase shifter in response to the step-like applied voltage. The rise time is defined as the duration during which the induced charge density in the MOS capacitor increases from 10% to 90% of its peak value. It is worth noting that as per measurement convention we have used the 20% and 80% points to calculate Tr and Tf for the eye diagram analysis presented above.

The 3-dB modulation bandwdith of a MOS capacitor phase shifter with the following dimensions is modeled: rib width is 1 µm, rib height is 1 µm, gate

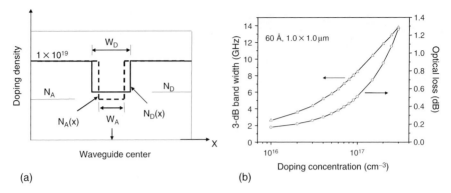

Fig. 12.9. (a) Doping profiles $N_D(x)$ and $N_A(x)$ for the thin (0.15 μm) n-type and p-type silicon layers in the MOS capacitor phase shifter. (b) Modeled 3-dB bandwidth and optical loss as a function of the doping concentration for a device with a 6-nm gate oxide. The optical loss is for a device length that leads to $\pi/2$ phase shift at a drive voltage of $V_D = 3$ V. The wavelength is $\lambda = 1.55$ μm

oxide thickness is 6 nm, and gate oxide is located vertically in the middle of the waveguide. It is assumed that the n-doping concentration is 5×10^{15} cm^{-3} except for the top thin (0.15 μm) layer with a doping profile $N_D(x)$ illustrated in Fig. 12.9a. For the p-type region, the doping concentration is N_A inside the waveguide and the top thin (0.15 μm) layer has a doping profile $N_A(x)$ shown in Fig. 12.9a. In the simulation, it is assumed that $N_A = N_D$ and the doping widths are $W_A = 1.0$ and $W_D = 2.0$ μm. The simulated results are shown in Fig. 12.9b; the 3-dB modulation bandwidth increases with an increase in the doping concentration. When the doping level reaches $\sim 1.5 \times 10^{17}$ cm^{-3}, a 10-GHz bandwidth is obtained. Figure 12.9b also shows that the modulation bandwidth can be scaled to even higher values with higher doping concentrations.

It is also worth noting again that increasing the modulation speed by increasing doping concentrations also increases the transmission loss due to FCA, as is evident from (12.3). To quantify such effects, phase shifter loss is also modeled as a function of the doping concentration. As the phase shifter length is only required to achieve $\pi/2$ phase shift for push–pull operation [27] in an MZI modulator, phase shifter loss is modeled for a device length of $L_{\pi/2}$. With a gate oxide of 6 nm and at $V_D = 3$ V, the modeled $L_{\pi/2} = 1.94$ mm. Figure 12.9b shows that the optical loss of the phase shifter increases with increasing the doping concentration. For a concentration of $\sim 1.5 \times 10^{17}$ cm^{-3}, which leads to 10-GHz modulation bandwidth, the modeled phase shifter loss is ~ 0.7 dB. Note that ~ 0.7 dB is the passive phase shifter loss without applied voltage. An additional loss of ~ 1 dB (for the required $\pi/2$ phase shift) due to voltage-induced FCA should be included in the MZI loss in the "on" state, in which the optical transmission of the modulator reaches the maximum value. In addition, the MZI modulator loss will include the Y-junction loss and passive waveguide loss due to light scattering, induced, for example, by sidewall roughness. With development of the advanced fabrication technology, it is expected that these losses could be minimal, as low transmission loss in small waveguides on SOI is well established. In this manner, modeling suggests that an optimized device can achieve 10 GHz modulation bandwidth with <2 dB optical loss.

12.4 Silicon Laser and Amplifier Based on Stimulated Raman Scattering

Coherent light generation and amplification in silicon has been one of the most challenging goals in silicon photonics [25, 33, 34]. Because of the indirect band gap, bulk silicon shows very inefficient band-to-band radiative electron–hole recombination. Light emission in silicon has thus focused on the use of silicon engineered materials such as nanocrystals [34–37], Si/SiO_2 superlattices [38], erbium-doped silicon-rich oxides [39–42], surface-textured bulk silicon [43], and Si/SiGe quantum cascade structures [44]. SRS has recently been demonstrated as a mechanism to generate optical gain in planar silicon waveguide structures [45–53]. However, due to the nonlinear optical loss associated with TPA-induced FCA [47, 49, 54], so far net optical gain and lasing has been limited to pulsed operation [55, 56]. To reduce FCA, we designed a silicon waveguide with a built-in p–i–n diode structure, which allows the free carriers generated in the TPA process be swept out of the waveguide by the electric field when the p–i–n diode is reverse biased. We experimentally prove that the nonlinear absorption is significantly reduced, enabling CW net optical gain [57]. With optical coatings applied to the waveguide facets, an optical cavity is formed, and we achieve CW lasing [58] in a silicon waveguide. The demonstration of Raman lasing in a compact, all-silicon, waveguide cavity on a single silicon chip represents an important milestone toward practical CW optical amplifiers and lasers that could be integrated with other optoelectronic components onto CMOS fabrication compatible silicon chips.

12.4.1 Device Description

The silicon rib waveguide is fabricated on the $\langle 100 \rangle$ surface of an undoped SOI substrate using standard photolithographic patterning and reactive ion etching techniques. The rib waveguide width (W), rib height (H), and etch depth (h) are 1.5, 1.55, and 0.7 μm, respectively. A cross-section SEM image of a typical p–i–n waveguide is shown in Fig. 12.10a. The effective core area [59] of the waveguide is calculated to be ~1.6 μm^2 by using a fully vectorial waveguide modal solver [60]. To increase the interaction length, the waveguide is formed in an S-shaped curve with a bend radius of 400 μm and a total length of 4.8 cm (Fig. 12.10b). The straight sections of the waveguide are oriented along the $\langle 011 \rangle$ crystallographic direction. To reduce the nonlinear optical loss due to the TPA induced FCA, a p–i–n diode structure is fabricated along the waveguide by implanting boron and phosphorus in the slab on either side

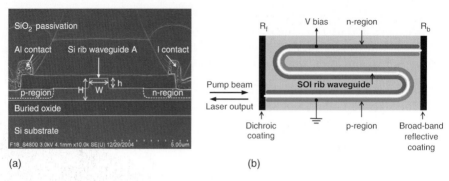

Fig. 12.10. (a) Scanning electron microscope image of a typical p–i–n waveguide cross-section. (b) S-shaped p–i–n waveguide with optical coatings on both facets to form a laser cavity. For amplifier application antireflection coatings are used

of the rib waveguide with a doping concentration of $\sim 1 \times 10^{20}\,\text{cm}^{-3}$. The separation between the p- and n-doped regions was $\sim 6\,\mu\text{m}$. Ohmic contacts were formed by depositing aluminum films on the surface of the p- and n-doped regions. This was followed by a SiO_2-passivation layer deposition. The doped regions and the metal contacts at the designed separation had negligible effect on the propagation loss of the waveguide because the optical mode is tightly confined in the waveguide. This was verified experimentally. The linear optical transmission loss of the S-bend waveguide was measured to be $0.4 \pm 0.1\,\text{dB}\,\text{cm}^{-1}$ based on Fabry–Pérot resonance measurements of a polished but uncoated waveguide [34].

When used as an amplifier, the waveguide facets are coated with antireflection films. To form a laser cavity the front facet is coated, with a dichroic film, having a reflectivity of $\sim 71\%$ for the Raman/Stokes wavelength of 1686 nm, and $\sim 24\%$ for the pump wavelength of 1550 nm. The back facet has a broadband high reflectivity coating of $\sim 90\%$ for both pump and Raman wavelengths (Fig. 12.10b).

12.4.2 Reduction of Nonlinear Loss with Reverse-Biased p–i–n Diode

When a reverse bias is applied to the p–i–n diode, the TPA-generated electron–hole pairs can be swept out of the silicon region between the p- and n-doped regions, where the optical mode is located. If the carrier transit time across the optical mode area is shorter than the free-carrier lifetime, then the transit time becomes the effective carrier lifetime. This effective carrier lifetime, which is associated with the free-carriers interaction with the optical mode, determines the TPA-induced carrier density. By varying the reverse bias voltage, the carrier transit time and hence effective carrier lifetime can be modified. In the CW pumping case, the free-carrier density induced by the TPA is given by [49, 50]:

$$N(z) = \frac{\tau\beta}{2h\nu}\frac{P^2(z)}{A_{\text{eff}}^2},$$ (12.6)

where β is the TPA coefficient, $P(z)$ is the pump power along the waveguide, A_{eff} is the effective core area of the waveguide, $h\nu$ is the one-photon energy of the pump beam, and τ is the effective carrier lifetime. Taking into account both the TPA and TPA-induced FCA, the pump power evolution along the waveguide is described by the equation [49, 50]:

$$\frac{dP(z)}{dz} = -\alpha P(z) - \frac{\beta}{A_{\text{eff}}}P^2(z) - \sigma N(z)P(z),$$ (12.7)

where α is the linear absorption coefficient and σ is the FCA cross-section. As evident from (12.6), the free-carrier density can be reduced if the carrier lifetime τ in the waveguide region is shortened. In turn, the FCA will also be reduced. To experimentally verify this, we measured the transmitted pump power vs. input power for a 4.8-cm long S-shaped silicon p–i–n diode waveguide with AR coatings applied to both facets for various reverse bias voltages. Results are presented in Fig. 12.11. The symbols represent the measured results and the solid curves are the modeled results by solving combined (12.6) and (12.7). Figure 12.11 indicates that there is a good agreement between modeled and measured results for various reverse bias voltages. At low input powers the output power is linearly dependent upon the input power. As the input power is increased above 200 mW, the output power saturates,

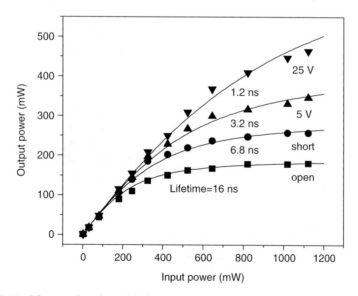

Fig. 12.11. Measured and modeled output power as a function of the input power for a 4.8-cm long waveguide containing a reverse-biased p–i–n diode with various bias voltages. Symbols represent the experimental data and curves are the modeled results. The carrier lifetime is used as a fitting parameter for each bias voltage

with the saturated output power depending upon the p–i–n diode bias. For an open circuited p–i–n waveguide, the output power saturates to a constant value of ~180 mW independent of the input power. Short circuiting and reverse biasing the p–i–n structure allows current to flow, which removes the free carriers from the optical mode and reduces the effective carrier lifetime, thus increasing the output intensity. This demonstrates that reduction in the FCA is achievable with this device structure.

By comparing the modeling and experiment, we can determine the effective carrier lifetime in the p–i–n diode waveguide. As the carrier lifetime depends on the reverse bias voltage, we use it as a fitting parameter. Other parameters used in the modeling are the linear absorption coefficient of $\alpha = 0.39\,\text{dB/cm}$, the previously reported TPA coefficient of $\beta = 0.5\,\text{cm/GW}$ [49, 50], and the FCA cross-section of $\sigma = 1.45 \times 10^{-17}\,\text{cm}^2$ [11] at a wavelength of 1.55 μm. When the p–i–n diode is open (no current flow is allowed), the modeled carrier lifetime is ~16 ns. When the p–i–n diode is short circuited (external bias is 0 but current flow is possible due to the internally built-in potential of the p–i–n), the carrier lifetime is much shorter ($\tau = 6.8\,\text{ns}$). By increasing the reverse bias voltage, the carrier lifetime is further reduced, and at a reverse bias of 25 V, an effective carrier lifetime of ~1 ns is obtained.

12.4.3 Measurement of CW Optical Gain

To measure the gain of the silicon waveguide containing a p–i–n diode, we performed a pump–probe experiment. The experimental setup is shown in Fig. 12.12. The pump and probe lasers are combined with a wavelength multiplexer into a lensed single-mode fiber whose output is used to couple to the waveguide under investigation. The output beam of the waveguide is collimated by a 50× objective lens, and a long-wavelength pass optical filter is

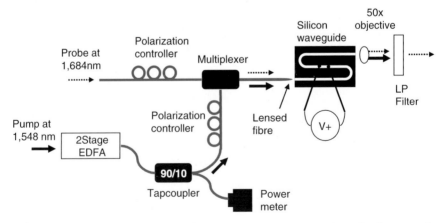

Fig. 12.12. Experimental setup of CW gain measurement, DUT is the device under test

used to separate the pump and probe beams. The probe beam passes through the filter and is detected with a photodetector while the pump beam is blocked by the filter. Fiber polarization controllers at the input to the device under test allow both probe and pump beam polarizations to be independently controlled. The device under test is mounted on a TEC and kept at a constant temperature of 25°C.

For the CW gain measurements, the pump laser is a CW-tunable external cavity laser emitting around 1,548.3 nm, which is amplified using two EDFAs to a maximum output power of 4 W. The probe laser is a 2-mW external cavity tunable diode laser operating at around 1,684 nm. Both the pump and probe laser sources have line widths of $\leq$100 MHz. The polarization of the probe and the pump beam is aligned with the TE mode of the waveguide.

To measure the SRS gain, the transmitted probe power is measured at the peak of the Raman gain profile (at the Stokes' wavelength) and compared to the input probe power. The input probe power is determined from the measured transmitted probe power without the pump beam by factorizing out the linear transmission loss of the waveguide.

Since the probe power is low, we ignore the pump depletion in the description of the pump-probe experiment. Taking into account the TPA and TPA-induced FCA at the Stokes' wavelength, the probe power $[P_s(z)]$ in the waveguide can be described by [50, 59]:

$$\frac{dP_s(z)}{dz} = -\alpha P_s(z) - \frac{2\beta - g_r}{A_{\text{eff}}} P(z)P_s(z) - \sigma N(z)P_s(z), \qquad (12.8)$$

where g_r is the Raman gain coefficient. Solving (12.8) using the pump power obtained from (12.7) and with probe input power $P_s(0)$ as an initial condition, one obtains the net optical gain of the waveguide via:

$$G = 10 \log \frac{P_s(L)}{P_s(0)}, \qquad (12.9)$$

where $P_s(L)$ is the probe output power obtained from (12.8) and L the waveguide length.

Figure 12.13 shows the net CW-Raman gain as a function of the pump power inside the waveguide for a 4.8-cm long waveguide at various bias voltages. The pump power is the power of the pump coupled into the waveguide and is determined by measuring the power exiting the lensed fiber and factorizing out coupling loss to the waveguide. The symbols represent the experimental data and the curves are the modeled results. In the modeling, we used a linear loss of 0.39 dB/cm. At the Stokes' wavelength, the FCA cross-section is 1.71×10^{-17} cm^2 [11]. Since the same waveguide was used for both the nonlinear transmission and pump-probe measurements, the corresponding carrier lifetime shown in Fig. 12.11 was used in the simulation of the net gain in Fig. 12.13. The only fitting parameter is the gain coefficient g_r. We see from Fig. 12.13 that there is no net gain when the p–i–n diode is open. This is

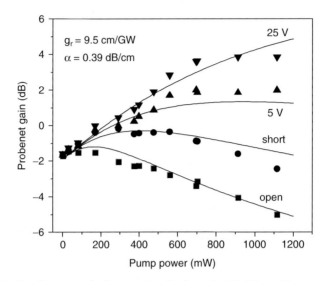

Fig. 12.13. Net Raman gain for a p–i–n diode embedded in a silicon waveguide as a function of the pump intensity for a 4.8 cm long silicon waveguide at different bias voltages. Symbols represent the experimental results and solid curve is the modeling result. The Raman gain coefficient used in the simulation is $g_r = 9.5$ cm/GW.

because, when the carrier lifetime is long, the Raman gain cannot compensate for the loss due to the linear waveguide scattering loss, TPA and TPA-induced FCA. When the diode is short circuited, the FCA is significantly reduced, but there is still no net gain observed. Once the p–i–n is reverse biased net gain is achieved. With a reverse bias of 5 V, we obtained a net gain of ∼2 dB with a pump power of ∼700 mW. Using a higher bias voltage of 25 V, a net gain of >3 dB is achieved for pump powers above 700 mW. From Fig. 12.13 we see that experimental data agree relatively well with modeling results based on a gain coefficient of $g_r = 9.5$ cm/GW.

Figure 12.14 shows the net Raman gain, at a probe wavelength of 1684 nm, as a function of the pump wavelength. The p–i–n bias was set at 25 V and the pump power at 511 mW. For this configuration net Raman gain is observed over 0.5 nm. Figure 12.14 also shows that the 3-dB linewidth of the SRS spectrum is ∼100 GHz.

12.4.4 CW Laser Experiment

The laser cavity is formed by coating the waveguide facets with multilayer films. The front facet coating is dichroic, having a reflectivity (R_f) of ∼71% for the Raman/Stokes wavelength of 1,686 nm, and ∼24% for the pump wavelength of 1,550 nm. The back facet has a broadband high reflectivity coating (R_b) of ∼90% for both pump and Raman wavelengths (Fig. 12.10b).

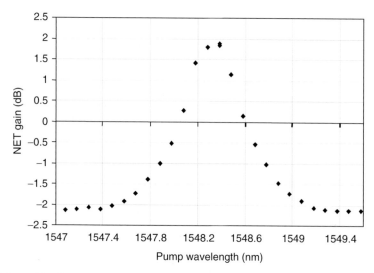

Fig. 12.14. Net Raman gain as a function of the pump wavelength for a 4.8-cm long silicon waveguide. The pump power is 511 mW, the probe wavelength is 1,684 nm, and the reverse bias on the p–i–n is 25 V

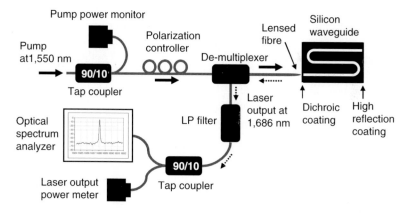

Fig. 12.15. Setup of the silicon Raman laser experiment. See main text for details

The waveguide loss was measured to be $0.35\,\mathrm{dB\,cm^{-1}}$ prior to applying the multilayer coating to the waveguide facet. Thus, the coating reflectivity can be determined by measuring the linear optical transmission spectrum of the same waveguides before and after application of the coating.

Figure 12.15 is a schematic of the Raman laser experiment. A CW external cavity diode laser (ECDL) at 1,550 nm is amplified by an erbium-doped fiber amplifier system to produce a pump beam of up to 3 W. The pump beam passes through a polarization controller followed by a thin-film-based wavelength demultiplexer and is coupled into the waveguide cavity by a lensed fiber through the dichroic-coated front facet. The Raman laser output and the reflected pump beam are coupled back into the lensed fiber and separated through the wavelength demultiplexer. The extracted laser output from the reflection port of the demultiplexer is further filtered by a long wavelength pass (LP) filter before being detected by a power meter or optical spectrum analyser. The coupling loss between the lensed fiber and the waveguide was measured to be ~4 dB and the insertion loss of the demultiplexer and the long-pass filter is ~0.6 dB. The silicon chip is mounted on a thermoelectric cooler and kept at a constant temperature of 25°C.

At the pump wavelength, a low finesse cavity is formed by the low reflectivity front facet and high reflectivity back facet. This configuration allows the cavity enhancement effect [62–64] of the pump power to be utilized to lower the lasing threshold. When the pump laser is tuned to the resonance of the cavity, the circulating power inside the waveguide cavity is enhanced, and the effective mean internal power (I_{eff}) inside the cavity can be expressed as [65]:

$$I_{\text{eff}} = I_i \frac{1 - e^{-\alpha L}}{\alpha L} \frac{(1 - R_f)(1 + R_b e^{-\alpha L})}{(1 - \sqrt{R_f R_b} e^{-\alpha L})^2}, \tag{12.10}$$

where I_i is the incident pump power (less coupling loss), R_f and R_b are the reflectivity of the front and back facets respectively, α is the absorption coefficient, and L is the waveguide length. At low power, we can estimate the power enhancement factor $M = I_{\text{eff}}/I_i$ to be ~2.2 using our experimental parameters. At high powers, however, the absorption coefficient α increases due to the TPA-induced nonlinear absorption, and M reduces accordingly.

Figure 12.16a plots the Raman laser output power vs. the input pump power (I_i) coupled into the laser cavity depicted in Fig. 12.10b, at two different reverse bias voltages applied to the p–i–n diode. In the experiment, the pump beam polarization is adjusted with a polarization controller and

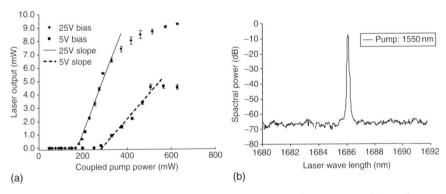

(a)   (b)

Fig. 12.16. (a) Laser output vs. pump power at two different reverse bias voltages. (b) Laser spectrum measured with optical spectrum analyzer

its wavelength is fine tuned to the cavity resonance to take advantage of the cavity enhancement of the pump and maximize the laser output. The Raman laser frequency is 15.6 THz red shifted from the pump laser. We see from Fig. 12.11 that the lasing threshold reduces with increasing reverse bias voltage. The lasing threshold is ~180 mW with a 25-V bias and ~280 mW with a 5-V bias. The lower threshold and higher laser output power with higher reverse bias voltage are expected because the effective carrier lifetime is shorter, resulting in lower nonlinear loss and higher gain [57]. For this waveguide cavity, the total loss of the feedback mirrors ($R_f = 71\%$ and $R_b = 90\%$) at lasing wavelength is ~2 dB, so ~1 dB single pass net gain is needed to reach lasing threshold. From previous measurements (Fig. 12.13), ~1 dB net CW gain is obtained at a pump power of ~400 mW with 25-V reverse bias and ~600 mW with 5-V reverse bias. Taking into account the cavity enhancement factor of ~2.2, the lasing threshold is expected to be ~182 and ~273 mW for a reverse bias of 25 and 5 V, respectively, which is consistent with our measurements. The slope efficiency (single side output) above threshold is ~4.3% with a reverse bias of 25 V, and 2% with 5-V reverse bias. Figure 12.16a also shows that the laser output power begins to saturate at a pump power >400 mW with 25-V bias and at a pump power >500 mW with 5-V bias. This is primarily due to the nonlinear loss caused by the TPA-induced FCA as a consequence of the nonzero carrier lifetime of the p–i–n waveguide, reducing the net gain at higher pump powers. In addition, the cavity enhancement factor M for the pump reduces with increasing nonlinear loss, lowering the effective pump power in the cavity.

Figure 12.16b is a plot of the laser output spectrum measured with a grating-based optical spectrum analyzer with 0.07 nm resolution. The input pump power was ~400 mW and a reverse bias of 25 V was applied to the p–i–n. The CW laser output has over 55-dB side mode suppression and its center wavelength corresponds to the appropriate Stokes' shift. The displayed linewidth is limited by the resolution of the spectrum analyser.

12.5 Conclusions

In this chapter we have presented optical characterization of a fast silicon modulator capable of transmitting up to 10 Gbps data stream with an extinction ratio of 3.8 dB. We showed that the phase efficiency of the MOS capacitor modulator can be significantly enhanced by reducing the waveguide dimension and properly placing the gate oxide in the waveguide. We demonstrated a CW silicon Raman laser using a low-loss silicon waveguide. We showed that the TPA-induced FCA in silicon can be significantly reduced by intruding a reverse-biased p-i-n diode in the waveguide. As a result, we achieved net gain in silicon large enough to compensate the cavity reflector loss. The demonstration of CW lasing in silicon represents an important milestone toward integration of silicon-based optoelectronics.

These recent breakthroughs in silicon photonics are important for both technical and psychological reasons. Technically, they demonstrate the existence of core capabilities in silicon that had not yet been explored, namely fast phase shifting and optical gain. These devices use the physical properties of silicon along with the ability to control free carriers with familiar semiconductor devices. They have applications beyond what has been presented here. Fast optical phase shifters have potential application in optical switching and dispersion compensation. Raman gain could also be used for modulation and wavelength conversion.

Psychologically these achievements, when added to worldwide efforts by researchers in silicon photonics, help to change silicon's reputation as an optical material. Recent advancements have been impressive, but it is essential that universities, companies, and government institutions increase funding and activities in silicon photonics R&D. The path is clear, but it must be cleared of several obstacles before the end can be reached.

Researchers should keep CMOS fabrication compatibility in mind when creating new technologies. If the intention is to benefit from the silicon manufacturing infrastructure, compatible materials and processing tools must be used. At the same time, the concept of CMOS compatibility extends beyond silicon. Materials such as optical polymers and inorganic dialectics can be straightforward to integrate into existing fabs. In fact these polymers are not so different from the photoresist used for lithography.

Several basic functions still need to be developed or extended in silicon. The most obvious is an efficient, compatible, room temperature light emitter. Modulation speeds must continue to accelerate beyond 10 Gbps. Wavelength conversion using the Raman scattering or other techniques should be explored. Integrated optical isolators will also be needed to stabilize silicon transmitters. In addition novel vertical couplers need to be developed to facilitate wafer level testing of planer devices and creative, automated techniques will be required for fiber coupling.

If researchers discover new capabilities and engineers develop methods to manufacture them in standard silicon fabs, silicon photonics will explode as a field and as a market. Today's strained copper interconnects already limit the potential of the world's information technology. Silicon photonics presents an opportunity to help create a world where computing, communication, and the potential of the human race become unlimited.

12.5.1 Acknowledgments

We thank R. Jones and D. Samara-Rubio for significant contributions to the silicon laser and modulator research, respectively. We also thank A. Barkai, O. Cohen, D. Rubin, S. Tubul, and D. Tran for technical assistance in device fabrication, U. Keil and T. Franck for modulator driver design and validation, D. Hodge and J. Tseng for sample preparation, and A. Alduino, A. Fang, H. Liu, M. Morse, and M. Salib for technical discussions.

References

1. M. Paniccia, M. Morse, and M. Salib, "Integrated photonics," Silicon Photonics in: L. Pavesi and D. J. Lockwood, Springer, Berlin Heidelberg New York (eds), (2004, Chap. 2)
2. E.L. Wooten et al. "A review of lithium niobate modulators for fiber-optic communications systems," IEEE J. Sel. Top. Quantum Electron. **6**, 69–82 (2000)
3. M.M. Howerton, R.P. Moeller, A.S. Greenblatt, and R. Krahenbuhl, "Fully packaged, broad-band LiNbO$_3$ modulator with low drive voltage," IEEE Photon. Technol. Lett. **12**, 792–794 (2000)
4. J.E. Zucker, K.L. Jones, B.I. Miller, and U. Koren, "Miniature Mach-Zehnder InGaAsP quantum well waveguide interferometers for 1.3 µm," IEEE Photon. Technol. Lett. **2**, 32–34 (1990)
5. J.S. Cites and P.R. Ashley, "High-performance Mach-Zehnder modulators in multiple quantum well GaAs/AlGaAs," J. Lightwave Technol. **12**, 1167–1173 (1992)
6. M. Fetterman, C.-P. Chao, and S.R. Forrest, "Fabrication and analysis of high-contrast InGaAsP–InP Mach-Zehnder modulators for use at 1.55 µm wavelength," IEEE Photon. Technol. Lett. **8**, 69–71 (1996)
7. T. Ido et al., "Ultra-high-speed multiple-quantum-well electro-absorption optical modulators with integrated waveguides," J. Lightwave Technol. **14**, 2026–2034 (1996)
8. R.A. Soref, "Silicon-based optoelectronics," Proc. IEEE **81**, 1687–1706 (1993)
9. C.K. Tang and G.T. Reed, "Highly efficient optical phase modulator in SOI waveguides," Electron. Lett. **31**, 451–452 (1995)
10. P. Dainesi et al. "CMOS compatible fully integrated Mach-Zehnder interferometer in SOI technology," IEEE Photon. Technol. Lett. **12**, 660–662 (2000)
11. R.A. Soref and B.R. Bennett, "Electrooptical effects in silicon," IEEE J. Quantum Electron. **QE-23**, 123–129 (1987)
12. C.E. Png, G.T. Reed, R.M. Atta, and G.J. Ensell, "Development of small silicon modulators in silicon-on-insulator (SOI)," Proc. SPIE **4997**, 190–197 (2003)
13. C.K. Tang and G.T. Reed, "Highly efficient optical phase modulator in SOI waveguides," Electron. Lett. **31**, 451–452 (1995)
14. P. Dainesi et al. "CMOS compatible fully integrated Mach-Zehnder interferometer in SOI technology," IEEE Photon. Technol. Lett. **12**, 660–662 (2000)
15. A. Liu, R. Jones, L. Liao, D. Samara-Rubio, D. Rubin, O. Cohen, R. Nicolaescu, and M. Paniccia, "A high-speed silicon optical modulator based on a metal-oxide-semiconductor capacitor," Nature **427**, 615–618 (2004)
16. L. Liao, A. Liu, R. Jones, D. Rubin, D. Samara-Rubio, O. Cohen, M. Salib, and M. Paniccia, "Phase modulation efficiency and transmission loss of silicon optical phase shifters," IEEE J. Quantum Electron. **QE-41**, 250–257 (2005)
17. L. Liao, D. Lim, A. Agarwal, X. Duan, K. Lee, and L. Kimerling, "Optical transmission losses in polycrystalline silicon strip waveguides: effects of waveguide dimensions, thermal treatment, hydrogen passivation, and wavelength," J. Electron. Mater. **29**, 1380–1386 (2000)
18. T. Kamins, Polycrystalline Silicon for Integrated Circuits and Displays, 2nd edn. (Kluwer Boston, MA 1998, Ch. 3)
19. E. Garmire and H. Stoll, "Propagation losses in metal–film–substrate optical waveguides," IEEE J. Quantum Electron. **QE-8**, 763–766 (1972)

20. S.M. Sze, Physics of Semiconductor Devices, 2nd edn. (Wiley, New York, 1981)
21. R.A. Soref and J.P. Larenzo, "All-silicon active and passive guided-wave components for $\lambda = 1.3$ and 1.6 µm," IEEE J. Quantum Electron. **QE-22**, 873–879 (1986)
22. R.A. Soref and B.R. Bennett, "Kramers–Kronig analysis of electro-optical switching in silicon," Proc. SPIE **704**, 32–37 (1986)
23. A.S. Sudbo, "Numerically stable formulation of the transverse resonance method for vector mode-field calculations in dielectric waveguides," IEEE Photon. Technol. Lett. **5**, 342–344 (1993)
24. H. Reisinger and R. Stengl, "Fundamental scaling laws of DRAM dielectrics," in: Proc. 2000 Third IEEE International Caracas Conference on Devices, Circuits and Systems, Piscataway, NJ, 2000, pp. D26/1–D26/4
25. G.T. Reed and A.P. Knights, Silicon Photonics: An Introduction (Wiley, West Sussex, 2004)
26. D. Samara-Rubio, U.D. Keil, L. Liao, T. Franck, A. Liu, D. Hodge, D. Rubin, and R. Cohen, "Customized drive electronics to extend silicon optical modulators to 4 Gbps," accepted for publication
27. J.C. Cartledge, "Performance of 10 Gbps lightwave systems based on lithium niobate Mach-Zehnder modulators with asymmetric Y-branch waveguides," IEEE Photon. Technol. Lett. **7**, 1090–1092 (1995)
28. S. Pae, T. Su, J.P. Denton, and G.W. Neudeck, "Multiple layers of silicon-on-insulator islands fabrication by selective epitaxial growth," IEEE Electron. Dev. Lett. **20**, 194–196 (1999)
29. L. Liao, D. Samara-Rubio, M. Morse, A. Liu, D. Hodge, D. Rubin, U.D. Keil, and T. Franck, "High speed silicon Mach-Zehnder modulator," Opt. Express Vol. 13, pp. 3129–3135, April 2005
30. A. Liu, D. Samara-Rubio, L. Liao, and M. Paniccia, "Scaling the modulation bandwidth and phase efficiency of a silicon optical modulator," IEEE J. Sel. Topics Quantum Electron. **11**, pp. 367–372 (March/April 2005)
31. R.E. Bank, D.J. Rose, and W. Fichtner, "Numerical methods for semiconductor device simulation," IEEE Trans. Electron. Dev. **ED-30**, 1031–1041 (1983)
32. G.P. Agrawal, Fiber-Optic Communication Systems 2nd edn. (Wiley, New York, 1997)
33. L. Pavesi, S. Gaponenko, and L. Negro (eds) Towards the First Silicon Laser (NATO Science Series, Kluwer, Dordrecht, 2003)
34. L. Pavesi, L.D. Negro, C. Mazzoleni, G. Franzo, and F. Priolo, "Optical gain in silicon nanocrystals," Nature **408**, 440–444 (2000)
35. T. Shimizu-Iwayama, K. Fujita, S. Nakao, K. Saitoh, T. Fujia, and N. Itoh, "Visible photoluminescence in Si^+-implanted silica glass," J. Appl. Phys. **75**, 7779–7783 (1994)
36. M.L. Brongersma, A. Polman, K.S. Min, T. Tambo, and H.A. Atwater. "Tuning the emission wavelength of Si nanocrystals in SiO_2 by oxidation." Appl. Phys. Phys. **72**, 2577–2579 (1998)
37. F. Iacona, G. Franzo, and C. Spinella, "Correlation between luminescence and structural properties of Si nanocrystals," J. Appl. Phys. **87**, 1295–1303 (2000)
38. D.J. Lockwood, Z. H. Lu, and J. M. Baribeau, "Quantum confined luminescence in Si/SiO_2 superlattices," Phys. Rev. Lett. **76**, 539–541 (1996)
39. S. Lombardo, S.U. Campisano, G.N. van den Hoven, A. Cacciato, and A. Polman, "Room-temperature luminescence from Er^{3+}-implanted semi-insulating polycrystalline silicon," Appl. Phys. Lett. **63**, 1942–1944 (1993)

40. M. Fujii, M. Yoshida, Y. Kanzawa, S. Hayashi, and K. Yamamoto, "1.54 μm photoluminescence of Er^{3+} doped into SiO_2 films containing Si nanocrytals: evidence for energy transfer from Si nanocrystals to Er^{3+}," Appl. Phys. Lett. **71**, 1198–1200 (1997)

41. P.G. Kik, M.L. Brongersma, and A. Polman, "Strong exciton-erbium coupling in Si nanocrystal-doped SiO_2," Appl. Phys. Lett. **76**, 2325–2327 (2000)

42. H.S. Han, S.Y. Seo, and J.H. Shin, "Optical gain at 1.54 μm in erbium-doped nanocluster sensitized waveguide," Appl. Phys. Lett. **79**, 4568–4570 (2001)

43. T. Trupke, J. Zhao, A. Wang, R. Corkish, and M. Green, "Very efficient light emission from bulk crystalline silicon," Appl. Phys. Lett. **82**, 2996–2998 (2003)

44. G. Dehlinger, L. Diehl, U. Gennser, H. Sigg, J. Faist, K. Ensslin, and D. Grutzmacher, "Intersubband electroluminescence from silicon-based quantum cascade structures," Science **290**, 2277–2280 (2000)

45. R. Claps, D. Dimitropoulos, Y. Han, and B. Jalali. "Observation of Raman emission in silicon waveguides at 1.54 μm," Opt. Express **10**, 1305–1313 (2002)

46. R. Claps, D. Dimitropoulos, V. Raghunathan, Y. Han, and B. Jalali, "Observation of stimulated Raman amplification in silicon waveguides," Opt. Express **11**, 1731–1739 (2003)

47. T.K. Liang and H.K. Tsang, "Role of free carriers from two-photon absorption in Raman amplification in silicon-on-insulator waveguides," Appl. Phys. Lett. **84**, 2745–2747 (2004)

48. R.L. Espinola, J.I. Dadap, R.M. Osgood, Jr., S.J. McNab, and Y.A. Vlasov, "Raman amplification in ultrasmall silicon-on-insulator wire waveguides," Opt. Express **12**, 3713–3718 (2004)

49. H. Rong, A. Liu, R. Nicolaescu, M. Paniccia, O. Cohen, and D. Hak, "Raman gain and nonlinear optical absorption measurement in a low loss silicon waveguide," Appl. Phys. Lett. **85**, 2196–2198 (2004)

50. A. Liu, H. Rong, M. Paniccia, O. Cohen, and D. Hak, "Net optical gain in a low loss silicon-on-insulator waveguide by stimulated Raman scattering," Opt. Express **12**, 4261–4267 (2004)

51. Q. Xu, V. Almeida, and M. Lipson. "Time-resolved study of Raman gain in highly confined silicon-on-insulator waveguides," Opt. Express **12**, 4437–4442 (2004)

52. T.K. Liang and H.K. Tsang, "Efficient Raman amplification in silicon-on-insulator waveguides," Appl. Phys. Lett. **85**, 3343–3345 (2004)

53. B. Jalali, V. Raghunathan, O. Boyraz, R. Claps, and D. Dimitropoulos, "Wavelength conversion and light amplification in silicon waveguides." In: Proceedings of Group IV Photonics Conference, September. 29–October. 1, 2004, Hong Kong

54. R. Claps, V. Raghunathan, D. Dimitropoulos, and B. Jalali, "Role of nonlinear absorption on Raman amplification in silicon waveguides," Opt. Express **12**, 2774–2780 (2004)

55. O. Boyraz and B. Jalali, "Demonstration of a silicon Raman laser," Opt. Express **12**, 5269–5273 (2004)

56. H. Rong, A. Liu, R. Jones, O. Cohen, D. Hak, R. Nicolaescu, A. Fang, and M. Paniccia, "An all-silicon Raman laser," Nature **433**, 292–294 (2005)

57. R. Jones, H. Rong, A. Liu, A. W. Fang, M. J. Paniccia, D. Hak, and O. Cohen, "Net continuous-wave optical gain in a low loss silicon-on-insulator waveguide by stimulated Raman scattering," Opt. Express **13**, 519–525 (2005)

58. H. Rong, R. Jones, A. Liu, O. Cohen, D. Hak, A. Fang, and M. Paniccia, "A continuous-wave Raman silicon laser," Nature **433**, 725–728 (2005)

59. G.P. Agrawal, Nonlinear Fiber Optics, 2nd edn (Academic, New York, 1995)
60. Details are available at http://www.photond.com
61. H.S. Han, S.Y. Seo, and J.H. Shin, "Optical gain at 1.54 µm in erbium-doped nanocluster sensitized waveguide," Appl. Phys. Lett. **79**, 4568–4570 (2001)
62. A.E. Siegman, Lasers (University Science Books, Mill Valley, 1986, p. 413-418)
63. K. Suto, T. Kimura, and J. Nishizawa. "Semiconductor Raman laser pumped with a fundamental mode," IEEE Proc. J **139**, 407–412 (1992)
64. J.K. Brasseur, K.S. Repasky, and J.L. Carlsten. "Continuous-wave Raman laser in H_2," Opt. Lett. **23**, 367–269 (1998)
65. E. Garmire, "Criteria for optical bistability in a lossy saturating Fabry–Perot," IEEE J. Quantum. Electron. **QE-25**, 289–295 (1989)

13

Free-Space Optical Interconnects

A.G. Kirk

Summary. Free-space optical interconnects (FSOI) have been widely studied as a possible solution to the interconnect bottleneck at several levels of high-speed digital systems. Here we review the underlying principles of the design of FSOIs and consider the physical limitations to interconnect density and distance. We discuss the issue of tolerance to misalignment and compare different packaging approaches. Throughout this chapter we illustrate the analysis with examples of recently implemented systems.

13.1 Introduction and Motivation

13.1.1 Background

Free-space optical interconnects (FSOI) provide an alternative to the guided wave approach to implement optical interconnects within high-performance electronic systems. In free-space systems light is routed from transmitter to receiver by way of discrete optical components (lenses and mirrors) rather than continuous waveguides. Although the term 'free-space' implies that light travels through air for most of its journey, in many cases the medium of propagation is actually glass or plastic. Most FSOI systems are designed to route several optical channels in parallel, and this can extend to more than 1,000 channels. The field of FSOIs emerged in the mid-1980s, having grown out of several different research strands. In the 1970s the limitations of electrical interconnects had already been identified, and proposals had been made for the optical distribution of clock signals across different sections of microprocessor chips [1]. At that time guided wave optics was in its infancy and a free-space implementation which used diffractive optical elements (DOEs) to replicate a single high-frequency clock signal seemed the most obvious choice. Also around this time the concept of 'optical computing' was introduced and was being used to refer to Fourier-plane optical signal processing techniques [2] and also to attempt to implement optical neural networks and other more general purpose computing systems. In all these cases free-space optics was used to relay light between different planes within the system, and the data often took the form of a continuous image that was being processed. In these systems the 'interconnect' was more than a simple point-to-point link and also performed part of the processing operation. In the 1980s researchers (many of them based in the AT&T laboratories but also elsewhere) were investigating techniques to implement optical switching operations for telecommunications networks and once again used free-space optics to interconnect 'smart pixel' arrays of optoelectronic devices [3,4]. However, by the late 1980s there was an increasing recognition that electronic components are superior to nonlinear optical components for information processing, and that in most applications a digital data representation results in higher performance than an analog representation. As a result there emerged the concept of 'optically interconnected electronics' in which the optical interconnects form simple point-to-point links between electronic devices. Figure 13.1 shows a generic free-space optical interconnect design in which a single lens performs an imaging operation between the optoelectronic device arrays. As it is immediately obvious from this figure, the attraction of the FSOI approach is that a multichannel interconnect can be implemented with only one optical component. However, there are equally obvious issues such as the degree of alignment precision required between transmitter and receiver modules, the performance requirements of the optical imaging element and the scalability of this approach that must be addressed. In this chapter we will focus on some of these issues and will show how the performance of FSOIs can be maximized. We will also provide examples of demonstration systems that have been implemented by various researchers.

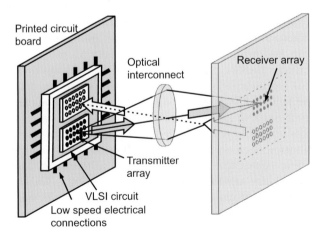

Fig. 13.1. Generic free-space optical interconnect

13.1.2 Requirements for Free-Space Optical Interconnects

Before proceeding to discuss FSOI in depth, it is worth considering the performance requirements for these systems. This will help us to evaluate the various design choices and trade-offs that must be made. Some of these requirements concern the capacity and format of the interconnect system (number of channels, interconnect distance, channel spacing) whilst others concern the degree of assembly precision required, the reliability and the robustness of the interconnect. Obviously, these will greatly depend on which level of system the interconnect is deployed. Free-space interconnects have been proposed for almost all system levels, from rack-to-rack applications to on-chip interconnects [5]. However, for a number of reasons they have been considered predominantly for chip-to-chip and board-to-board links. They are generally considered to be unsuitable for rack-to-rack applications due to the need for techniques to compensate for misalignment between racks and by the difficulty in relaying a large number of parallel channels across relatively large distances (of up to several metres). For this interconnect level, parallel fibre ribbons have already been shown to be more suitable and 2-D fibre array systems have also emerged [6]. For the on-chip interconnect level there is still some discussion as to whether optics can actually help to reduce the interconnect bottleneck or not [7]. It is possible that there is a role for optics in interconnecting different sections of large integrated circuits (ICs) but this remains to be proven. Therefore we will principally consider the requirements of chip-to-chip and board-to-board interconnects and try to identify the design space in which FSOIs must operate. Since this technology is aimed at future high-performance electronic systems, we refer to the International Technology Road Map for semiconductors [8]. Figure 13.2 shows the projected increase in VLSI transistor on-chip clock speed, off-chip clock speed, number of high-speed off-chip clock lines and total off-chip I/O capacity as a function of time for high-performance systems, taken from [9]. It can be seen that by 2014 the off-chip clock speed is projected to reach 1.8 GHz and the width of the off-chip bus is also projected to increase to 3,000 high-speed lines, with a total projected off-chip I/O capacity of 5 Tb s^{-1}. The on-chip clock speed is projected to reach 13.5 Gb s^{-1}. From this data, we can obtain a view of the design space for off-chip interconnects by the year 2014, under the assumption that the necessary off-chip bandwidth will be 5 Tb s^{-1}. This is shown in Fig. 13.3, where the data rate per channel necessary to achieve 5 Tb s^{-1} is plotted as a function of the number of parallel channels [9]. We have shaded different regions corresponding to different possible optical interconnect formats. It seems apparent that 1-D fibre ribbons will not be capable of delivering the necessary bandwidth as data rates of more than 200 Gb s^{-1} per channel would be necessary when, for example, a 24 fibre ribbon is used. The step up in density is provided by fibre arrays. Recently 2-D fibre arrays have been developed for large-scale micro-electromechanica-systems (MEMS) switch arrays and arrays from 8×8 to 36×36 have been demonstrated [10]. However, volume fabrication of these

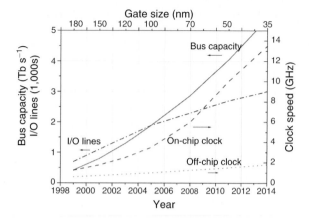

Fig. 13.2. Projected evolution of on-chip clock speed (*dashed line, right axis*), off-chip clock speed (*dotted line, right axis*), number of high-speed I/O lines (*dash-dot line, left axis*) and total bus capacity (*solid line, left axis*) as a function of time, using data from [8]

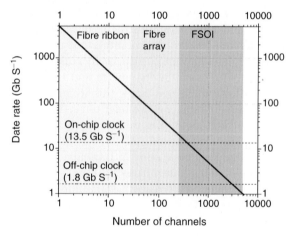

Fig. 13.3. Data rates per channel necessary to achieve $5\,\mathrm{Tb\ s^{-1}}$

elements remains to be properly developed. A 1,296 channel fibre array would require per channel data rates of $3.85\,\mathrm{Gb\ s^{-1}}$, which is reasonable but for the very short distances that we consider here, the limited bending radius and high fabrication cost of fibre arrays may render them impractical. From Fig. 13.3 it is apparent that, if we assume that the optical interconnect channel data rate is not to exceed the projected on-chip data rate of $13.5\,\mathrm{Gb\ s^{-1}}$ (which would otherwise require the use of serialization/deserialization circuits) then a $5\,\mathrm{Tb\ s^{-1}}$ aggregate data rate implies the presence of 370 optical lines. Alternatively if we assume that the off-chip optical links run at the projected electrical off-chip clock speed of $1.8\,\mathrm{Gb\ s^{-1}}$ then this implies approximately

2,700 optical lines. We suggest that these two values represent the boundaries for optical solutions to the off-chip interconnect problem, and that they also represent the domain in which FSOI represents a possible solution. Several experimental free-space interconnects have achieved the lower end of the parallelism range [11–14] and we have recently reported an Optoelectronic-VLSI (OE-VLSI) ASIC with 1,080 optical I/O that approaches the middle of the required parallelism range [15]. Other researchers have demonstrated matrix addressed vertical cavity surface emitting laser (VCSEL) arrays with 4,096 outputs [16,17].

Fibre image guides and fibre image conduits [18,19] represent another possible alternative high density interconnection medium. However, fibre image guides suffer from transmission nonuniformity unless large channel spacings are used, and questions also exist as to their potential for low cost fabrication. Image conduits are closer to rivaling free-space optics in terms of performance. Since they are also a rigid technology, a comparison should be made in terms of cost and performance, but we will not consider them further here. It has also been suggested that wavelength division multiplexing could be used as an alternative to spatial multiplexing, but we will also not consider that option [20].

In addition to the number of channels, we also need to consider channel density and interconnect distance. Future chip high-performance chip areas are predicted to be $140\,\text{mm}^2$ [8]. Using the channel counts estimated above (370–2,700 channels), and assuming the entire surface of the die can be used for surface-emitting light sources, then this implies a channel pitch of between 615 and $227\,\mu\text{m}$. For chip-to-chip interconnects, the maximum distance is the size of a board, i.e. approximately 50 cm. For board-to-board interconnects, the distance may be as short as 2.5 cm (the typical interboard spacing) to 1 m (if the signal is to run the length of a backplane). We will consider the practicality of this later in this chapter.

If any interconnect technology is to be adopted by system designers and manufacturers then there are also many other important issues to consider, including cost, reliability, ease of manufacture, ease of assembly, durability and whether its adoption will place constraints on other aspects of system design (for example cooling or electrical interconnections). We will consider these issues too in a later section.

13.1.3 System Example

An example of an experimentally realized point-to-point FSOI is shown in Fig. 13.4 [12, 21]. In this system two OE-VLSI [15] chips are bi-directionally interconnected over a distance of 86 mm. Each chip contains 256 VCSELs and 256 photodetectors and operates at a wavelength of 850 nm. This interconnection distance was selected such that by inserting two additional prisms into the beam path it could interconnect two boards in a bookshelf configuration [12]. The relay optics are based on clustered diffractive lens and the array density that was achieved was 28 channels per mm^2. This design was chosen to simultaneously maximize the interconnection distance and the channel density.

Fig. 13.4. Experimentally realized 512 channel chip-to-chip optical interconnect [12, 21]

In the following sections, we will explain how this design was carried out and compare it with other approaches.

13.2 Basic Principles of Free-Space Optical Interconnects

13.2.1 Light Propagation in Free-Space

Light is an electromagnetic wave and so will propagate through space. There are many different methods that can be used to calculate the propagation of light, depending on the degree of precision required, the propagation distance, the ratio of any material dimensions to the wavelength of light and whether a scalar or a vectorial solution is required. We will not attempt to provide a full treatment of light propagation here but instead refer the interested reader to the large number of texts that exist on this subject [22–24]. In many cases of interest to FSOIs the scalar Helmholtz equation is sufficient to model the propagation of light, which states that all allowed solutions must be of the form $\nabla^2 U + k^2 U = 0$, where $U(x, y, z)$ is the light amplitude and $k = \omega/v$ is the wave number for light of frequency ω and phase velocity v. For some simple classes of source amplitude distributions, such as the point source and the plane wave, we can calculate the propagation of the wave analytically. For an arbitrary source distribution it is not possible to do this and either numerical methods or approximations must be employed. One of the most widely used approximations is the Fresnel–Kirchhoff diffraction integral which is reasonably accurate for regions which are many wavelengths from the source region [22]. One important conclusion from this analysis is that any wavefront of finite extent will diffract and spread out as it travels through space. Therefore all FSOIs, except those over very trivial distances, require lenses, mirrors or other optical components to prevent the propagating waves from spreading. In most cases of interest to FSOI designers the Fresnel–Kirchhoff integral can be further simplified to obtain the Fresnel diffraction integral.

13.2.2 Gaussian Beams

One other source amplitude distribution for which we can find an analytical model for propagation is the Gaussian distribution [23] (in fact this model makes use of a paraxial approximation to the scalar Helmholtz equation but we will ignore that here). This is important because optical fibres and many classes of lasers emit beams that are more or less Gaussian, and we will be using the Gaussian beam extensively in designing FSOI systems. A Gaussian beam is circularly symmetrical and has an intensity profile

$$I\left(r\right) = I_0 \exp\left(\frac{2r^2}{\omega^2}\right) \tag{13.1}$$

(see Fig. 13.5a) where ω is a parameter known as the beam radius and represents the radius at which the beam amplitude falls to $1/e^2$ (13.5%) of its peak value. As the beam travels through space (along the z-axis) the beam radius expands according to

$$\omega\left(z\right) = \omega_0 \left(1 + \frac{z^2}{z_R^2}\right)^{\frac{1}{2}}, \tag{13.2}$$

where ω_0 is the *initial beam waist* and z_R is a parameter known as the Rayleigh range, given by

$$z_R = \frac{\pi\omega_0^2}{\lambda}, \tag{13.3}$$

where λ is the wavelength of light (see Fig. 13.5b). From this we can see that the behaviour of the Gaussian beam is entirely determined by the initial

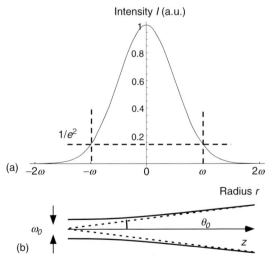

Fig. 13.5. Gaussian beam intensity profile (**a**) and evolution (**b**)

beam waist and the wavelength. The more tightly confined the initial beam, the more rapidly it will diverge. When the axial distance is several times the Rayleigh range we approach an asymptotic regime where the beam appears to diverge linearly, with divergence angle $\theta_0 = \lambda/\pi w_0$, as shown in Fig. 13.5(b). We can also see from this that beams that have a small beam waist will diverge rapidly whilst those with a large beam waist will diverge slowly.

We can use lenses and mirrors to modify the divergence of Gaussian beams. If a lens of focal length f is positioned so that the initial beam waist w_0 is a distance x_1 from the object focal point then a new beam waist w_1 is obtained at position x_2 from the image focal point, according to the equation

$$x_2 = \frac{x_1 f^2}{x_1^2 + z_R^2}. \tag{13.4}$$

Here the signs of x_1 and x_2 are defined according to the Newtonian convention [23] so that x_1 is positive when it is to the left of the object focal point and x_2 is positive when it is to the right of the image focal point. The new beam waist is given by

$$w_1^2 = \frac{f^2 w_0^2}{x_1^2 + z_R^2} \tag{13.5}$$

from which we can see that depending on the relative magnitude of the initial Rayleigh range z_R and the focal length we can either focus the beam to a smaller beam waist (and increase its divergence) or expand it (and reduce its divergence). If the initial beam waist is placed at the object focal point (so that $x_1 = 0$) then we obtain the simple result that the second beam waist is given by $w_1 = f\lambda/\pi w_0 = f\theta_0$, which is the result that one would expect from geometric optics. The new beam waist occurs at the image focus. If we wish to place the new beam waist as far as possible from the lens (i.e. to obtain the maximum possible optical throw) then study of (13.4) shows that this will be achieved when $x_1 = z_R$ (i.e. the source is one Rayleigh range in front of the object focus) and that the new beam waist is found at a distance $x_2 = f^2/2z_R$ from the image focal point. This is known as the maximum lens-to-waist configuration and is often used when long relays are required. However, as we will see in a later section it is also the least stable configuration in terms of alignment. To give an idea of the difference this makes in terms of interconnect distance, suppose that we wish to relay 633 nm light emitted from an optical fibre with an initial beam-waist radius of 5 μm (Rayleigh range=124 μm) using a pair of 50 mm focal length lens then the maximum possible distance to the next 5 μm beam waist w_2 is 200 mm (four times the focal length, hence this is described as a 4-f system). However if we use the maximum lens-to-waist configuration, and place the source 124 μm to the left of the object focus then $x_2 = 10.1$ m and so the next 5 μm beam waist is obtained at a distance of 20.2 m from the first, a 100-fold increase in interconnect distance.

One last issue that must be considered for Gaussian beams is the aperture size necessary to transmit a beam without clipping it. Since the Gaussian beam intensity profile only goes exponentially to zero, any aperture will cause some clipping. However, as a rule of thumb the aperture should be sufficiently large to transmit at least 99% of the power in the beam. By integrating (13.1) with respect to radius we obtain an expression for the fraction of power in a Gaussian beam that is transmitted through an aperture of radius a, which is

$$P(a) = P_0 \left[1 - \exp\left(-\frac{2a^2}{\omega^2} \right) \right]. \tag{13.6}$$

The 99% radius is approximately 1.5 times the beam radius. If the beam is clipped more severely than this then the departure from the Gaussian intensity profile will be too great and it will no longer be possible to use the Gaussian beam model to predict the propagation of the beam.

In practice, the emission characteristics of real lasers will only be approximately Gaussian. Real laser beams diverge faster than would be predicted by a Gaussian beam analysis due to the departure of the wavefront from the ideal Gaussian. In some cases it may be possible to model the beam as a superposition of higher order Gaussian modes (either Gauss–Hermite or Gauss–Laguerre depending on the resonator symmetry). Although in principle the amplitude of each mode can be obtained from experimental measurements this is not straightforward or accurate. A simpler approach is to determine the M^2 parameter for the laser beam. We can write the divergence of such a beam as

$$\omega(z) = \omega_0 \left[1 + \left(\frac{z M^2}{z_R} \right)^2 \right]^{1/2}. \tag{13.7}$$

This is obviously a modification of (13.2), where the parameter M^2 has been introduced. For an ideal Gaussian beam $M^2 = 1$, whereas for real lasers M^2 may be of the order of 2 to 3. This can be readily be determined experimentally.

13.2.3 Refractive and Diffractive Micro-Optical Elements

We obviously need suitable optical components in order to control the propagation of light. Some of the earliest FSOI systems made use of conventional lenses. However, most more recent systems have employed micro-optical elements. There are two main reasons for this. First most FSOI systems relay multiple optical channels in parallel, and as we will see conventional (bulk) optics is not necessarily well-suited for this purpose. Second the demands of integration have required optical elements that can be easily aligned to each other. This again has motivated the use of micro-optics. We can divide micro-optical components into refractive and diffractive elements. We will briefly consider each of these.

Refractive Micro-Optical Elements

As the name implies, refractive micro-optical elements make use of refraction to control the propagation of light. The most widely used of these are surface-relief microlenses, which have curved surfaces for focusing and collimation. An excellent review of the design and fabrication of refractive microlenses is found in [25] but we will provide a brief summary. Microlenses behave in exactly the same way as conventional lenses, in that the curved surface profile will modify the phase of the light waves which pass through them. An appropriate curved surface profile can be obtained in a variety of ways, but one of the simplest is the thermal reflow technique [26]. In this approach a thick layer of photoresist is spun onto a glass or quartz substrate and then an array of cylinders is formed in the photoresist using optical lithography with a circular mask array. After development the substrate is placed in an oven and the temperature is brought up to the melting point of the resist. As a result of surface tension the melted resist cylinders will develop a spherical surface profile. The photoresist thickness T necessary to obtain a lens of sag h and aperture radius r is given by [26]

$$T = \frac{h}{6}\left(3 + \frac{h^2}{r^2}\right) \tag{13.8}$$

and surface sag and surface radius of curvature R are related by

$$h = R - \sqrt{R^2 - r^2} \tag{13.9}$$

from which we can obtain the paraxial focal length of the lens $f = R/(n-1)$ where n is the refractive index of the photoresist. Lens diameters of 5–800 μm can be obtained in this way with f-numbers ranging from $f/2$ to $f/0.8$ [26]. As the lens diameter increases the surface curvature departs from spherical and becomes flatter, resulting in aberrations. It is more difficult to obtain large f-numbers by this process. This approach is not limited to circular lens profiles. Square packing can also be accomplished, as is shown in Fig. 13.6(c). Once the resist has been melted it is hardened by baking it. The photoresist lenses can then be used directly as optical components, but they tend to have

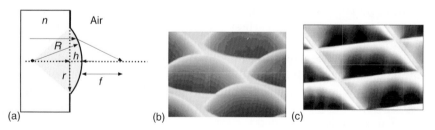

Fig. 13.6. Microlens parameters (**a**); circular and square-packed microlenses (**b**) and (**c**)

relatively high absorption and poor durability. More often these are then used as a mask in a reactive ion etching (RIE) process in which the lens profile is transferred into the substrate. This also provides another step in which the surface profile can be modified, since the final shape will be a function of the selectivity of the etchant for the photoresist and the substrate material. Lenses of this type can also be mass produced by molding into a suitable transparent plastic and are used to increase the light collecting ability of digital detector arrays.

As is well-known, spherical surfaces are only ideal for focusing paraxially, and display spherical aberration for larger aperture sizes. However, it has been shown that through careful fabrication it is possible to obtain the ideal hyperboloidal profile that can transform between spherical and plane waves [27].

There are other types of refractive microlenses of interest. Rather than making use of surface curvature it is possible to induce a suitable refractive index gradient in a glass or plastic substrate, and so obtain a gradient index microlens array. This can be achieved through an ion exchange process in glass [28] or a polymer indiffusion process [5]. If the indiffused species have a larger atomic weight than the species they replace then this process will also be accompanied by surface swelling, which will increase the power of the lenses and provides an extra degree of freedom (DOF) for designers.

Probably the greatest flexibility in the fabrication of microlenses can be obtained through the use of 'grey scale' mask techniques, such as high energy beam sensitivity (HEBS) glass [29]. Masks of this type have a graded optical transmittance, allowing for a varying photoresist exposure during optical lithography. As with the thermal reflow approach, the photoresist is then used as a mask for a subsequent etching process. This makes fabrication of larger f-number microlenses more practical and also allows arbitrary aspheric surfaces to be generated [30].

In addition to microlenses, other types of refractive micro-optical elements may be of interest. These include microprisms, which may be of use for beam deflection [25] and which can be fabricated using grey scale processes, thermal imprinting or by an LIGA (Lithographic Galvonoformung Abformung, or electroplating and molding) deep lithography process. We will return to some of these when we consider integration processes.

Diffractive Micro-Optical Elements

Diffractive optics is the other main class of micro-optical elements. Whereas refractive elements operate on continuous wavefronts (and thus obey Fermat's principle), diffractive elements operate by creating interference between different sections of successive wavefronts. Any refractive component can have a diffractive equivalent, and diffractive elements can also carry out some tasks, such as array generation, which are not as easy to achieve with refractive elements. Probably the two most common diffractive elements are Fresnel microlenses and linear gratings. As is shown in Fig. 13.7, a Fresnel microlens

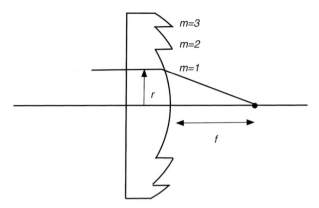

Fig. 13.7. Profile of Fresnel diffractive lens

operates by retarding the phase of a plane wave such that the optical path length from any location in the aperture to the focus of the lens is a constant plus an integer number of wavelengths. The phase $\phi(r)$ imparted to the wavefront is given by

$$\phi(r) = \text{mod}\left[\phi_0 + \frac{2\pi}{\lambda}\left(f - \sqrt{f^2 + r^2}\right),\ 2\pi\right],\qquad(13.10)$$

where f is the focal length of the lens and ϕ_0 is a constant. Thus in contrast to refractive lenses, a Fresnel lens has zones whose radius is given by $r_m^2 = 2mf\lambda + m^2\lambda^2$ (where m is an integer) where the surface height returns to zero. The surface height $d(r)$ is given by $d(r) = \lambda\phi(r)/2\pi(n-1)$, where n is the refractive index of the material. Although a surface profile of this form can be fabricated using grey scale masks, these lenses are often also fabricated using a multilevel process. The surface profile can be quantized, thus obtaining a discrete number of phase levels that can be fabricated using the 'binary optics' process [31]. In this fabrication process, a succession of etching steps is carried out (typically using reactive ion etching), each removing one half the depth of the previous etch. In this way 2^N etch depths can be obtained using N masks. Fewer phase levels means simpler fabrication, but also results in loss of efficiency as power is diffracted into unwanted orders. The dependence of diffraction efficiency on the number of phase levels is given by

$$\eta = \left[\frac{\sin(\pi/N)}{\pi/N}\right]^2.\qquad(13.11)$$

This is shown in Table 13.1, where we can see that a binary element has a diffraction efficiency of only 40.5% whereas a 16 level element has a theoretical efficiency of 98.7%.

Because these lenses do not require a large amount of material to be etched away they are often more suitable for large diameter lenses than are refractive lenses. Furthermore, an arbitrary phase profile can be obtained and so they

Table 13.1. Diffraction efficiency of multilevel diffractive lenses as a function of the number of phase levels N (and the corresponding number of masks required)

N	masks	$h(\%)$
2	1	40.5
4	2	81.0
8	3	95.0
16	4	98.7

can be used for aberration compensation. However, there are a number of important restrictions. First the numerical aperture is limited by the maximum diffraction angle that can be achieved, which is determined by the size of the smallest features, around the outer edge of the lens. It can be shown [31] that if the minimum feature size that can be etched is of width w then the $f/\#$ is given by

$$f/\# = \frac{wN}{2\lambda} \qquad (13.12)$$

Therefore a lens operated at a wavelength of 850 nm, with an eight level etching process that has a minimum feature size of $2\,\mu$m will have a minimum $f/\#$ of 9.4 (corresponding to a numerical aperture of 0.21). Some designers have attempted to circumvent this by reducing the number of phase levels towards the edge of the lens. Moreover the diffraction efficiency that can be obtained is also a function of the minimum feature size. All the above equation assume that scalar diffraction theory holds. However, as the feature size is reduced this is no longer true and it is necessary to consider rigorous diffraction calculations [32]. Figure 13.8 shows the calculated diffraction efficiency (including reflection losses) of an 8.5 mm focal length continuous profile (grey scale) Fresnel lens as a function of lens radius [33]. It can be seen that there is a significant reduction in efficiency. The 1.5 mm radius lens corresponds to an $f/2.8$ lens, and the graph shows that at this aperture size the diffraction efficiency has fallen by 15%. The polarization dependence is minimal. It is possible to optimize the grating profile to increase the diffraction efficiency [34] but this does not result in 100% efficiency and often increases the demands on fabrication. Due to their nature diffractive lenses are also very wavelength sensitive, although this is not usually a problem in optical interconnect applications where single wavelength sources are used.

Diffractive elements can also be used to split wavefronts into multiple paths. By using suitable computer optimization techniques it is possible to obtain a 1-D or 2-D array of equal amplitude beams [35].

13.3 Design Principles and Approaches

13.3.1 Overview

Having discussed some of the components that will be used for optical interconnects, and also having defined the properties of optical beams, we can now

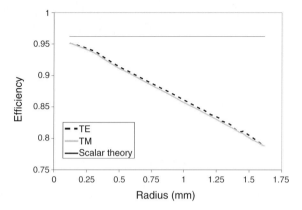

Fig. 13.8. Diffraction efficiency of an $f = 8.5\,\text{mm}$ focal length grey scale Fresnel lens as a function of radius [33]

investigate suitable design principles for free-space interconnects. In contrast to guided wave optical interconnects where waveguide loss, cross-section and minimum bend radius dominate the design process, free-space interconnect design is less clear-cut and has a different set of trade-offs. These include the way in which the optical aperture is partitioned, the number of optical relay stages and the degree of tolerance to misalignment. We will consider these below.

13.3.2 Aperture Division Choices

The first choice that must be made is in regard to the way the system aperture will be divided. Figure 13.9 shows the three basic choices: macrolens, microlens and clustered [36, 37]. The macrolens approach has the simplest design; the entire system has a single aperture stop. The two planes are mutually imaged onto each other with unit magnification and inversion. As is shown in the diagram, since typical optical sources such as VCSELs have a relatively large divergence, microlenses are usually used to collimate them. This system is telecentric, which has several advantages. One of the most important of these is that small axial misalignments of the system components result in defocus but there is no change in the magnification of the system. This is a significant advantage as it ensures that light transmitted from the sources will be coincident with the receivers (assuming no other misalignment). The ease of construction of this system has resulted in the development of many examples [11, 38, 39]. Although simple to design and construct with off-the-shelf components, macro-optical systems have several disadvantages. The first of these is lack of scalability [36, 37, 40]. The separation of the two device planes is normally determined by the requirements of the electrical system in which the interconnect is deployed. This is typically 5–10 cm for a board-to-board

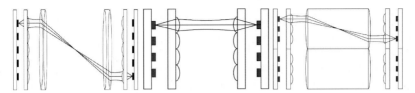

Fig. 13.9. Aperture division choices: (**a**) macrolens; (**b**) microlens; (**c**) clustered design

interconnect. If we wish to image a chip which is $1\,cm^2$ with a telecentric optical system over a $10\,cm$ distance then this implies that we need a lens with a half field of view in excess of $20°$. Such a lens will require multiple elements (in order to correct aberrations) [41] and will not fit within the required space. There is therefore a trade-off between the size of the array that can be imaged and the optical throw (or board spacing). One alternative which partially circumvents this problem is to use gradient refractive index lenses as relay elements [42, 43], however, these also become excessively long when used for array sizes of more than $2\,mm$.

One possible solution to this problem is to use a microchannel, or microoptical system [44]. In this paradigm one lens aperture is employed for each channel. The advantage of this approach is that each lens operates with a field of view of a single source, rather than the entire array. The array size can be increased simply by adding extra microchannels, which does not require a modification of the lens design. However, this system also has two major disadvantages. First the interconnection distance is limited by diffraction to very short distances, due to the small aperture size (even when the maximum lens-to-waist configuration is employed). We will calculate the limits to interconnection distance in Sect. 13.3.3. The second disadvantage of a microchannel system is the poor tolerance to misalignment. Therefore although microchannel systems are suitable for very short interconnection distances ($<20\,mm$) they are not practical for board-to-board interconnects.

An alternative and scalable approach to designing parallel optical relays is to employ a clustered optical system [12, 14, 36, 45–47] such as the one shown in Fig. 13.9c. In systems of this type, transmitter and receiver windows are arranged in clusters, each of which is imaged by a single lens. These lenses are often called minilenses since they have a size between a microlens and a macro(bulk)lens. This system design seeks to combine the relatively long optical throw and misalignment tolerance of macro-optical systems with the scalability and moderate field-of-view requirements of microchannel systems. This permits simple spherical lenses to be employed (which may be either diffractive or refractive) rather than the multielement lenses required in macro-optical systems. Several studies describe in detail the design of clustered optical systems for particular applications [47, 48]. In the design process the fixed constraints are typically the spacing of the two device planes (i.e. the optical throw) and the maximum size of the device plane. Within these constraints the designer balances various parameters such as device density (which will determine the number of parallel channels), lens $f/\#$ (which will determine the cost and practicality of the lenses) and alignment tolerance (which will determine the degree of difficulty in assembling the system). In Sect. 13.3.3 we will determine the optimum cluster size for a given interconnect distance and device plane size.

13.3.3 Optimal System Design

There are three main factors which influence the performance and scalability of the clustered FSOI. These are diffraction (which places an upper limit on lens spacing), geometric aberrations (which places an upper limit on the field size) and the speed of the relay lens (which will also limit field size for a given focal length) [48]. Figure 13.10 represents the important parameters of a clustered interconnect. The optical sources are arranged, in a square $N \times N$ array of pitch p. The distance from the centre of the array to the outermost line of sources is h (i.e. this is the distance to the edge of the optical field of the cluster lens, measured along a vertex). The light from the sources is collimated by a microlens which has an aperture equal to the pitch of the channels and a focal length f_μ which for a given pitch is determined by the divergence of the VCSELs and the degree of clipping that can be tolerated). The clustered relay lenses are confocal with the microlenses and are assumed to have square apertures (side length D) and focal length f. They are attached at either end of a glass block, of index n, which has a total length of $2fn$ (in order to maintain telecentricity). The optical system is symmetrical in that the microlenses at the detector end are identically located relative to the detectors. This ensures that the system is bi-directional. The optical throw L (or interconnect length) is defined as the distance between the focal planes

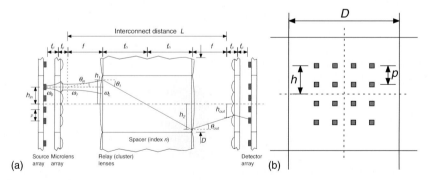

Fig. 13.10. Optical interconnect parameters; (**a**) array layout and (**b**) optical system.

of the clustered relay lenses (i.e. $L = 2f(n+1)$. We can now determine the maximum field size h for a given cluster lens size D, optical throw L and a set of source properties. From this we can calculate the number of channels that is supported by the interconnect and hence the channel density (defined as N^2/D^2 since the relay lenses are assumed to have a 100% fill factor).

We have determined the maximum off-axis source position h_{in} as limited by the speed of the relay lens and the aberrations of the relay system. We treat the system by modelling the light as Gaussian beams at the input and output side and via ray-tracing in the relay block. It can be shown [48] that for a ray that enters the relay block with angle θ_{in}, originating at height h_{in}, the third-order expansion of the ray intercept error ε at the output is given by

$$\varepsilon = \frac{3h_{in}^2 \sin\theta_{in}}{f\,n^2}.$$
(13.13)

Here the angle θ_{in} is the angle of the ray that corresponds to the 99% power asymptote of the incident multimode Gaussian beam after collimation by the microlens array. This assumes that the diffractive lenses are thin elements with a quadratic phase profile [48]. From this we can obtain an approximate expression for the increased clipping loss γ that will be undergone at the output microlens. For a given degree of tolerable clipping loss, we can obtain an allowable aberration ε_{max}. This is given by

$$\varepsilon_{max} = p\left(\frac{3}{4}\sqrt{\frac{2}{-\log(\gamma)}} - \frac{1}{2}\right)$$
(13.14)

from which we can obtain an expression for the maximum possible source height, as limited by aberrations. This is given by

$$h_{ab} = n\sqrt{\frac{f\,\varepsilon_{max}}{6\sin\theta_{in}}}.$$
(13.15)

Note that an additional factor of $\sqrt{2}$ has been incorporated since h measures distances along the axis rather than along the diagonal (which will have the most aberrated beams).

The second limit to the source height is the aperture of the relay lens. If we assume that the $f/\#$ of the lens is defined as the ratio between the focal length and the diagonal aperture size, and that a 1% clipping condition again obtains, the maximum source height in this case is given by

$$h_{ap} = f\left(\frac{1}{2\sqrt{2}f/\#} - \tan(\theta_{in})\right).$$
(13.16)

The maximum source height h_{max} will therefore be determined by the minimum of (13.15) and (13.16). Therefore the maximum number of channels (in one-dimension) is given by

$$N = 2\frac{h_{\max}}{p} + 1. \tag{13.17}$$

Finally we need to determine the area of the relay lens, in order to calculate the channel density. The relay lens aperture is given by

$$D = 2\left[h_{\max} + f\tan\left(\theta_{\text{in}}\right)\right] \tag{13.18}$$

We can now calculate the channel density ρ from

$$\rho = \frac{N^2}{D^2}. \tag{13.19}$$

Figure 13.11 shows the calculated density as a function of interconnection distance for an interconnect which uses 850 nm multimode VCSELs with $M^2 = 2$ and $\omega_0 = 3\,\mu\text{m}$ [21]. The array pitch is 125 μm and the loss tolerance γ is 5%. It can be seen that at very short distances and for fast lenses a density of almost 64 channels per mm^2 can be achieved (which is the limit for devices on a 125 μm pitch). As the distance increases, however, the achievable density decreases. Slow lenses offer a lower density than fast lenses at short distances, but eventually the performance is dominated by aberrations (as occurs at a distance of 120 mm for $f/3$ lenses) and so the relay lens speed is no longer an advantage.

We can also use this model to calculate the impact of VCSEL divergence on performance. It has been found that for devices on a 125 μm pitch at distances above 100 mm a single mode VCSEL (with $\omega_0 = 3\,\mu\text{m}$, $M^2 = 1$, $1/e^2$ half-beam divergence $= 5.2°$) allows a 50% increase in channel density when compared to a multimode VCSEL which has twice the divergence (i.e. $\omega_0 = 3\,\mu\text{m}$, $M^2 = 2$ and $1/e^2$ half-beam divergence $= 10.4°$).

This model can be used to design an interconnect system that will maximize density for a given interconnect distance. As an example, Fig. 13.12

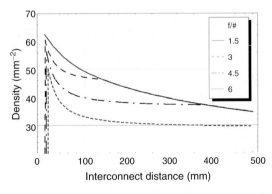

Fig. 13.11. Channel density as a function of interconnect distance [48]

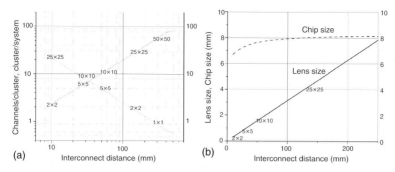

Fig. 13.12. System design example for a 50×50 $f/4.5$ system on a $125\,\mu$m pitch; (**a**) channels/cluster (*solid line*) and clusters/system (*dashed line*); (**b**) lens size (*solid line*) and chip size (*dashed line*) [48]

shows the variation in the number of channels per cluster and clusters per system for an interconnect that requires 50×50 channels, with multimode ($M^2 = 2$) VCSELs on a $125\,\mu$m pitch with $f/4.5$ relay lenses, as a function of interconnect distance. It can be seen that for a short distance ($10\,$mm) there will be a small number of channels per cluster (2×2) and hence many clusters per system (25×25). As the interconnect distance increases it eventually becomes most efficient to use just one relay lens (i.e. a macrolens) with all 50×50 channels traversing it. This holds for distances above $250\,$mm. The relay lens size increases almost linearly with distance and the chip size (determined as the product of the relay lens size and the number of clusters per system) remains almost constant at $8\,$mm. Obviously there exist intermediate distances where it would not be possible to have an integer number of clusters per system. This indicates that at these distances it is not possible to obtain the maximum density. However, by selecting a different source array pitch it may be possible to improve the density which can be obtained.

We can also make a comparison between the clustered relay system and a microchannel interconnect. As we have seen in Fig. 13.12, as the interconnection distance decreases the number of channels per cluster also decreases until there is eventually only one channel per cluster. However the clustered interconnect remains telecentric, and so does not provide the same interconnection distance (as a function of relay lens focal length) as can be achieved by a microlens relay in the maximum lens-to-waist configuration [44]. This effectively means that at very short interconnection distances the microchannel interconnect should offer better densities. This is shown in Fig. 13.13 for a multimode ($M^2 = 2$) system for which the clustered system has a device pitch of $125\,\mu$m (the pitch of the microchannel relay is a function of interconnection distance [5]). As can be seen, for distances below 5–$15\,$mm (as a function of relay lens $f/\#$) the microchannel relay offers a better density. However, for inter-chip distances the clustered interconnect is preferred.

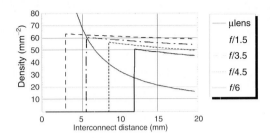

Fig. 13.13. Comparison of interconnect density for microlens relay (lens) and clustered relay, for a range of f-numbers. This assumes the clustered relay has sources with a pitch of 125 µm and that the VCSEL M parameter is 2 for all systems [48]

13.3.4 Misalignment and Modularization

In general misalignment is one of the most critical limitations of FSOI. Most free-space interconnects are multielement systems that require the mutual alignment of many different components in different planes. In designing the clustered interconnect above, we saw that it maximized the interconnection density but we did not consider to what precision the components must be aligned. A proper study of misalignment is intimately connected to the packaging approach that will be used for the system. In general, we seek designs that can be readily partitioned into modules that contain groups of components. Because modules can be assembled to quite high precision (often using lithographic techniques), elements within modules can have a relatively low tolerance to misalignment. However, the tolerance to misalignment between different modules should be as large as possible as these are often assembled using mechanical techniques [49]. To illustrate this concept, consider Fig. 13.14, which shows two possible approaches to modularizing the clustered interconnect. In Fig. 13.14a we have packaged the microlenses to the optoelectronic devices, whereas in b we have packaged the microlenses to the relay lenses. In this context 'packaging' implies that two elements are aligned using some sort of high precision process and then bonded together permanently. Modules are then assembled using an integration process.

This then brings us to the question of system partitioning: how should a system be broken down into modules such that it has the maximum possible tolerance to misalignment?

To understand why there may be a difference we can refer to Fig. 13.15 (after [50]). In part (a), the microlens is not packaged with the detectors. Since detectors are usually small (to obtain high speed), a small lateral misalignment of the detector versus the microlens Δx results in the beam rapidly shifting from the detector. In contrast, in part (b), the microlens is aligned to the detector as a single module. Now when the beam is misaligned by the same amount relative to the module, the beam is focused onto the centre of the detector, although some light will be lost due to clipping at the microlens and

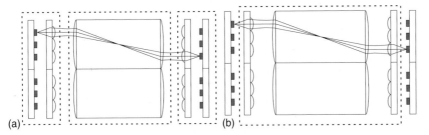

Fig. 13.14. Partitioning of clustered interconnect: (**a**) with device-microlens modules and (**b**) with optoelectronic devices separated from the optics

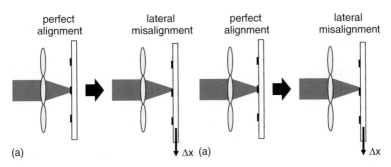

Fig. 13.15. Lateral misalignment tolerance for microlens-detector combination: (**a**) microlens not integrated with detector; (**b**) microlens integrated with detector (after [50])

may appear as crosstalk at an adjacent detector. If we do a similar analysis for angular misalignment $\Delta\theta$, of the detector array and microlens array in the first instance and of the beam and the microlens-array and detector module in the second instance, then we find that the situation is reversed and that design (a) is more tolerant of misalignment than design (b). We can calculate the misalignment product $\Delta x \Delta\theta$ for the system, which is a product of the angular and lateral misalignment that will result in a 50% drop in power at the detector. In [9] this product is calculated for a range of different interconnect designs. Clustered designs in which the relay lenses are packaged into a single module, as in Fig. 13.14a are shown to be much more tolerant to misalignment than simple microchannel designs, although microchannel designs can be improved by placing a field lens in front of the detector.

This analysis can be extended further to consider why this should be so. Consider Fig. 13.16 which is a module that contains a single detector and a lens. The lens produces an image of the detector in space (which is essentially the entrance pupil of the system). It is assumed that the magnification of the system is such that the size of the detector image is smaller than the lens aperture. The alignment tolerance of the module is given by the product of

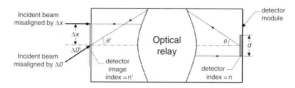

Fig. 13.16. Relationship between the optical invariant and the misalignment product

Δx and $\Delta \theta$. The image of the detector is drawn using one oblique ray and one axial ray and has a total size of d'. The angle the oblique ray makes with the optical axis is θ' and θ in the image plane and object plane, respectively. These angles correspond to the entrance and exit numerical apertures (NA) of the module. Within the laws of geometrical optics it can be shown that the following relationship is true at all points within an optical system [41]:

$$\frac{d}{2}\theta = \frac{d'}{2}\theta'. \tag{13.20}$$

This is referred to as the optical invariant. We can now relate this to the misalignment product. If we assume a 50% loss tolerance, then we can argue that the lateral tolerance Δx is equal to $d'/2$ and that the angular tolerance $\Delta \theta$ is equal to the NA of the module θ'. Therefore from this and (13.10) we can write

$$\Delta x \Delta \theta \propto \frac{d}{2}\theta. \tag{13.21}$$

Therefore the alignment product is directly related to the optical invariant, which is a product of the detector size and the NA of receiving optics. This is an invariant of the system, regardless of the complexity of a module. We can therefore conclude that we can maximize the alignment product first by increasing d, which means using large detectors (up to the limit imposed by capacitance and bit rate) and second by maximizing the NA of the receiving optics. This provides a quick and easy way to determine the alignment tolerance of an entire interconnect system. One simply needs to determine the value of the optical invariant at the detector plane; if the invariant is small then the design is inherently difficult to align. This approach also explains why Gaussian beam relays, in which the detector is imaged in the aperture of the second relay microlens [51, 52] have a good misalignment tolerance. Similarly field lenses increase the apparent size of the detector in object space and so also improve the tolerance to misalignment [9].

Misalignment tolerance can also be traded off for channel density and interconnect distance. For example in the system design shown in Fig. 13.10 the effect of misalignment can be reduced by increasing the size of the microlenses (in effect oversizing them). However, this obviously reduces the maximum channel density and so there is a trade-off to be made between misalignment tolerance and channel density. Without going into details, it has been shown that for the system described above, reducing the channel density from 100 channels per mm^2 to 30 channels per mm^2 increases the lateral misalignment tolerance from 5 to 20 µm [48]. This assumes that the microlenses are perfectly aligned to the VCSELs and to the detector arrays. In the Gaussian beam relay design [52] misalignment tolerance is increased by reducing the space between relay lenses.

13.3.5 Sensitivity Analysis

In Sect. 13.3.4 we were interested in identifying the principles that lead to designs that have a high tolerance to misalignment. The optical invariant approach represents a powerful way of understanding the misalignment tolerance of an optical interconnect. However, it does not necessarily tell us everything we need to know about the ease or difficulty of assembly a particular system when many individual components are randomly misaligned. It is particularly important to determine the way in which misalignments of individual components stack-up to determine the performance of the system as a whole. In particular, we need to be able to answer the following question: how precisely must each individual component be aligned in order to obtain a high probability that the system as a whole will work. The starting point is a sensitivity analysis in which each component is misaligned in each DOF in a ray-tracing or Gaussian beam simulation and the impact on throughput is calculated [53]. The degree of misalignment that can be tolerated for a given throughput reduction is described as the sensitivity of the component. The selection of the tolerance metric is a critical part of the sensitivity analysis. For example a lens in an optical system may be found to have a lateral misalignment tolerance of 5 µm at the 10% power loss level, indicating that a 5 µm misalignment will result in a 10% throughput reduction. Systems assembled to a relaxed metric might be easier and cheaper to assemble but might also provide lower performance in the assembled system due to accumulated losses. A more severe metric will mean that the tolerances are tighter, providing a lower loss but also increasing the fabrication cost. Once the individual tolerances have been calculated, it is then necessary to determine the way in which individual component misalignments interact and stack up. There are a variety of ways by which this can be done. The most direct approach is to obtain an expression for throughput as a function of the positions of all of the individual components. For systems with more than a few components this rapidly becomes an intractable calculation. The simplest alternative is a worse-case misalignment tolerance calculation in which all components are assumed to be maximally misaligned. Another technique that is commonly used is a root-sum-of-square (RSS) analysis [52, 54] in which each component is assumed to be misaligned such that it causes a given reduction in throughput and the total throughput reduction is then calculated as the RSS of the individual values. RSS analysis is fast but it is known that this method is not valid as error functions for optical systems are not additive and tolerances are not linearly independent. Finally it is possible to obtain an accurate determination of misalignment tolerance by performing a Monte Carlo analysis [55]. Many simulated systems are generated in which all components are assumed to be misaligned with some suitable probability distribution and the final throughput in each case is calculated. In this way a probability distribution of throughputs can be obtained and the expected throughput (for an average system) can be predicted [56]. The sensitivity of clustered interconnect designs to misalignment was investigated in [55]. In this paper, a variety of lens speeds and relay stages were considered, with up to 50,000 samples for each system. It was found that increasing the number of relay stages did not greatly decrease the tolerance to misalignment for each component, except for tolerance to source tilt. However, increasing the number of components did reduce the overall probability of success, and resulted in dramatic departures from RSS predictions. Thus a 4-f relay system, assembled with components that are all individually aligned to a 1% tolerance (i.e. 99%) transmittance resulted in an expected system throughput of only 69%, whereas an RSS approach would have predicted a throughput of 96%. Obviously one key aspect of this analysis is to make realistic predictions of misalignment distributions, which is a function of the packaging techniques used. Software tools that can calculate the effect of misalignment have also been developed [57]. This approach can be extended further. Once a Monte Carlo analysis has been performed, it is then possible to perform a regression analysis in order to obtain a relationship between the system variables and the performance metric, as described in [58]. This can provide more insight into the sensitivity of the system in each DOF and thus informs the system designer as to the areas in which the greatest care is required.

13.3.6 Other Aspects of Design

In Sect. 13.3.5 we considered chiefly centro-symmetric telecentric systems. Telecentric systems are inherently distortion free, although they may have other aberrations. A wide variety of different types of optical interconnect systems have been introduced, including off-axis planar optics designs [59] and systems that contain three levels of lenses [60] rather than two as in the clustered system design, and reflective designs [13]. These require more advanced analysis of optical aberrations and techniques such as eikonal analysis [61] can be very powerful in developing designs that have low distortion and other aberrations. It is also possible to incorporate misalignment analysis into the standard optical ray-matrix equations [62] by extending them from 2×2 to 3×3 matrices, which can simplify analysis. It is also possible to increase misalignment tolerance through redundancy. For example [63] describes the use of nine redundant spatial channels to achieve a misalignment tolerance of ± 1 mm and $\pm 1°$ over an interconnect distance of 20 cm.

13.4 Fabrication, Packaging and Assembly

13.4.1 Introduction

Once a system has been designed and the components (micro-optical and optoelectrinic) have been designed, it must be packaged and assembled. Because FSOI have not yet been deployed in commercial systems, there are few standards in this area, although the recent wide scale developments of optical MEMS for optical fibre networks has resulted in a significant progress. However, there are some important differences between the requirements of FSOI and components for optical fibre networks. The alignment tolerances required for optical fibres are typically much more demanding than for FSOI. This is because in typical FSOI systems the final collecting device is a detector that may have a cross-section of 20–$100\,\mu m$, whereas a beam focused onto the end facet of a single-mode fibre must be aligned to better than $1\,\mu m$. This lateral tolerance can be relaxed by using a collimating lens, but only at the expense of a reduced angular tolerance. As a result optical fibre systems are often assembled using active alignment techniques, where components are assembled with a 6 DOF stage while light is propagating through the system and only bonded once maximum throughput has been achieved. This results in relatively expensive systems. In FSOI systems typical lateral tolerances for interconnect distances of 10–$100\,cm$ are 5–$100\,\mu m$, with angular tolerances of 10–500 arc minutes [9]. However, FSOI systems will only be adopted if costs can be minimized The ITRS calls out for costs around \$0.01 per pin for high-performance systems, so that a 1,000 channel optical interconnect should cost less than \$10.

13.4.2 Module Fabrication and Packaging

These cost concerns explain why it is so critical to have carried out a careful analysis of misalignment tolerance. With proper design it is possible to arrange for components that have a high sensitivity to mutual misalignment to be fabricated in a single monolithic component or module. As we have seen in Sect 13.4.1, it is often good practice to package relay lenses together. The simplest way to achieve this is fabricate them within the same lithographic process. This has been accomplished in several ways by a variety of researchers. Reference [5] describes an optical bridge which interconnects two 2×8 arrays of optical channels over a distance of $5\,mm$ (see Fig. 13.17). The element is approximately $5 \times 6 \times 2\,mm$ in size and is fabricated in polymethylmethacrylate (PMMA) which is exposed to a proton beam. As the piece is moved through the beam in three-dimensions the necessary surfaces are formed as the proton beam modifies the cross-linking of the polymer. Exposed regions are removed in a subsequent development process. This approach is very powerful in that all of the optical surfaces are aligned to each other with an accuracy that is limited only by the precision of the positioning system that is used to move

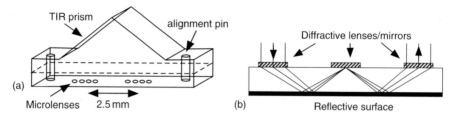

Fig. 13.17. Examples of optical integration strategies: (**a**) optical interconnect system fabricated via deep proton lithography; (**b**) planar optics approach

the work piece through the beam. In a second fabrication process a pair of total-internal prisms is also fabricated in PMMA with corresponding features that lock to the lens substrate. Typically, this process also yields optically smooth surfaces (down to 20 nm [5]). However, there are limits in terms of the thickness of material that can be machined this way (due to absorption of the beam), the size of the total piece and also to the types of shapes that can be fabricated.

A second way to make use of lithographic precision in the assembly of optical interconnects is to use the 'planar optics' approach [64,65]. Here a thick (1–10 mm) transparent substrate is patterned with diffractive or refractive optical components on the top and the bottom, using standard microlithographic procedures and with a high precision mask aligner to ensure that features on the top surface are aligned to features on the bottom surface (see Fig. 13.17b). This is a powerful approach that enables quite complex systems to be developed. Reference [59] describes a system which interconnects 96 optical fibres to an optoelectronic integrated circuit using this approach. Some of the disadvantages of this approach are the increased aberrations due to the off-axis propagation (although this can be mitigated by careful design), the loss caused by multiple reflections and the fact that the maximum interconnect distance is limited by the substrate size and the microlithographic system that is employed.

A related approach is to fabricate the various optical components separately and then use optical alignment techniques to ensure precise assembly. Because FSOI systems require the alignment of 2-D arrays this is more complex than the process required for most optical fibre systems which have cylindrical symmetry. The first example of this technique was the use of interferometric alignment lenses for the alignment of optical relays [66] (see Fig. 13.18a). The diffractive lenses at the edges of the relay lenses create interference between the zero-order and first-order beams, resulting in fringes that are only flattened when the lenses are precisely aligned. The lenses are glued to a spacer after alignment. The assembled relay block (which forms part of the system shown in Fig. 13.4) is shown in Fig. 13.18b. This approach can deliver alignment tolerances of 1–5 microns for device plane separations of

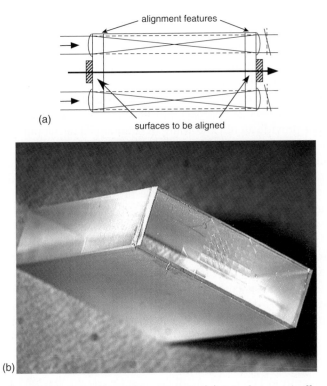

Fig. 13.18. Interferometric alignment approach: (**a**) interferometrically aligned relay lenses (after [24]); (**b**) aligned lens block [46])

30 mm [66]. This concept has been extended to include off-axis alignment gratings and lenses which can provide alignment information about all 6 DOF [67]
and most recently to designs that provide diagnostic information on the state
of alignment (via the creation of focused beams on targets) which could be
used for automated assembly [67,68]. If we design elements that have optical
surfaces only on the outside then we can replicate them at low cost via plastic
moulding [69].

Another approach that must be mentioned is the use of integrated micromechanical elements to provide realignment after assembly. This approach is
already used in the assembly of some commercial telecommunications switches
which are based on beam-steering micromirrors. Obviously, this approach is
less attractive if the mirrors are not already required as part of the system.
However, a similar approach is to incorporate built-in micromachined actuators to reposition components such as optical fibres [70], microlenses [71]
or micromirrors [72] after assembly in order to minimize insertion loss. Once
alignment has been achieved the moving components can then be locked into
place with a suitable adhesive.

13.4.3 System Assembly

Once optical and optoelectronic modules have been assembled they must be incorporated into a system. In a free-space optical system a rigid support structure is required to act as a mechanical reference. Early FSOI employed high precision commercial optical positioning equipment to hold each part or module. This is a flexible approach but limits the system size and complexity and does not provide an obvious route to commercial development. The development of the slot baseplate approach [38] permitted the rapid prototyping of FSOI whilst allowing complex systems with many axes to be demonstrated. Slot baseplates can permit very high alignment tolerances ($\pm 12\,\mu$m with careful machining) although maintaining this precision over the entire baseplate is not easy. However, slot baseplates are massive components and are not suitable for commercial devices. Once the system has been demonstrated the slot plate may be replaced by a moulded housing into which the optical components can be placed. Another example of a precision module assembly is the system described in [73]. In this system a series of L-frames are used, together with ceramic spacers which define the relative postion of each module. In some cases, it is desirable to be able to insert and remove modules from the system, in order to permit defective parts to be replaced, for routine servicing or upgrades. In order to be able to remove and replace a component in a parallel free-space optical system without realignment it is important that that component is kinematically attached to the system. Various researchers have demonstrated kinematic modules in this context [74–76]. These are often inspired by the techniques employed in fiber ribbon connectors in which pins are mated with high precision holes. Again a sensitivity analysis is essential in order to understand which DOI require critical alignment and which are relaxed. As an example of system performance, the 512 channel interconnect described in [12], which made use of a combination of interferometrically aligned modules and kinematic assembly achieved an end to end average loss of $-10.4\,$dB and a crosstalk of $-9.6\,$dB. However, most of the loss in that case was due to the fact that the VCSEL divergence was larger than the value for which the system was designed, resulting in large losses at the first microlens array. However, with careful design, fabrication and assembly there is no intrinsic reason why FSOI cannot achieve low losses.

13.5 Discussion and Conclusions

In this chapter, we have described some of the concepts that underlie the design and operation of FSOI. As we have shown, FSOI design is intrinsically bound up with alignment tolerance, channel density and interconnect distance, and also by the components, assembly and integration processes that are available. There are a large number of parameters that may be traded off with each other and so in general it is difficult to draw general conclusions. Although there has been much progress in the design and fabrication of components for FSOI (including sources, detectors and optics) there are no commercial FSOI systems, with the exception of some specialized image processing and filtering systems. We may ask ourselves why this is the case. However, before considering FSOI in particular, it is worth noting that optical interconnects of any kind have not yet got 'into the box' and that some researchers in the microelectronics industry [77] have pointed out that it may be a long time before optics is used at the chip-to-chip or board-to-board level. Optical interconnects are so disruptive that manufacturers will push electrical interconnects to the limit in order to avoid making such huge changes. So long as optics is limited to the interconnection of widely separated systems (such as racks and chassis) then free-space optics has no role to play. However, if we assume that eventually some sort of optical interconnect is required at the board or chip level, then we can compare FSOI with guided wave interconnects. Guided wave interconnects, and in particular waveguides embedded in PCBs 'feel' much more like conventional electrical interconnects than do free-space interconnects, and therefore are more likely to appeal to electrical system designers. Furthermore, guided wave interconnects can be laid out like wires, with arbitrary lengths and with the possibility to make bends (within the limits of the waveguide technology). In contrast, as we have seen, FSOI systems have a channel density that is a function of length. Therefore unless we wish to sacrifice channel density, or develop standard interconnect distances, each FSOI system design will be unique. The one significant advantage of FSOI designs is that they can achieve a high degree of parallelism, since they allow 2-D device arrays to be interconnected to each other. However, if the devices to be interconnected are chips on conventional PCBS, then, as is shown in Fig. 13.4, the interconnect has to rise up off the board. The larger the device array, the larger the stand-off height required. This may prevent other boards or components being packaged as closely. Furthermore, designers of guided wave interconnects have also come up with clever approaches to getting a 1-D ribbon connected to a 2-D device array by rotating the device array by 45°, which tends to reduce this advantage. In addition, any technology which makes use of surface-normal optical emission from a chip needs to take cooling requirements into account. At present most high-performance systems remove heat from one side of a chip. In many FSOI designs, such as the schematic of Fig. 13.1, light is emitted from one side and electrical connections are made to the other. This prevents a heat sink from being attached. Some possible solutions include using a transparent substrate and emitting light downwards [78]. As point-to-point interconnects then, free-space optics have yet to prove themselves. However, FSOI do have one big advantage that is hard match with other electrical or optical approaches, and that is the possibility of implementing fan-out from one source to many receivers. The use of free-space optics for clock distribution was one of the first inspirations for the field [1] and thus they may have uses in applications such as parallel computing where multicast buses may be important [79]. The growth of free-space optics in optical fibre networks and the development of new low cost laser sources and fabrication techniques mean that there are still many possibilities to explore.

Acknowledgments

This research represents the work of many people and I would like to thank all of the current and former members of the McGill Photonics Systems Group for their efforts. In particular, I am very grateful to my colleague Professor David Plant for his enthusiastic collaboration on this topic.

References

1. Goodman, J. W., Leonberger, F. J., Kung, S. C., and Athale, R. A., "Optical interconnections for VLSI systems," *Proceedings of the IEEE*, vol. 72, no. 7, pp. 850–866, July 1984
2. Van der Lugt, A., "Review of optical processing," *Journal of the Optical Society of America*, vol. 63, no. 10, p. 1302, October 1973
3. Hinton, H. S., *An Introduction to Photonic Switching Fabrics* New York: Plenum, 2003
4. Jahns, J. and Lee, S. H., *Optical Computing Hardware* Boston: Academic, 1993
5. Thienpont, H., Debaes, C., Baukens, V., Ottevaere, H., Vynck, P., Tuteleers, P., Verschaffelt, G., Volckaerts, B., Hermanne, A., and Hanney, M., "Plastic microoptical interconnection modules for parallel free-space inter- and intra-MCM data communication," *Proceedings of the IEEE*, vol. 88, no. 6, pp. 769–779, June 2000
6. Cryan, C. V., "Two-dimensional multimode fibre array for optical interconnects," *Electronics Letters*, vol. 34, no. 6, pp. 586-587, March 1998
7. Collet, J. H., Caignet, F., Sellaye, F., and Litaize, D., "Performance constraints for onchip optical interconnects," *Selected Topics in Quantum Electronics, IEEE Journal on*, vol. 9, no. 2, pp. 425–432, 2003
8. Semiconductor Industry Association. *International Technology Roadmap for Semiconductors: 2004 edition.* Austin, Texas, 2004
9. Kirk, A. G., Plant, D. V., Ayliffe, M. H., Chateauneuf, M., and Lacroix, F., "Design rules for highly parallel free-space optical interconnects," *IEEE Journal of Selected Topics in Quantum Electronics*, vol. 9, no. 2, pp. 531–547, March 2003
10. Kim, J., Nuzman, C. J., Kumar, B., Lieuwen, D. F., Kraus, J. S., Weiss, A., Lichtenwalner, C. P., Papazian, A. R., Frahm, R. E., Basavanhally, N. R., Ramsey, D. A., Aksyuk, V. A., Pardo, F., Simon, M. E., Lifton, V., Chan, H. B., Haueis, M., Gasparyan, A., Shea, H. R., Arney, S., Bolle, C. A., Kolodner, P. R., Ryf, R., Neilson, D. T., and Gates, J. V., "1100 × 1100 port MEMS-based optical crossconnect with 4-dB maximum loss," *IEEE Photonics Technology Letters*, vol. 15, no. 11, pp. 1537–1539, November 2003

11. Berger, C., Xiaoqing, W., Ekman, J. T., Marchand, P. J., Spaanenburg, H., Wang, M. M., Kiamilev, F., and Esener, S. C., "Parallel processing demonstrator with plug-on-top free-space interconnect optics," *Proceedings of the SPIE - The International Society for Optical Engineering*, vol. 4292 pp. 105–116, 2001

12. Chateauneuf, M., Kirk, A. G., Plant, D. V., Yamamoto, T., and Ahearn, J. D., "512-channel vertical-cavity surface-emitting laser based free-space optical link," *Applied Optics*, vol. 41, no. 26, pp. 5552–5561, September 2002

13. Haney, M. W., Christensen, M. P., Milojkovic, P., Fokken, G. J., Vickberg, M., Gilbert, B. K., Rieve, J., Ekman, J., Chandramani, P., and Kiamilev, F., "Description and evaluation of the FAST-Net smart pixel-based optical interconnection prototype," *Proceedings of the IEEE*, vol. 88, no. 6, pp. 819–828, June 2000

14. Kirk, A. G., Plant, D. V., Szymanski, T. H., Vranesic, Z. G., Tooley, F. A. P., Rolston, D. R., Ayliffe, M. H., Lacroix, F. K., Robertson, B., Bernier, E., and Brosseau, D. F., "Design and implementation of a modulator-based free-space optical backplane for multiprocessor applications," *Applied Optics*, vol. 42, no. 14, pp. 2465–2481, May 2003

15. Venditti, M. B., Laprise, E., Faucher, J., Laprise, P. O., Eduardo, J., Lugo, A., and Plant, D. V., "Design and test of an optoelectronic-VLSI chip with 540-element receiver-transmitter arrays using differential optical signaling," *IEEE Journal of Selected Topics in Quantum Electronics*, vol. 9, no. 2, pp. 361–379, March 2003

16. Geib, K. M., Choquette, K. D., Serkland, D. K., Allerman, A. A., and Hargett, T. W., "Fabrication and performance of two-dimensional matrix addressable arrays of integrated vertical-cavity lasers and resonant cavity photodetectors," *IEEE Journal of Selected Topics in Quantum Electronics*, vol. 8, no. 4, pp. 943–947, July 2002

17. Geib, K. M., Serkland, D. K., Allerman, A. A., Hargett, T. W., and Choquette, K. D., "High-density interleaved VCSEL-RCPD arrays for optical information processing," *Proceedings of the SPIE - The International Society for Optical Engineering*, vol. 4942, pp. 207–212, 2003

18. Chiarulli, D. M., Levitan, S. P., Hansson, J., and Weisser, M. Chip-to-chip multipoint optoelectronic interconnections. 111-113. 2003. USA, Optical Society of America. Optics in Computing (Trends in Optics and Photonics Series vol. 90)

19. Maj, T., Kirk, A. G., Plant, D. V., Ahadian, J. F., Fonstad, C. G., Lear, K. L., Tatah, K., Robinson, M. S., and Trezza, J. A., "Interconnection of a two-dimensional array of vertical-cavity surface-emitting lasers to a receiver array by means of a fiber image guide," *Applied Optics*, vol. 39, no. 5, pp. 683–689, Febuary 2000

20. Willner, A. E., Chang-Hasnain, C. J., and Leight, J. E., "2-D WDM optical interconnections using multiple-wavelength VCSEL's for simultaneous and reconfigurable communication among many planes," *IEEE Photonics Technology Letters*, vol. 5, no. 7, pp. 838–841, 1993

21. Plant, D. V., Venditti, M. B., Laprise, E., Faucher, J., Razavi, K., Chateauneuf, M., Kirk, A. G., and Ahearn, J. S., "256-channel bidirectional optical interconnect using VCSELs and photodiodes on CMOS," *Journal of Lightwave Technology*, vol. 19, no. 8, pp. 1093–1103, August 2001

22. Born, M. and Wolf, E., *Principals of Optics*, 6 ed. Cambridge: Cambridge University Press, 1997

23. Hecht, E., *Optics*, 4 ed. Reading, MA: Addison-Wesley, 2002
24. Goodman, J. W., *Introduction to Fourier Optics*, 2 ed. Boston, MA: McGraw-Hill, 1996
25. Herzig, H. P., *Micro-optics* London: Taylor and Francis, 1997
26. Daly, D., Stevens, R. F., Hutley, M. C., and Davies, N., "The manufacture of microlenses by melting photoresist," *Measurement Science & Technology*, vol. 1, no. 8, pp. 759–766, August 1990
27. He, M., Yuan, X., and Bu, J., "Sample-inverted reflow technique for fabrication of a revolved-hyperboloid microlens array in hybrid solgel glass," *Optics Letters*, vol. 29, no. 17, pp. 2004–2006, 2004
28. Oikawa, M. and Hamanaka, K., "The physics of planar microlenses," in Jahns, J. and Lee, S. H. (Eds.) *Optical Computing Hardware* San Diego: Academic, 1994
29. Wu, C.-K. High energy beam colored glasses exhibiting insensitivity to actinic radiation. Canyon Materials Research & Engineering. U.S. Patent 4,567,104. 1986. CA. 1983
30. Wang, G. J., Wang, S. Y., and Chin, C. H., "Fabrication and modeling of the gray-scale mask based aspheric refraction microlens array," *JSME International Journal, Series C: Mechanical Systems, Machine Elements and Manufacturing*, vol. 46, no. 4, pp. 1598–1603, 2003
31. Jahns, J., "Diffractive optical elements for optical computers," in Jahns, J. and Lee, S. H. (eds.) *Optical computing hardware* Boston: Academic, 1994, pp. 137–167
32. Prather, D. W., Pustai, D., and Shouyuan, S., "Performance of multilevel diffractive lenses as a function of f-number," *Applied Optics*, vol. 40, no. 2, pp. 207–210, January 2001
33. Alleyne, C. J. and Kirk, A. G. "Rigorous coupled wave analysis applied to transmission efficiency of diffractive beam array relays for free-space optical interconnects", Applied Optics, vol. 44, pp. 1200–1206, March 2005
34. Noponen, E., Turunen, J., and Vasara, A., "Electromagnetic theory and design of diffractive-lens arrays," *Journal of the Optical Society of America A (Optics and Image Science)*, vol. 10, no. 3, pp. 434–443, March 1993
35. Kress, B. and Meyrueis, P., *Digital Diffractive Optics* Chichester: John New York, 2000
36. Lohmann, A. W., "Image-Formation of Dilute Arrays for Optical Information-Processing," *Optics Communications*, vol. 86, no. 5, pp. 365–370, December 1991
37. Tooley, F. A. P., "Challenges in optically interconnecting electronics," *IEEE Journal of Selected Topics in Quantum Electronics*, vol. 2, no. 1, pp. 3–13, April 1996
38. Hinton, H. S., Cloonan, T. J., Mccormick, F. B., Lentine, A. L., and Tooley, F. A. P., "Free-Space Digital Optical-Systems," *Proceedings of the IEEE*, vol. 82, no. 11, pp. 1632–1649, November 1994
39. Walker, A. C., Desmulliez, M. P. Y., Forbes, M. G., Fancey, S. J., Buller, G. S., Taghizadeh, M. R., Dines, J. A. B., Stanley, C. R., Pennelli, G., Boyd, A. R., Horan, P., Byrne, D., Hegarty, J., Eitel, S., Gauggel, H. P., Gulden, K. H., Gauthier, A., Benabes, P., Gutzwiller, J. L., and Goetz, M., "Design and construction of an optoelectronic crossbar switch containing a terabit per second free-space optical interconnect," *IEEE Journal of Selected Topics in Quantum Electronics*, vol. 5, no. 2, pp. 236–249, March 1999

40. Neilson, D. T. and Barrett, C. P., "Performance trade-offs for conventional lenses for free-space digital optics," *Applied Optics*, vol. 35, no. 8, pp. 1240–1248, March 1996

41. Smith, W. J., *Modern Optical Engineering: The Design of Optical Systems*, 2 ed. New York: McGraw-Hill, 1990

42. Hamanaka, K., "Optical bus interconnection system using Selfoc lenses," *Optics Letters*, vol. 16, no. 16, pp. 1222–1224, August 1991

43. Kirk, A. G., Thienpont, H., Goulet, A., Heremans, P., Borghs, G., Vounckx, R., Kuijk, M., and Veretennicoff, I., "Demonstration of optoelectronic logic operations with differential pairs of optical thyristors," *IEEE Photonics Technology Letters*, vol. 8, no. 3, pp. 467–469, March 1996

44. McCormick, F. B., Tooley, F. A. P., Cloonan, T. J., Sasian, J. M., Hinton, H. S., Mersereau, K. O., and Feldblum, A. Y., "Optical interconnections using microlens arrays," *Optical and Quantum Electronics*, vol. 24, no. 4, pp. 465–477, April 1992

45. Haney, M. W., Christensen, M. P., Milojkovic, P., Ekman, J., Chandramani, P., Rozier, R., Kiamilev, F., Liu, Y., Hibbs-Brenner, M., Nohava, J., Kalweit, E., Bounnak, S., Marta, T., and Walterson, B., "FAST-Net optical interconnection prototype demonstration," *Journal of Optics A-Pure and Applied Optics*, vol. 1, no. 2, pp. 228–232, March 1999

46. Lacroix, F., Bernier, E., Ayliffe, M. H., Tooley, F. A. P., Plant, D. V., and Kirk, A. G., "Implementation of a compact, four-stage, scalable optical interconnect for photonic backplane applications," *Applied Optics*, vol. 41, no. 8, pp. 1541–1555, March 2002

47. Rolston, D. R., Robertson, B., Hinton, H. S., and Plant, D. V., "Analysis of a microchannel interconnect based on the clustering of smart-pixel-device windows," *Applied Optics*, vol. 35, no. 8, pp. 1220–1233, March 1996

48. Chateauneuf, M. and Kirk, A. G., "Determination of the optimum cluster parameters in a clustered free-space optical interconnect," *Applied Optics*, vol. 42, no. 29, pp. 5906–5917, October 2003

49. Plant, D. V. and Kirk, A. G., "Optical interconnects at the chip and board level: Challenges and solutions," *Proceedings of the IEEE*, vol. 88, no. 6, pp. 806–818, June 2000

50. Ayliffe, M. H., "Alignment and packaging techniques for two-dimensional free-space optical interconnects." Ph.D. Thesis, McGill University, 2001

51. Neilson, D. T., "Tolerance of optical interconnections to misalignment," *Applied Optics*, vol. 38, no. 11, pp. 2282–2290, April 1999

52. Neilson, D. T. and Schenfeld, E., "Free-space optical relay for the interconnection of multimode fibers," *Applied Optics*, vol. 38, no. 11, pp. 2291–2296, April 1999

53. Lacroix, F., Chateauneuf, M., Xue, X., and Kirk, A. G., "Experimental and numerical analyses of misalignment tolerances in free-space optical interconnects," *Applied Optics*, vol. 39, no. 5, pp. 704–713, Febuary 2000

54. Zaleta, D., Patra, S., Ozguz, V., Ma, J., and Lee, S. H., "Tolerancing of board-level-free-space optical interconnects," *Applied Optics*, vol. 35, no. 8, pp. 1317, 1996

55. Lacroix, F. and Kirk, A. G., "Tolerance stackup effects in free-space optical interconnects," *Applied Optics*, vol. 40, no. 29, pp. 5240–5247, October 2001

56. Nigam, S. D. and Turner, J. U., "Review of statistical approaches to tolerance analysis," *Computer Aided Design*, vol. 27, no. 1, pp. 6–15, January 1995

57. Levitan, S. P., Kurzweg, T. P., Marchand, P. J., Rempel, M. A., Chiarulli, D. M., Martinez, J. A., Bridgen, J. M., Chi, F., and Mccormick, F. B., "Chatoyant: a computer-aided-design tool for free-space optoelectronic systems," *Applied Optics*, vol. 37, no. 26, pp. 6078–6092, September 1998

58. Ozkan, N. S. F., Marchand, P. J., Esener, S. C., and Lee Hendrick, W., "Misalignment tolerance analysis of free-space optical interconnects via statistical methods," *Applied Optics*, vol. 41, no. 14, pp. 2686–2694, 2002

59. Gruber, M., Jahns, J., El Joudi, E. M., and Sinzinger, S., "Practical realization of massively parallel fiber-free-space optical interconnects," *Applied Optics*, vol. 40, no. 17, pp. 2902–2908, June 2001

60. Christensen, M. P., Milojkovic, P., McFadden, M. J., and Haney, M. W., "Multiscale optical design for global chip-to-chip optical interconnections and misalignment tolerant packaging," *IEEE Journal of Selected Topics in Quantum Electronics*, vol. 9, no. 2, pp. 548–556, March 2003

61. Christensen, M. P., Milojkovic, P., and Haney, M. W., "Analysis of a hybrid micro/macro-optical method for distortion removal in free-space optical interconnections," *Journal of the Optical Society of America A (Optics, Image Science and Vision)*, vol. 19, no. 12, pp. 2473–2478, December 2002

62. Johnstone, R. W. and Parameswaran, M., "Determination of optical alignment in free-space micro-optical-bench systems using paraxial ray-transfer matrices," *Journal of Microlithography, Microfabrication, and Microsystems*, vol. 2, no. 1, pp. 49–53, January 2003

63. Bisaillon, E., Brosseau, D. F., Yamamoto, T., Mony, A., Bernier, E., Goodwill, D., Plant, D. V., and Kirk, A. G., "Free-space optical link with spatial redundancy for misalignment tolerance," *IEEE Photonics Technology Letters*, vol. 14, no. 2, pp. 242–244, Febuary 2002

64. Jahns, J., "Planar Packaging of Free-Space Optical Interconnections," *Proceedings of the IEEE*, vol. 82, no. 11, pp. 1623–1631, November 1994

65. Gruber, M., Kerssenfischer, R., and Jahns, J., "Planar-integrated free-space optical fan-out module for MT-connected fiber ribbons," *Journal of Lightwave Technology*, vol. 22, no. 9, pp. 2218–2222, September 2004

66. Robertson, B., Liu, Y. S., Boisset, G. C., Tagizadeh, M. R., and Plant, D. V., "In situ interferometric alignment systems for the assembly of microchannel relay systems," *Applied Optics*, vol. 36, no. 35, pp. 9253–9260, December 1997

67. Chateauneuf, M. and Kirk, A. G., "Six-degrees-of-freedom alignment technique that provides diagnostic misalignment information," *Applied Optics*, vol. 43, no. 13, pp. 2689–2694, May 2004

68. Chateauneuf, M., Ayliffe, M. H., and Kirk, A. G., "In situ technique for measuring the orthogonality of a plane wave to a substrate," *Optics Letters*, vol. 28, no. 9, pp. 677–679, May 2003

69. Debaes, C., Vervaeke, M., Ottevaere, H., Meeus, W., Tuteleers, P., Brunfaut, M., Baukens, V., Van Campenhout, J., and Thienpont, H. Demonstration of manufacturable free-space modules for multi-channel intra-chip optical interconnects. vol.1, 63–64. 2002. Glasgow, UK, IEEE. LEOS 2002. 2002 IEEE/LEOS Annual Meeting Conference Proceedings. 15th Annual Meeting of the IEEE Lasers and Electro-Optics Society (Cat. No. 02CH37369)

70. Haake, J. M. Microactuator for precisely aligning an optical fiber and an associated fabrication method. McDonnell Douglas Corporation. US Patent 5,602,955. 1997. MO, USA. 7–6–1995

71. Tuantranont, A., Bright, V. M., Zhang, J., Zhang, W., Neff, J. A., and Lee, Y. C., "Optical beam steering using MEMS-controllable microlens array," *Sensors and Actuators A-Physical*, vol. 91, no. 3, pp. 363–372, July 2001

72. Ishikawa, K., Zhang, J. L., Tuantranont, A., Bright, V. M., and Lee, Y. C., "An integrated micro-optical system for VCSEL-to-fiber active alignment," *Sensors and Actuators A-Physical*, vol. 103, no. 1–2, pp. 109–115, January 2003

73. Yamaguchi, M., Yamamoto, T., Hirabayashi, K., Matsuo, S., and Koyabu, K., "High-density digital free-space photonic-switching fabrics using exciton absorption reflection-switch (EARS) arrays and microbeam optical interconnections," *IEEE Journal of Selected Topics in Quantum Electronics*, vol. 2, no. 1, pp. 47–54, April 1996

74. Ayliffe, M. H., Kabal, D., Lacroix, F., Bernier, E., Khurana, P., Kirk, A. G., Tooley, F. A. P., and Plant, D. V., "Electrical, thermal and optomechanical packaging of large 2D optoelectronic device arrays for free-space optical interconnects," *Journal of Optics A-Pure and Applied Optics*, vol. 1, no. 2, pp. 267–271, March 1999

75. Ayliffe, M. H., Chateauneuf, M., Rolston, D. R., Kirk, A. G., and Plant, D. V., "Six-degrees-of-freedom alignment of two-dimensional arrays components by use of off-axis linear Fresnel zone plates," *Applied Optics*, vol. 40, no. 35, pp. 6515–6526, December 2001

76. Brosseau, D. F., Lacroix, F., Ayliffe, M. H., Bernier, E., Robertson, B., Tooley, F. A. P., Plant, D. V., and Kirk, A. G., "Design, implementation, and characterization of a kinematically aligned, cascaded spot-array generator for a modulator-based free-space optical interconnect," *Applied Optics*, vol. 39, no. 5, pp. 733–745, Febuary 2000

77. Dawei, H., Sze, T., Landin, A., Lytel, R., and Davidson, H. L., "Optical interconnects: out of the box forever?," *IEEE Journal of Selected Topics in Quantum Electronics*, vol. 9, no. 2, pp. 614–623, March 2003

78. Thai, S., Kuznia, C., Divakar, M. P., Albares, D., Pendleton, M., Le, T., Bachta, P., Hagan, R., Pommer, D., Cable, J., and Reedy, R. Parallel interconnect components for optical modules utilizing flip-chip VCSEL on ultra-thin silicon-on-sapphire substrate. 4649, 230–235. 2002. San Jose, CA, United States, The International Society for Optical Engineering. *Proceedings of SPIE - The International Society for Optical Engineering*

79. Lukowicz, P., Jahns, J., Barbieri, R., Benabes, P., Bierhoff, T., Gauthier, A., Jarczynski, M., Russell, G. A., Schrage, J., Sullau, W., Snowdon, J. F., Wirz, M., and Troster, G., "Optoelectronic interconnection technology in the HOLMS system," *IEEE Journal of Selected Topics in Quantum Electronics*, vol. 9, no. 2, pp. 624–635, March 2003

Index

Springer Series in
OPTICAL SCIENCES

Springer Series in
OPTICAL SCIENCES